Environment and Development

Antonio Augusto Rossotto Ioris
Editor

Environment and Development

Challenges, Policies and Practices

palgrave
macmillan

Editor
Antonio Augusto Rossotto Ioris
School of Geography and Planning
Cardiff University
Cardiff, UK

ISBN 978-3-030-55415-6 ISBN 978-3-030-55416-3 (eBook)
https://doi.org/10.1007/978-3-030-55416-3

This Palgrave Macmillan imprint is published by the registered company Springer Nature Switzerland AG.
The registered company address is: Gewerbestrasse 11, 6330 Cham, Switzerland

This book is dedicated to the Guarani-Kaiowa and the many other indigenous peoples of South America horribly affected by the genocidal violence of mainstream development, which is anti-environment because it is anti-human.

Contents

List of Contributors

Mariana D. Barbieri University of Campinas (UNICAMP), Campinas, Brazil

Fabiana Barbi Centre for Environmental Studies and Research (NEPAM), University of Campinas (UNICAMP), Campinas, Brazil

Gilberto F. Barroso Department of Oceanography and Ecology, Federal University of Espírito Santo (UFES), Vitória, Brazil

Felipe A. Bastos State Institute of Environment and Water Resources of Espírito Santo (IEMA), Vitória, Brazil

Elisabeth Cremona University of Malta, Msida, Malta

Leila da Costa Ferreira Institute of Philosophy and Human Sciences (IFCH), University of Campinas (UNICAMP), Campinas, Brazil

Marcelo Urbano Ferreira Department of Parasitology, Institute of Biomedical Sciences, University of São Paulo (USP), São Paulo, Brazil

Daniela Maimoni Figueiredo Federal University of Mato Grosso (UFMT), Cuiabá, Brazil

Helena I. Gomes Food, Water, Waste Research Group, Faculty of Engineering, University of Nottingham, Nottingham, UK

Adrian Gonzalez School of Geography and Planning, Cardiff University, Cardiff, UK

Geoff Goodwin Oxford Department of International Development, Queen Elizabeth House, University of Oxford, Oxford, UK

Edson Grandisoli Institute of Advanced Studies, University of São Paulo (USP), São Paulo, Brazil

Regina Beatriz Guimarães Neto Federal University of Pernambuco (UFPE), Recife, Brazil

Antonio Augusto Rossotto Ioris School of Geography and Planning, Cardiff University, Cardiff, UK

Rafael R. Ioris History Department, University of Denver, Denver, CO, USA

Pedro Roberto Jacobi Institute of Energy and the Environment, University of São Paulo (USP), São Paulo, Brazil

Vitale Joanoni Neto Federal University of Mato Grosso (UFMT), Cuiabá, Brazil

Igor Cavallini Johansen Department of Parasitology, Institute of Biomedical Sciences, University of São Paulo (USP), São Paulo, Brazil

Nazli Koseoglu Social, Economic and Geographical Sciences, The James Hutton Institute, Aberdeen, UK

Claire Lagier Rachel Carson Centre for Environment and Society, Ludwig-Maximilians-Universität München, Munich, Germany

Ana T. Lima Department of Oceanography and Ecology, Federal University of Espírito Santo (UFES), Vitória, Brazil

Paul E. Little University of Brasília (UnB), Brasília, Brazil

Renato Rodrigues Neto Department of Oceanography and Ecology, Federal University of Espírito Santo (UFES), Vitória, Brazil

Louis Rice World Health Organisation Collaborating Centre for Healthy Urban Environments, University of the West of England, Bristol, UK

Daniele Tubino P. de Souza Institute of Energy and the Environment, University of São Paulo (USP), São Paulo, Brazil

Fernando Jakes Teubner Junior Brazilian Institute of Geography and Statistics (IBGE), Vitória, Brazil

Stefano Varese Department of Native American Studies, University of California-Davis, Davis, CA, USA

Arjen E. J. Wals Wageningen University, Wageningen, The Netherlands

List of Figures

List of Graphs

List of Tables

Part I

Introduction

1

Environmental Roots of Development Problems

Antonio Augusto Rossotto Ioris

Environment and Development: The Turbulent Journey of Modernity

This book is about the complex and highly contested nexus between the goals and promises of development, and the frustrating rise of environmental problems and tensions. In the past half century, environmental questions, multiple and broadly defined, have increasingly occupied centre stage in global debates about the maelstrom of present-day socio-economic tendencies and uncertain future prospects. Despite many attempts to elucidate its meaning, development remains an elusive concept, fraught with contradictions, associated with ethereal promises and recurrent frustrations. Likewise, the definition of environment, as well as associated terms such as nature and ecology, depends on the moral, religious and technological basis of each society and its time-space circumstances. It is possible to argue, nonetheless, that most of the controversies here are related to the reduction of development to economic growth, as

A. A. R. Ioris (✉)
School of Geography and Planning, Cardiff University, Cardiff, UK
e-mail: IorisA@cardiff.ac.uk

A. A. R. Ioris (ed.), *Environment and Development*,
https://doi.org/10.1007/978-3-030-55416-3_1

well as the threat of environmental problems as mere calls for improved management strategies (increasingly through market-based strategies that transfer the rationality of economic growth to the supposed mitigation of environmental impacts). In effect, environment-development dilemmas are actually much broader and more complicated than suggested by simplistic narratives of progress and conservation. All major challenges around the world are directly associated with controversial processes of environmental change and landscape degradation, as in the case of mass migration, staggering levels of energy and material waste, food insecurity, widespread water and air pollution, obscene levels of socio-spatial inequality and genocidal practices against indigenous peoples and other minority groups. It has also to do with the historical agency of nature and its impact on socio-political relations, as in the case of the 2020 Coronavirus outbreak, the burning of the Amazon in recent years and the escalation of anthropogenic global warming.

The difficulty to properly understand and effectively deal with the socio-ecological consequences of development are basically derived from the uneasiness to challenge the advance of capitalist-based modernity. That is, the normativity of development has been translated into the affirmation of Western modernity as the unescapable model for the rest of the planet, and, in that way, the self-granted permission to conquer and explore nature has also become a global phenomenon. Development intrinsically requires the mobilisation of nature and its conversion into economic resources, but it is a process that 'exists through the actions that it legitimates, through the institutions it keeps alive and the signs that testify its presence' (Rist 2008: 10). From the period of European colonisation to the phase of national independence and international development in the post-war decades and, finally, the more recent adjustments of public policies under the influence of neoliberalising ideologies, the pursuit of capitalist production and reproduction institutions has transformed local, national and global spheres of socio-ecological interaction. Since the golden age of Portuguese navigation in the fifteenth century, capitalism has systematically and purposefully expanded out of Europe Modernity as an unstoppable project with global repercussions. Capitalist activity has taken over continents, locations, cultures and practices throughout the world, which was translated in the twentieth century in the agenda and process of development. In this intense geographical process, the production of new spatial settings has become a central pillar of the circulation and accumulation of capital (Ioris 2020). Socio-spatial frontiers of development function as a mirror, where the bare and explicit features of capitalism are exposed through the formation of site-specific mechanisms of resource extraction, economic production and political justification. Because of these interdependent relations between 'core' and 'periphery' through joint processes of exploitation, realignment and reinforcement in both old and new areas, it can be argued that capitalist development also largely depends on accumulation by frontier making and socio-ecological change (Ioris 2018a).

Our basic aim in this introductory chapter is to examine the meaning and fundamental features of the advance of development, what is later detailed and exemplified in the other chapters of this book. The main purpose is to suggest a new conceptual approach to frontier-making dynamics that underpin the socio-ecology of development. This involves reflecting on the necessity, the configuration and the contestation of socio-spatial frontiers, beyond the conventional descriptive and quantitative assessments of land-use change, resource extraction or commodity production. The examination of the environmental roots of development and its relentless expiation to all corners of the planet is extremely relevant because, as pointed out by Sachs (2010: vii), in an age of globalisation 'politics is compelled to push either equity without ecology, or ecology without equity. It is hard to see how this dilemma can be resolved unless the belief in 'development' is dismantled'. The uneven and ferocious development of capitalist modernity and the production of economic frontiers has been the object of a long debate and occupied critical scholars working in several different disciplines. The spaces of capitalist frontiers have been repeatedly studied by many researchers, and interpretations range from neoclassical enthusiasm about the economic outcomes of new frontiers to critical voices discussing growing proletarianisation and acts of resistance following the penetration of capitalism. Although often used interchangeably, Watts (2018) differentiates between 'frontier' as a zone of socio-economic advance and 'border' as a line of demarcation between national territories or administrative units. Lund and Rachman (2018) also make a distinction between frontier dynamics (the frontier is a free resource zone, where social order is eliminated, property is disrupted and social contracts dissolved) and territorialisation (when spaces acquire new systems of authority and regulation). Departing from such literature, our focus on these pages is on the political ontology of frontspaces (i.e. the spaces of the frontier of development), that is, the locales, landscapes and places undergoing changes because of the advance of new socio-economic processes that are dialectically connected with wider spatial trends at broader scales.

We are primarily concerned with the socio-spatial frontier as an area that is undergoing a rapid transformation because of the migration of people and the opening of new economic opportunities, and where authority and governance are significantly diluted and transformed. Moreover, frontiers are more than 'liminal spaces', characterised by a state of ambiguity or in-between-ness fraught with possibilities and opportunities to start over (McDowell and Crooke 2019) because the room for innovation is actually limited because of the normativity of capitalism and the power asymmetries between central, relatively consolidated areas and the frontier. Frontspaces are projections of the institutions and relations of central politico-economic areas; in other words, they are the centre being reasserted and gradually incorporate the margins in the same developmental project. Marx argued that capitalism is not merely the movement of exchange values because circulation alone can never realise capital (considering that the mere exchange of equivalents extinguishes value, and the circulation of money and commodity cannot lead to self-renewal). The success of circulation requires mediation of the total economic process, including geographical connections and socio-ideological aspects, as in the case of new sites for extracting resources and producing commodities. 'Commodities constantly have to be thrown into it anew from the outside, like fuel into a fire' (Marx 1973: 255). New development frontiers, as in the case of Africa, the Amazon and in Southeast Asia, today, are therefore required to accommodate economic and social demands and divert attention away from home-grown problems. There is always a demand for materials and resources, new markets and business opportunities, for compensating socio-ecological degradation and reducing socio-political tensions through migration. Marx (1976: 794) observes that some workers emigrate, but 'in fact they are merely following capital, which has itself emigrated'. The next section summarises our five main points (or conceptual claims) about the advance of capitalist development and its socio-ecological outcomes.

Theorising the Time-Spaces of Economic Development and Frontier Making

The socio-spatial frontiers of capitalist development exist not only because they provide economic and social opportunities that may be more difficult to access elsewhere but because problems accumulated in central areas are responsible for frontier making (insofar as this constitutes a fundamental endeavour to renovate the whole economic system). At the frontier, the new remains tamed, subordinate and relativised as genuine economic creativity because the ontic reality of the frontier is fundamentally shaped by the transplantation of exogenous socio-economic relations. The above-mentioned empirical discussion helps to illustrate the projection of the situation of the politico-economic centre on the frontier. For instance, for more than 400 years, there were limited and only sporadic attempts to develop productive activities in the Amazon (Théry 1997), due to the more compelling possibility of collecting and exploiting the region's abundant resources, resulting in its perennial frontier status (Bunker 1985). In the middle of the last century, due to the complexification of the South American economy and mounting tensions in the eastern core areas, the Amazon entered a new phase of economic development, which, nevertheless, did not change its status as a frontier maker. On the contrary, the new phase of frontier making was predicated upon the old: the time was ripe for a focus on production (at least the image of production) and for a new language of modernity (Ioris 2015). Since the 1960s, the region has been a target for the construction of hydropower schemes, roads, mega timber processing plants, navigation infrastructure, the industrial pole of Manaus (Brazil) and, most importantly, for the advance of agribusiness. The region is still a frontier, despite looking ever more like the politico-economic centres, and even more of a frontier precisely because of this attempt to imitate the centres. (and even more of a frontier precisely because of that).

Consequently, in order to interrogate the frontier, one has to simultaneously comprehend the achievements and contradictions of the most developed (central) economic areas. We can rely on the politico-economic methodology adopted by Marx to describe the evolution of capitalism and from his claim that the bourgeois economy supplies the key to the ancient economies that preceded it. Marx (1973: 105) argues that the anatomy of humans 'contains a key to the anatomy of the ape', that is, 'the intimations of higher development in the less developed systems can be understood only after the higher development is already known'. To understand the interpenetration between past and present, Marx specifically warns that one must progress carefully because similarities and differences must be examined in detail, avoiding simplistic, non-critical associations. That is certainly useful, but we need first to reverse the chronological direction of Marx's analysis, from the centre to the (more recent) socio-spatial frontier. The reason is that, although the frontier obviously emerged after the economy evolved in central areas, it, nonetheless, always revisits the basic elements of the centre's economic past. This is the case regarding, for example, processes of primitive accumulation, dispossession, unregulated appropriation of resources and labour exploitation, ideological argumentation of progress and the imposition of a bourgeois order and associated values. Starting from the realisation that development frontiers are predicated and constrained by the lingering failures of the areas where migrants and business people originally came from, the controversial and politicised dynamics that emerge can be schematically summarised in five main points, which are relevant for the examination of other contemporary frontiers around the world.

The *first main conceptual claim is that development frontier making may appear as a social and spatial dislocation, as distanced from the centre, while, in effect, the centre is being projected, restated and restored.* There is no essential disconnection between the socio-economy in the consolidated (central) and the new (frontier) areas, but actually a coherent continuity between the centre and the frontier. Frontiers are therefore more than 'zones of incorporation' of an expanding world system (cf. Wallerstein 1974); they play a key role in the reorganisation and revitalisation of the centre. There is no spatial contradiction between the centre and the new areas, but in reality, the frontier is presupposed in the contradictions of the centre itself and functions as a mechanism to mitigate those tensions and prolong the existence of the centre. In this way, the areas where capitalism keeps expanding are, from the outset, loaded with the practices, realisations and vices of the old areas. The European colonisation of the Americas transferred to the new continent socio-political hierarchies, values and ideologies that underpinned metropolitan societies in Europe. The Americas contained not only riches and resources but became catalysts for controlled, top-down reforms in the centre. Similar processes were present in the eighteenth century when colonisation expanded in India, and in the nineteenth century during the Scramble for Africa and the partition of the continent among European colonial powers. Simultaneously, other parts of Asia and Oceania were turned into spatial frontiers. In more recent decades, several of these former colonies have been recolonised, this time particularly under the rhetoric of market globalisation, foreign investment (often in the form of land, water and nature grabbing) and international development.

In this way, the social, economic and spatial configuration of frontiers fundamentally replicates the mechanisms of exploitation and exclusion that define capitalist nations (with all their contradictions and frictions). The development frontier is more than a spatial fix, as argued by Harvey (2006), that is, a location where capital can be diverted and invested in infrastructure and real estate, in order to respond to problems of overaccumulation. Frontiers are areas where time and space are reconfigured and that in theory could result in something new but because of their subordinate status rapidly assume a configuration that largely mimics the core areas. That is, time and space are transfigured but retain the properties they had in the original, central areas. The frontier is where the trajectory of time and space (as established in the central area) is disrupted, creating a realm of potentialities, but then, due to the dynamics of frontier making, the new area is retained within the sphere of influence of the centre. The geography of the frontier fundamentally unfolds around the troublesome gap between the possibility of the new and the concrete reproduction of key features of the (older) centre. As theorised in Aristotle's *Physics*, change requires the existence of potentiality, which is actualised and realised according to specific circumstances but without full independence. The end-state of the process of change depends on the specific properties of the system and its potentiality for change. The frontier is less fixed, more tentative and, to some extent, open, but the previous order is only superficially and temporarily interrupted, and then rapidly and effectively reconfigured according to what existed before in the centre. The reasons for the stymied potentiality of frontier activity are located in the controlling power of the processes of expropriation, enrichment and authority, which empower some sections of society and establish a new spatial order that only partially, asymmetrically incorporates the majority of the frontier population.

The centre must be upheld (to safeguard dominant interests and political power) through the production of new economic and social peripheries, but only to the extent that the periphery mirrors the decisive politico-economic features of the centre. (That is the crucial difference between frontiers in a capitalist economy and refugees from the advance of capitalism, as in the case of alternative communities and isolated Indigenous tribes). In all social spaces defined by market-centred relations, the 'critical economic decisions are often far removed from where the primary effects will be felt and without the input from those who will be directly affected. That makes it especially difficult to fully grasp and combat the extent of environmental degradation' (Magdoff and Williams 2017: 77). Although the mindset of frontier people may still not be strictly capitalist (in the sense of efficient realisation of profit and maximisation of economic outputs), the logic of the frontier is indirectly associated with capitalist tendencies operating at different scales. While Harvey (2006) did not pay special attention to the lived processes at the frontiers of capital expansion, his observation that location is an active element in the overall circulation and accumulation of capital is spot on: frontier still has a fluid and unsettled organisation, but which from the outset has very favourable conditions for supporting the centre and creating new spaces for capital to flourish. The new socio-spatial frontiers emerge out of the contradictions and constraints of consolidated areas but also with limited freedom to reconstitute social and economic relations much beyond the given hegemonic conditions. The strategic role of frontier situations is not simply to recirculate capital, but to mitigate mounting social tensions (such as unemployment and political unrest, e.g. Scottish and Irish migrants to British colonies or ex-colonies), release the value of natural resources hitherto beyond economic use (minerals, water, land, labour power, etc.) and pave the way for new cycles of investment and circulation. Frontiers may be located in either urban or rural areas; in some cases, the movement is from conflict-prone rural areas towards the periphery of large cities, as has typically happened in Latin America (the case of Lima's *barriadas* is paradigmatic, exacerbated by Peru's civil war in the 1980s and the liberalising reforms in the 1990s).

Our *second main claim is about the production of development frontiers through the reinforcement of the multiple and interconnected dualisms* that characterise Western culture and the Westernised pattern of economic development. Many authors have identified the origins of Western dualisms in Hebrew theology and Greek philosophy, sustained through the Christian polarity of good-evil and the Cartesian mind-body dichotomy that permeate Western patterns of thought and associated scientific knowledge. The result is a situation in which interpersonal dualisms (male/female, white/non-white, learned/traditional, archaic/modern, unproductive/productive, etc.) are deeply rooted in, and help, to reinforce, Western culture and its socio-economic structures (Mellor 2000). These ingrained dualisms are forced upon spatial frontiers by dominant interest groups as part of the attempt to consolidate new relations of production and reproduction. The long chain of dualisms derives from the fundamental dichotomy between core and frontier, which is nurtured by the supposed superiority of the centre and the alleged deficiencies of the frontier. The fabricated contrast between 'superior' people in central areas and 'second-class' frontier inhabitants is instrumental both for the institutionalisation of the new frontier and for upholding the authority of the centre. The typical narrative of frontier making replicates the hierarchical differentiation between core and frontier, and reinforces the message that activities at the frontier must reproduce social roles and institutions imported from consolidated areas.

Dualistic thinking has been put to work to further the advance and legitimisation of frontier making in different parts of the world. Australian settler culture vividly illustrates the superposition of various dualisms; this society was highly racist, anti-native and hierarchical, permeated by a masculine discourse about the supremacy of the free white settler (Woollacott 2015). Australian frontier making has also been denounced for the recurrent practice of genocide. As described by Rogers and Bain (2016), between 1788 and 1928, extreme brutality was rationalised through the melding of Darwinian ideas about the survival of the fittest (i.e. the white, male settlers) with notions of inferior races that would inevitably die out (i.e. the aborigines). That led to a perverse combination of extinction and extermination, due to the impossibility of pastoralism coexisting with indigenous prairie management. Despite this tragic history, so far, there has been limited academic interest in frontier genocide, which Staner (1968, mentioned by Rogers and Bain 2016) calls the 'great Australian silence'. There are still unresolved questions about who should be considered responsible for genocidal colonisation, considering the impact of white settlers, local authorities and, ultimately, the British colonial masters in London. Evans (2007) details the full extent of this genocide, the result of a coordinated onslaught on lives, land and culture, which was central to the evolution of capitalism from the age of mercantilist colonisation to the time of industrialisation. Of course, local aborigines tried to resist, sometimes violently, but this 'does not change the fact that genocide occurred' (Rogers and Bain 2016: 90).

The self-professed supremacy of white, Western society, and its self-granted permission to conquer and exploit, were potentialised during colonisation by strategic scientific developments (navigation, firearms, production tools, etc.) and the application of scientific and religious knowledge (for instance, geography to support imperialist projects; biology and geology to identify valuable resources; and Christian morality to disempower and subjugate the locals). However, old dualisms have continued to reverberate, resulting in accumulated dualities, long after the end of colonialism. The tension between the advance of novel social and economic relations and the persistence of the old values and institutions is one of the main characteristics of frontier making. The range of interconnected dualisms was felt intensely in the Amazon, where the culture, knowledge and skills of traditional peoples, including those who migrated to the region after the 1970s and their families, were systematically devalued to pave the way for the commodification of production and consumption practices. Barbier (2012) asserts that the expansion of the Amazon frontier was in itself dualist, split between agribusiness farms and family agricultural units, while, in fact, the agricultural frontier is multiple, and the different categories of farming are materially and socio-culturally interdependent. Nonetheless, the promise of a better life for the large majority of impoverished migrants was never fulfilled in the new reality dominated by large-scale farmers and transnational corporations (Ioris 2017). The appeal of the modern world, at the expense of social traditions and community life, is also illustrated by the ongoing advance of processed, frozen food into the most remote corners of the region (with all the associated problems for health and the local economy), as in the case of the upper Negro River Basin on the border between Brazil and Colombia.

The *third ontic feature of socio-spatial frontiers is dialectically related to the previous two: time and space at the frontier are compressed, reconfigured and launched in different directions*. Spatial and temporal changes do not necessarily progress in the linear and sequential manner typical of core areas; at the frontier, the basic mechanisms of expropriation, commodification, proleterianisation and so on will follow unique patterns (obviously connected with the wider socio-economic trends and structures). The frontier has different phases, which normally begin slowly and then, when circumstances are favourable, accelerate rapidly. The frontier's very existence is never guaranteed, but one frontier can open and close several times on different occasions. A particular area that was considered a functional frontier for the purpose of capitalist relations can suddenly lose that status, for instance, due to competition from new products or production areas. Then, after some time, what had become an obsolete frontier can be re-created and incorporated into new rounds of migration and production, as new opportunities and additional technologies become available (e.g. the handling of new products and goods in spaces previously used for others). This means that old frontiers are excavated through the redeployment of knowledge and practices that, once again, are externally imposed from the centre—if the frontier could re-emerge independently of the centre, it would no longer be a frontier. Not only can old frontiers be supplanted by new ones, but different 'frontier moments' can be both imposed and superposed on previous socio-spatial experiences. Ultimately, 'frontier is not space itself. It is something that happens *in* and *to* space. Frontiers *take* pace. Literally' (Rasmussen and Lund 2018: 388).

New socio-spatial relations are built upon past experiences, not necessarily improving practices or procedures; in fact, a spatial frontier may be new in historical terms, while still replicating some of the oldest and vilest relations and institutions. At the frontier, capitalism is renewed through novel technologies and productive platforms, but is also virtually free to reinstate elements of violence, exploitation, dispossession, racism and other injustices that characterised previous historical periods. This has been the case with the Amazon region, which was a frontier of biodiversity and mineral extraction during colonial times, then the main source of plant latex at the time of the Second Industrial Revolution and, more recently, due to the demand for agribusiness goods, has become a dynamic frontier for plantation production and export. Violence was employed as a central element of colonisation strategy, and the expanding frontiers pushed forward by the invading Europeans did not, in fact, advance civilisation, but rather destroyed social groups and their sophisticated knowledge and art (Hemming 1987). One of the most notorious examples was the legislation introduced by the Marquis of Pombal, prime minister of Portugal, in 1757 (called *Diretório*), which forced indigenous groups in the Amazon to move to settlements managed by a 'director' [*diretor*], where racial assimilation was encouraged and cultural and linguistic identity subsumed. During the rubber extraction period in the nineteen century, the existence of indigenous groups was ignored, and contingents of very poor migrants were attracted to the region to collect latex, which in the end served to enrich a very small elite in Manaus and Belém, while satisfying the growing industrial demand for natural rubber. This exemplifies how genocide, slavery and violence were not sporadic incidents, but constituted an ongoing, systematic and transnational phenomenon underpinning frontier making.

The *fourth element of our conceptualisation is the mystification of the benefits and opportunities available at the frontier of capitalist development.* The condition of the frontier is always highly hierarchical and often manipulated to serve mainly the interests of those in more favourable positions (which include land speculators, rural development companies, intermediaries and fixers, opportunistic investors, traders who receive and export goods produced by a large number of individual agents, etc.), but these asymmetries are disguised by the appearance of accessibility and better prospects for earning a living. The mystification of what frontier areas are really offering is based on deliberate misrepresentations or omissions. There exists a fetishism of the frontier that is nourished by ambiguous evidence of success and vague stories about people who thrived. Turner (1920) misrepresented the frontier as a conduit of democracy and equality, and his account exemplifies the positive narrative constructed by those who gain from the frontier. Along similar lines, Bowman (1927: 64) argued in his work on 'pioneer fringes' that a 'changing environment breeds liberalism if the resources are abundant enough to support close settlements and the development of independent social and political institutions'. However, instead of a political vacuum, the frontier is a space of social control, where autonomy was a clear strategy for governing the territory (Hogan 1985). The mystification of the frontier also determines that success is measured according to the values of central areas and Western standards. Failures are seldom attributed to frontier conditions; rather, the blame is placed on the incompetence of migrants and pioneers who failed to take advantage of the opportunities presented to them.

The memories of those who migrated from the south of Brazil to the Amazon were populated by images of courage associated with their Italian and German ancestors (who moved to South America in the nineteenth century) and the mythology of bravery related to the consolidation of international borders with the Spaniards, and later the Argentineans. Wealthy landowners, subsistence farmers and workers were all products of the same agrarian past who lived in the south of the country, but the symbolism of the frontier was appropriated differently by different social groups. The rhetoric of victory and anticipated success was repeatedly invoked by those who led the opening of new agricultural areas in the Amazon in the 1970s, which helped to downplay the obstacles faced by the newcomers and the crude reality of socio-ecological exploitation. The plan of the military dictatorship was to allocate land to impoverished peasants and landless people (considered troublemakers in their areas of origin), but it was a monumental disappointment as it largely failed to foster agrarian capitalism in the region (Rivière d'Arc and Apestéguy 1978). The majority of those who migrated to the Amazon did not have the means to secure or maintain land and ended up as proletarians in urban or rural areas (Ioris 2019). In practice, the frontier was less epic and more a daily fight for survival (a significant proportion of migrants did not find success and returned, even more impoverished, to their areas of origin in the south, where many took part in other forms of protests and land occupation).

The Australian frontier-making experience, which has significant differences from that of the Amazon and other parts of South America, is frequently associated with an image of conquest and triumph, but this is largely explained by the construction of a new society that mirrored British values and social hierarchies. Australia in the early nineteenth century was a major destination for convicts and ex-convicts (who the authorities wished to remove from the motherland) and then, after the 1840s, a target for free settlers, encouraged to move to the colony to take control of 'free land', in fact, aboriginal land (Woollacott 2015). As far back as 1834, an Australian settler argued that those who colonise a new area 'are sure to enjoy a greater degree of consideration and importance among their companions than they could reasonably have hopped to attain in the older society' (Wakefield 1967: 327). Inequalities were not only found between settlers and aborigines; a small squatter elite (described as a 'squattocracy', and including, among others, members of the Melbourne Club, established in 1839) controlled most of the land and limited access to new migrant contingents. This created serious resentments and pressure for land reform. New legislation was introduced, such as the 1860 Nicholson Land Act, but its effectiveness was limited, as the powerful squatter elite could still purchase whatever land they required through the use of 'dummy bidders'. Wealthy squatters also used their knowledge of the land to buy up the best locations, leaving only the least fertile ground for less privileged settlers.

The mystification of frontier opportunities was also an important feature of the conquest of the American Midwest around Chicago. After 1833, the regional population and the economy grew rapidly due to the activity of settlers, investors and speculators. These so-called boosters advanced economic theories and well-crafted rhetoric about the natural endowments of the Chicago area (minimising the obvious need to invest in infrastructure and logistics to make the frontier really flourish). Their actions were typically loaded with enthusiastic exaggeration and self-interested promotion. Due to improvements in rail and boat transportation, their promotion of the frontier became a self-fulfilling prophecy, and during the nineteenth century, Chicago became a main trade centre for grain, timber and meat (Cronon 1992). The accomplishments of the frontier were internationally celebrated, culminating with the staging of the 1893 World's Columbian Exposition in Chicago to applaud the progress of American civilisation and the fulfilment of Columbus' dreams four centuries earlier (obviously discounting the immense social and environmental impacts). It is interesting that, while mystification and ideological pressures were extremely powerful, these phenomena also flourished because people in their daily struggle for survival are often led to express conservative views and put up with the current state of affairs. The most vulnerable and disorganised groups at the frontier tend to have serious difficulty developing coordinated opposition, despite their actual level of consciousness.[1]

Fifth, even in a globalised and highly interconnected world, frontier dynamics will not die out. On the contrary, *frontier making will continue to expand and flourish around the world either through the incorporation of new areas hitherto subject to less capitalist influence, or with the replacement of previous frontier making activity with novel rounds of capitalist relations of production and reproduction.* Accelerated market fluxes and population mobility do not dispense with spatial frontiers because the modern Western world persistently strives for new places and landscapes to conquer (Ioris 2018b). One of the decisive features of capitalist modernity is how it aims to standardise location-specific processes and incorporate them into the same market-centred rationality. Frontier situations favour the reaffirmation of Western modernity because existing socio-ecological elements of their reality are typically disorganised and thus easily replaced with new features associated with the Global North. However, it is important to realise also that capitalism does not need frontiers merely to renew itself, but, on the contrary, frontier making helps the centre to

[1] In his study on territorial conquest and European border disputes, Namier (1942: 69–70) perspicaciously observed that: 'One would expect people to remember the past and to imagine the future. But in fact, when discussing or writing about history, they imagine it in terms of their own experience, and when trying to gauge the future they cite supposed analogies from the past: till, by a double process of repetition, they imagine the past and remember the future'.

remain largely as it has always been. At the spatial frontiers, capitalism can be more capitalistic, in the sense that it is less constrained and more potentialised by the unique conditions of frontiers.

There exists a necessity for new socio-spatial frontiers to work as opportunities to try to erase socio-ecologies and produce novel spaces or new socio-economic relations under the influence of capitalist modernity. Frontier making is intrinsic in the peculiar trajectory of capitalism that combines dualisms and accumulates tensions between old and new, exploitation and production, particularities and universalisms. But the necessity for spatial frontiers needs to be understood in dialectical and non-prescriptive terms. Žižek (2011) claims that historical necessity is really a convergence of contingencies. He argues that the Hegelian notions of totality and historical necessity are, in fact, elements of flexible reasoning that imply a radically open contingency of history. The relation between contingency and necessity is dialectical, in the sense that there is a necessity for contingencies and, more radically, a contingency of necessities (i.e. things became necessary only in a contingent way). The relation between past and present is also dialectical, as the present is obviously influenced by the past, but the past is also reinterpreted and reconstructed by the present. As also observed by Bukharin (1929), necessity is really a chain of historical events that connect cause and effect. Rather than the trends of history being determined a priori by some overpowering force, historical necessities can only be explained retrospectively. In this way, Hegelian necessity should be seen not as a cause, but as the central property of the process of change (Mann 2008). The notions of historical necessity and dialectics are particularly relevant to understanding the Hegelian theorisation of global trends and the interventions of the state apparatus (Ioris 2014), as much as to the search for alternatives to capitalist frontiers, as examined next.

Perspectives of Development, Space and the Environment

The five main ontic features of the trajectory of the establishment of new development spaces discussed earlier constitute an attempt to identify, if only schematically, the basis for the insertion of new areas and reinsertion of old ones into the sphere of influence of Western, capitalist modernity. Socio-spatial frontiers continue to expand, including processes of production, extraction and politico-ideological containment, not only because of favourable economic opportunities but mainly due to the need to stabilise and invigorate core economic and political trends. At its frontiers capitalism can reassert hegemony with much lower costs and fewer restrictions. At the frontier, order, authority and convention are suspended; time and space acquire new meanings; and excesses are committed—Martins (2009: 09) describes the frontier expanding into the Amazon as the 'scenery of intolerance, ambition and death'—but the compelling symbolism of abundance, potential wealth and a bright future represents a powerful legitimisation tool. Such process has not been interrupted by market-based globalisation because frontier making continues to be predicated in rising tensions in central economic areas and their unstoppable demand for goods and services. This inexorable interdependency between centre and frontier is clearly a process with major ideological and political significance. The evolution of the frontier could hypothetically take any direction and lead to different social arrangements, but, in practice, there is a great deal of constraint due to the hegemony of capitalist relations. It is a self-fulfilling prophecy normally recounted only by the winners. The success of frontier making, from the perspective of Western economists and policymakers, depends on the consolidation of private property, the ability to exploit socionature, incentives for the circulation of capital and widespread commodification, and financialisation. Experiences that deviate from this model are considered anomalies and curiosities rather than genuine, viable frontiers.

The organisation of capitalist frontiers effects the maintenance of what Rancière describes as the police order, a symbolic constitution of the social that both fragments and incorporates, insists on homogenisation and pushes for consensus. The individual must passively comply, circulate in a space emptied of politics, as 'the space of circulation is nothing other than the space of circulation' [*l'espace de la circulation n'est que l'espace de la circulation*] (Rancière 1998: 242). In agribusiness frontiers such as the Amazon, crop monoculture has disturbing material and symbolic parallels with political and cultural monocultures that permeate highly hierarchical and exclusionary societies. Inequalities have become so widespread and structural that Gross (2019: 14) affirms that it is practically obsolete to mention it and now 'we now in the age of indecency' [*Nous sommes à l'âge de l'indécence*]. If the frontier spatial relations are unique in terms of potentiality and transformation being contained by the hegemony of capitalist prerogatives, these mechanisms of control and legitimisation are not absolute. On the contrary, because of the fluid boundaries and high mobility of newcomers, the frontier contains interstices in which political consciousness and reactions can emerge. Resistance to the capitalist logic of the frontier is not something that flares up by chance, but is located in the very constitution of the frontier that nurtures those interstices. The frontier has multiple time-space discontinuities as much as it shows rugosity (which are leftover characteristics from prior periods, according to Santos 1985), and it is in this context that opposition, almost always silent, but sometimes intense, erupts.

Resistance is therefore also related to the complex translocation of economic and social patterns from the centre to the frontier, considering that both migrants and established residents retain complex memories and experiences from pre-frontier times that allow them, depending on circumstances, the possibility to imagine some form of alternative. Durkheim (*On Suicide*) demonstrated that the most personal problems have sociological bases and spatial expressions; in our case, the frontier is a sociological process subject to significant individual forces. Resistance and reactions are informed and fuelled by past individual and collective experiences, such as previous conflicts or repression elsewhere. Depending on how people became involved and were relocated to the frontier (spontaneously or via government agencies), they will be more or less willing to question authority and risk whatever they have. The past is mobilised and influences the present because of cultural proclivities and subjective attitudes related to particular experiences, which ensure that particular kinds of historical consciousness become meaningful (Whitehead 2003). Frontier making is a cross-scale process that arises from national and international pressures, but the interplay between culture and history (beyond any false dichotomy between these) is resolved at the level of landscape change. The crafting of landscapes encapsulates historical and political consciousness that help to shape group identity. In the case of the Amazon, the landscape of large agribusiness areas is the consequence of multiple agencies that both converge and diverge, as the heavy machinery of wealthy landowners contrasts with pockets of family agriculture and those living along roads and in marginalised areas.

The socio-cultural construction of the development frontier as a space of opportunity and likely rewards for those who persevere helps to maintain social inequalities, under the assumption that social mobility merely depends on hard work (and a bit of luck). Although open rebellion and confrontation are rare, the main form of resistance is a silent process that happens through various forms of practices and positions (Scott 1985). Ultimately, the prospects of consciousness and resistance based on daily survival through the production of new landscapes at the frontier have major implications for a critical research agenda. Interrogating the frontier is a formidable challenge for critical, left-wing thinking (primarily concentrated on justice and equality), considering that frontier making is by definition a generation and perpetuation of inequalities. Likewise, critical scholars need to develop the ability to work through the political, apparently chaotic process of landscape change and silent resistance through the interstices of the established, taken-for-granted foundations of the frontier. All this requires a serious reflexive commitment and rejection of positivistic, politically void accounts of frontier making. As advocated by Lacoste (1973, in Quaini 1982), we need to reflect in order to measure, and not measure first to reflect later. That does not mean looking for facts to fit the conceptual model, but rather a firm investigative effort that amalgamates comprehensive empirical data and constant critical thinking. A development frontier may be a tentative, uncertain space where capitalist relations more easily prevail, but it can also be seen as a frontier of resistance and of overcoming the perverse socio-spatial consequences of capital's ascendancy. Before we could progress further, it is necessary to understand the structure of the book and how the different parts and chapters relate and complement each other.

Book Structure and Summary of the Chapters

This collective book is the result of extensive and continuous North-South collaboration between academics at different stages of their career and based in various universities and research centres. What has brought all together and in constant dialogue with non-academics is their common interest in questioning development trends and its multiple socio-ecological contradictions. The authors have worked together over the years in several research projects, publications and scientific events, which provided the common ground for different perspectives and disciplinary approaches to meet. Its first part includes mainly authors who took part in the workshop 'Environment and Development: Shared Twenty-First Century Sustainability Challenges', chaired by the editor, and that took place in the State University of Campinas (Unicamp, Brazil). This initiative aimed to provide new insights and methodological approaches to understanding the multiple challenges related to the interface between development and environmental sustainability. These are crucial challenges, of greater importance today, as the world is increasingly interconnected, with growing rates of production and trade but also with serious levels of inequality, environmental degradation and mounting socio-ecological risks (for instance, due to climate change, soil erosion, water scarcity, biodiversity loss and social inequality). One of the main policy contradictions explored during the workshop in Campinas was the limits of supposedly sustainable and integrated responses, as well as the reductionism of technical assessments and the narrow consideration of alternatives. These dilemmas were examined and the theoretical and lived experiences of development were critically addressed, paying attention to multiple scales, local realities and economic frontiers. This workshop was funded by the UK-Newton Fund, facilitated by the British Council and managed by Cardiff University and Unicamp's Institute of Geosciences.

The second part of the book is directly associated with the series of events and other activities of the Agrocultures International Network between 2018 and 2020 that investigated socio-spatial interactions and historico-geographical transformations of the agricultural frontiers in the Amazon Region (more information at www.agrocultures.org). The initiative created a privileged space for critical discussion and interaction between academics and wider civil society focused on local knowledge, habits, language and subjectivities in the context of a perverse process of market-centred regional development. The Agrocultures Network produced a number of technical publications and specific recommendations for development and policymaking. The initiative was helpful in proposing alternatives to the existing socio-economic trends and to the mounting failures of politico-economic decision-making. It has also enabled knowledge transfer, research networking and cross-agency learning for the actors involved at the local, national and international levels. The series of workshops were directly informed by the input of the publics who live, produce and consume in agricultural frontier areas, and the organisations that advocate for change on their behalf. A vital part of resolving academic debates, and making these relevant to real-world change, was to present practitioners' views at events where general members of the public do not normally contribute. The activities benefited from the visualisation and representation of the agricultural frontier through the use and recording of images, symbols, maps and artefacts, which informed the understanding of the spatial history of the agricultural advance in the Amazon. The Agrocultures Network workshop was funded by the Arts and Humanities Research Council (AHRC) and managed by the editor, based at Cardiff University, and his colleague, the historian Vitale Joanoni Neto, of the Federal University of Mato Grosso (UFMT). Financial and institutional support from the Newton Fund, AHRC and the respective universities are warmly acknowledged and much appreciated.

After this introductory pages, in Chap. 2, Rice discusses growing urban health issues around the world, whereby social and environmental living conditions in cities and towns are contributing to poor health outcomes. Empirical evidence and theoretical narratives point to these new urban centres, which have contributed to the proliferation of ill health. As a result, there is growing awareness and urgency of the need for 'healthy cities' that support well-being and improve health outcomes for inhabitants. The chapter specifically deals with the emergence of 'diseasogenic cities' and the implementation of Healthy City strategies in the Global South, ranging from embracing alternative ontologies and epistemologies, local knowledge and societal change, through to practical solutions for delivering context-specific sanitation and healthcare. Chapter 3, by Souza and collaborators, examines the trend of urban water problems and river basin degradation, what is based on a case study conducted in Porto Alegre, Brazil. Those tendencies are associated with challenging issues such as poverty, unplanned settlements, social exclusion, lack of basic sanitation infrastructure, poor management of water resources and inefficiencies in environmental governance systems. The responses should necessarily involve the participation of diverse stakeholders in a complex process of dialogue and learning that must deal with the multifaceted ecological and social issues affecting urban water systems and populations inhabiting their surroundings. In this context, social learning is considered a relevant approach to inform collective action towards the sustainability of these socio-ecological systems.

Chapter 4, by Koseoglu, provides a discussion of the importance of water management for development, and vice versa. An effective coordination of water should consider factors beyond economics such as political, social and ecological dynamics in complex interactions between stakeholders to prevent present and future conflicts related to water. In that context, systems analysis (SA) models have emerged in response to this need and allowed for a holistic approach to address a variety of water management issues from regional planning and river basin management to water quality, flooding and draught management, and sectorial water allocation. The chapter offers an introduction to the systems analysis modelling approach and its evolution, as decision support tool since the 1950s, in response to contemporary water challenges. Following a brief conceptualisation, an extensive literature review showcases the major methodological trajectory and practical contributions of the SA field of investigation, particularly in the Global South. In Chap. 5, Lima and colleagues examine the relevance of environmental risk and catastrophe for conservation and development. The analysis is based on large-scale disaster that occurred in the Doce River Basin, in the southeast of Brazil, that led to massive destruction and losses of human life. Following the tragedy, a series of political and management decisions were taken, impairing ecosystem recovery. The creation of a new governance structure that could potentially deal with river basin reclamation is assessed according to each stakeholder's decision-making power. Results show that there remain many biases towards funds management, diminishing effective institutional diversity in the decision-making process. Despite the efforts in creating new environmental governance systems to address post-disaster ruptures, the lack of equanimity in the governance structure, combined with frail law enforcement, is still a primary obstacle.

In Chap. 6, Leila Ferreira and colleagues discuss how climate change represents a multilevel challenge, simultaneously relating to the local and global scales. National governments are key actors in developing climate change mitigation and adaptation strategies. Brazil, China and Mozambique are emblematic cases where climate risks may deepen vulnerabilities and undermine development prospects. The analysis is focused on the institutional arrangements and political strategies that have emerged since then in those three countries and how climate issues were considered in their political processes. Empirical results suggest that both Brazil and China have focused much of their efforts on mitigating actions, that is, the reduction of greenhouse gas emissions. On the other hand, in Mozambique, political action on climate change is largely based on the need to respond to the risks and disasters that are intensified by these changes. In Chap. 7, Rafael Ioris and his collaborator revisit the momentous occasion when, on 12 November 2016, the Colombian president Juan Manuel Santos optimistically proclaimed the reaching of a final accord that would put an end to the longest military conflict in modern history. The leader of this violence-ridden nation sought to assuage the stalemate created by the defeat of the government's position in the referendum held on 2 October, in that same year. The rejection of what had become known as the Havana Accord, signed between the Colombian government and FARC (*Fuerzas Armadas Revolucionarias de Colombia*, that is, the Revolutionary Armed Forces of Colombia), that same month represented an unmistakable and serious warning that, though essential, formal peace negotiations were not all the country needed. Despite the feeling of achievement and the optimism conveyed by the agreement, its implementation has been impacted by the legacy of past development, growing environmental problems and the many disputes between sectors, regions and groups of interest.

In Chap. 8, Goodwin reflects on the politics of water scarcity and community water associations in rural Highland Ecuador. The author argues that the water that now flows from the tap is the fruit of a long-term process of collective organisation that has united rural families and communities in a struggle to overcome socially constructed inequalities. Through this ongoing struggle, community water associations have developed the autonomous capacity to deliver water services and manage water supplies. Not only has this given them the capacity to supply potable and irrigation water to rural communities but the space to create new relations and practices and transform relations with the state. Community water associations there have confronted multiple obstacles in the construction of autonomy, but the process performed a fundamental role in challenging water scarcity, reconfiguring water politics and reshaping capitalist development. The last chapter (Chap. 9) of the first part, by Cremona and her colleague examines a gap in the ecosocialist literature related to the subject of transformative agency, exactly those actors with the capacity and power to realise an ecosocialist vision. The authors explore several emblematic publications to show that five agents feature prominently: social movements, indigenous groups, the affluent capitalist class, left-wing political parties and the state. Their specific capabilities and power resources seem to determine the different roles they could assume in a transition from a capitalist present to an ecosocialist future.

Opening the second part of the book, which is dedicated to environment and development issues in the Amazon Region, Varese (Chap. 10) examines four parallel stories that reveal the relative success of indigenous communities in defending their autonomy and examines how their territory is linked to their ability to mobilise external solidarity around the use of national and international legal instruments as effective shields to protect their rights. Considering the past decades of ecological history, it seems that of all the victims of this violence exercised on the Amazon region by economic sectors in collusion with the national state, many indigenous peoples have been able to endure the aggression and secure some circumstantial victories. Through their political and cultural struggle, these groups have brought to the centre of national and international politics the ethical and environmental principles that should inform the relation of the rest of the country with the Amazon. In Chap. 11, Gonzalez shows that neoliberal extractivism remains an integral part of Latin America's economic agenda, particularly in the expanding new frontiers of development. That happens despite widespread societal opposition and adverse socio-environmental impacts. It, therefore, presents significant challenges not only for sustainable development but the relationship between state, society and business. The chapter is focused on Peru, one country which has seen significant neoliberal expansion, particularly in its rainforest Loreto Region. It outlines the challenges facing Peruvian society in holding oil companies accountable for pollution through the political ecology of voice (PEV) theoretical framework. Empirical findings, examined against that conceptual framework, indicate that Peruvian citizens, particularly indigenous people, face major difficulties in holding oil companies accountable for pollution in an environment where inaccessibility, injustice and inequality prevail.

In Chap. 12, Little makes use of ethnohistorical studies to review diverse processes of ethnogenesis in the Americas, involving indigenous peoples and maroon societies, in which the turbulent, often complementary, relationship between frontier processes of ethnocide and subsequent processes of ethnocide take place. The concept of 'cycles of ethnogenesis' is applied to the contemporary Brazilian Amazonia where, beginning in the 1980s, a new set of social and ethnic groups emerged in the wake of a previous developmentalist frontier. In this process, the global environmental movement, which emerged as a powerful force in its own right in the debates and decisions regarding the development of the Amazon region, played an important role in assisting emerging socio-environmental and ethnic groups in gaining visibility. The chapter also considers the political and theoretical implications of incorporating a dynamic concept of ethnogenesis into socio-cultural analysis. Chapter 13, by Joanoni Neto and Guimarães Neto, contains a reflection on the interventions of the Brazilian State, especially after the 1964 civil-military coup d'état, which transformed the Amazon into a strategic priority for conservative national development. The authors analyse the general power strategy of the dictatorship and demonstrate how the Amazon became the political and economic axis in this governmental experience, materialising in several public policies. The text includes a review of the governmental discourses and practices that underpinned power relations, especially by focusing on the assumptions behind laws, decrees, plans, institutional programmes and administrative reforms. The main conclusion is that the business logic that permeated the administrative practice during this period persisted after the end of the dictatorial regime.

Chapter 14, a collaboration between Johansen and Marcelo Ferreira, demonstrates how development policies by different spheres of the Brazilian government (federal, state and municipal) have contributed to the occurrence of malaria in parts of the Amazon. A detailed case study is provided, focused on Mâncio Lima, a municipality in the state of Acre. A state policy in the early 2000s encouraged fish farming by opening fish ponds. The initiative, however, eventually led to massive production of breeding sites for the mosquito that transmits malaria in the Brazilian Amazon. Malaria transmission exploded in the municipality and produced the largest number of cases per inhabitant in Brazil. The chapter finishes with an analysis of the case in relation to development and post-development theories. The authors argue that any intervention in the area must consider social participation, environmental preservation and defence against local health risks. Chapter 15, by Antonio Ioris, interrogates the construction of dams and other related water infrastructure on the Brazilian section of the Amazon. Water infrastructure has been used to propel and celebrate the modernisation of the regional economy and also to reveal how modernity is always partial, fraught with gaps and contradictions. The politics of water management is analysed from a cultural, political and economic perspective that aims to contribute to an understanding of the politics of distribution, recognition and resignification that play an important role in the evolution of water modernity in the region. The chapter proposes a periodisation of regional water management, divided in three moments of intense socio-ecological change and political and economic influences. That is intended to facilitate the understanding of the impact of hegemonic pro-development ideologies and discourses, as well as to comprehend the interface between economic and more-than-economic practices.

Expanding the immediately preceding case study on water infrastructure, Chap. 16, by Figueiredo and her colleague, show that over the past three decades, a series of global initiatives have promoted changes in water governance adopted by many countries and, in Brazil, resulted in the approval in 1997 of the Water Resources National Policy Act. Some tangible progress has been made ever since, but there is still a clear lack of effective implementation, particularly in the Amazon Hydrographic Region. This authors critically review the process of water governance and the hydrosocial territory of the Teles Pires River Basin (TPRB) in the state of Mato Grosso. Together with the analysis of documents and reports, semi-structured interviews were carried out with members of three river basin committees installed in the TPRB. Results reveal that the water governance model remains centralised, and water management model has been only partially implemented. In Chap. 17, Lagier identifies significant gaps in knowledge within the literature on rural social movements and local ecological transitions in the Amazon. The text is centred on the social movements affiliated with La Via Campesina, the world's largest international peasants' organisation, which has adopted 'agroecology' and 'food sovereignty' as common political principles. Empirical data show that contrary to what is often taken for granted by the more celebratory literature on rural social movements and agroecological transitions, there are complex and contradictory effects on both local ecological transitions and young activists' capacity to promote agroecology as an alternative to agribusiness. The author argues that studying agroecology's social and intergenerational sustainability is crucial to countering current devastating ecological and land-use patterns, which means paying attention to Amazonian youth and women's demands.

The book finishes with the *Oxford Letter for the Amazon*, edited by Antonio Ioris and approved during the fifth workshop of the Agrocultures network, which took place in Oxford in 2020. The event started on 31 January, the same day that the UK departed from the European Union and a period of growing uncertainties related to Coronavirus disease 2019 (Covid-19) outbreak. It was a moment when the Amazon was facing increasing levels of destruction, aggression and nothing short of a genocide. The workshop at Oxford University was planned three years earlier, when the organisers could have hardly imagined how relevant and opportune that date was going to be, academically and politically. One fundamental reason, for instance, is that the governments of the nine countries that share the Amazon have preferred to invest energy and resources on the perpetuation of troubles, instead of trying to understand and address the fundamental causes and the major consequences of impacts and growing inequalities. After years of intense debate and rich conversations stimulated by the Agrocultures initiative, something was very clear for those attending the workshop: it was necessary to say 'no!', 'enough', 'stop!' Things had to change significantly because the Amazon was much bigger, more complex, more beautiful and more important than the political and economic rulers seem to believe. In many ways, the Amazon has been the most crucial contemporary experiment of environment and development, where irrational, anti-people, anti-nature and anti-intelligence measures are being adopted, regardless of the will and the needs of most of its population. Old and perverse mechanisms of conquest and colonisation have been revitalised in the name of economic growth and national integration. That resulted in a trail of violence, stupidity and systemic abuses associated with new roads, dam construction, mining, timber extraction and, more than anything else, export-oriented agribusiness.

In more recent years, a new round of projects and market-friendly conservation initiatives under the hegemony of global financial capital is threatening even the small concessions secured through a difficult struggle since the 1980s, including the establishment of a few national parks and the recognition of some indigenous lands and peasant areas. Another thorny question was debated in Oxford, considering that the meeting took place in the oldest English-speaking university in the world, considered the best academic centre in the planet, a place that has produced many Nobel Prize winners and for more than a thousand years has offered a monumental contribution in fields such as philosophy, anthropology, biology, geography, history and so on. However, those walls and colleges are also very guilty of the continuous misrepresentation and mismanagement of the Amazon. The treasured Oxford University, as well as all the other universities represented there, encapsulate the failures of Western science and the incompetence of mainstream academics when they insist in maintaining an arrogant position of superiority and decide to ignore the political core of the problems we are going to discuss during this event. That is exactly why the moment of reflection was needed: to go beyond the conventional, money-driven type of research and teaching that prevails in the universities today. To identify the hidden connections between the past and present of economic frontier making, the links between attacks on local communities in the Amazon and global climatic catastrophe, the synergies between reactionary ideologies, political dishonesty, the reproduction of poverty and the neglect of the Amazon peoples.

The departure point of the workshop was the recognition of the inequalities produced by development-as-economic-growth and the associated asymmetry of gains and losses, as well as an appreciation of the creativity, inventiveness and courage of marginalised, subaltern groups. It was remembered that in October 2019, the Amazon Synod took place, under the leadership of Pope Francis, and ascertained that religion is a collective project radically and truthfully in favour of life, love, tolerance and openness to the other. During the Agrocultures event in 2020, the suffering and the wrong which has been inflicted on the poor, exploited Amazon worker, on women and children, peasant, riparian and indigenous communities; on their ecosystems; and livelihoods were, likewise, discussed, as well as appreciating their knowledge, skills, practices and ability to resist and to dream. As previously noticed by Latour (1996: 5), 'the presumed vanishing cultures are very much present, they are active, vibrant, inventive, proliferating in all directions, reinventing their past, subverting their own exoticism and removing the monotonous homogeneity of a global market and deterritorialized capitalism'. Despite their apparent material poverty, these groups have a lot to offer and are ready to take on any challenge. Just a few days before the workshop, the indigenous peoples of Brazil had met in Piaraçu (in the Capoto Jarina Indigenous Land), under the guidance of Chief Raoni Metuktire—who was himself the main speaker of the Oxford meeting, considered the moral Head of State of Brazil and a key intellectual and political reference in the world today—to reiterate their readiness to fight and to demonstrate that they are on the right size of history, decency and justice. These are some of the visible invisible affected by the advance of development and environmental degradation, the voices of the silent majority and the many oppressed minorities who represent what is colourful, valuable and alive in the Amazon, the living proof that the region had a long, miserable past, has a turbulent present, but, because of its people, it can also nourish a much better future.

References

Barbier, E. B. (2012). Scarcity, Frontiers and Development. *The Geographical Journal, 178*(2), 110–122.

Bowman, I. (1927). The Pioneer Fringe. *Foreign Affairs, 6*(1), 49–66.

Bukharin, N. (1929 [1917]). *Imperialism and World Economy*. London: Martin Lawrence.

Bunker, S. G. (1985). *Underdeveloping the Amazon: Extraction, Unequal Exchange and the Failure of the Modern State*. Chicago/London: University of Chicago Press.

Cronon, W. (1992). *Nature's Metropolis: Chicago and the Great West*. New York: Norton and Company.

Durkheim, É. (2002 [1897]). *On Suicide: A Study on Sociology*. Trans. J.A. Spaulding and G. Simpson. London: Routledge.

Evans, R. (2007). *A History of Queensland*. Cambridge: Cambridge University Press.

Gros, F. (2019). *Désobéir*. Paris: Flammarion.

Harvey, D. (2006 [1982]). *The Limits to Capital*. New Edition. London/New York: Verso.

Hemming, J. (1987). *Amazon Frontier: The Defeat of the Brazilian Indians*. London: Papermac.

Hogan, R. (1985). The Frontier as Social Control. *Theory and Society, 14*(1), 35–51.

Ioris, A. A. R. (2014). *The Political Ecology of the State: The Basis and the Evolution of Environmental Statehood*. London: Routledge.

Ioris, A. A. R. (2015). The Production of Poverty and the Poverty of Production in the Amazon: Reflections from those at the Sharp End of Development. *Capitalism Nature Socialism, 26*(4), 176–192.

Ioris, A. A. R. (2017). *Agribusiness and the Neoliberal Food System in Brazil: Frontiers and Fissures of Agro-neoliberalism*. London: Routledge.

Ioris, A. A. R. (2018a). Amazon's Dead Ends: Frontier-Making the Centre. *Political Geography, 65*, 98–106.

Ioris, A. A. R. (2018b). Place-Making at the Frontier of Brazilian Agribusiness. *GeoJournal, 83*, 61–72.

Ioris, A. A. R. (2019). Peasant Farming in the Southern Tracts of the Amazon: The Reluctant Alterity of Agribusiness. *Perspectives on Global Development and Technology, 18*, 375–400.
Ioris, A. A. R. (2020). *Frontier Making in the Amazon: Economic, Political and Socioecological Conversion* (Key Challenges in Geography: EUROGEO Book Series). Cham: Springer.
Latour, B. (1996). Not the Question. *Anthropology Newsletter, 37*(3), 4–5.
Lund, C., & Rachman, N. F. (2018). Indirect Recognition. Frontiers and Territorialization around Mount Halimun-Salak National Park, Indonesia. *World Development, 101*, 417–428.
Magdoff, F., & Williams, C. (2017). *Creating an Ecological Society: Toward a Revolutionary Transformation*. New York: Monthly Review Press.
Mann, G. (2008). A Negative Geography of Necessity. *Antipode, 40*(5), 921–934.
Martins, J. S. (2009). *Fronteira: A Degradação do Outro nos Confins do Humano*. São Paulo: Contexto.
Marx, K. (1973). *Grundrisse*. Trans. M. Nicolaus. London: Penguin.
Marx, K. (1976). *Capital, Volume I*. Trans. B. Fowkes. London: Penguin.
McDowell, S., & Crooke, E. (2019). Creating Liminal Spaces of Collective Possibility in Divided Societies: Building and Burning the Temple. *Cultural Geographies, 26*(3), 323–339.
Mellor, M. (2000). Feminism and Environmental Ethics: A Materialist Perspective. *Ethics and the Environment, 5*(1), 107–123.
Namier, L. B. (1942). *Conflicts*. New York: Books for Libraries Press.
Quaini, M. (1982 [1974]). *Geography and Marxism*. Trans. A. Braley. Oxford: Basil Blackwell.
Rancière, J. (1998). *Aux Bords du Politique*. Paris: Gallimard.
Rasmussen, M. B., & Lund, C. (2018). Reconfiguring Frontier Spaces: The Territorialization of Resource Control. *World Development, 101*, 388–399.
Rist, G. (2008). *The History of Development*. London: Zed Books.
Rivière d'Arc, H., & Apestéguy, C. (1978). Les Nouvelles Franges Pionnières en Amazonie Brésilienne. *Études Rurales, 69*, 81–100.
Rogers, T. J., & Bain, S. (2016). Genocide and Frontier Violence in Australia. *Journal of Genocide Research, 18*(1), 83–100.
Sachs, W. (Ed.). (2010). *The Development Dictionary. A Guide to Knowledge as Power* (2nd ed.). London: Zed Books.
Santos, M. (1985). *Espaço e Método*. São Paulo: Nobel.
Scott, J. C. (1985). *Weapons of the Weak: Everyday Forms of Peasant Resistance*. New Heaven/London: Yale University Press.

Théry, H. (Ed.). (1997). *Environnement et Développement en Amazonie Brésilienne*. Paris: Belin.
Turner, F. J. (1920). *The Frontier in American History*. New York: H. Holt and Company.
Wakefield, E. G. (1967 [1834]). *England and America*. New York: Augustus M. Kelley.
Wallerstein, I. (1974). *The Modern World-System I*. New York: Academic Press.
Watts, M. J. (2018). Frontiers: Authority, Precarity, and Insurgency at the Edge of the State. *World Development, 101*, 477–488.
Whitehead, N. (Ed.). (2003). *Histories and Historicities in the Amazon*. Lincoln/London: University of Nebraska Pres.
Woollacott, A. (2015). *Settler Society in the Australian Colonies*. Oxford: Oxford University Press.
Žižek, S. (2011). Hegel's Century. In S. Žižek, C. Crockett, & C. Davis (Eds.), *Hegel and the Infinite* (pp. ix–xi). New York: Columbia University Press.

Part II

Understanding the Environment and Development Nexus

2

Healthy Cities, Diseasogenic Cities and the Global South

Louis Rice

Introduction

This chapter explores the contested realm of 'healthy cities' in the Global South. The notion of a 'healthy city' concept emerged in the Global North and much of the subsequent debate, best practice exemplars, theoretical and critical framework has mostly remained a Global North phenomenon. Furthermore, attempts to contextualise and situate the Health City concept in the Global South have been critiqued according to three principal agendas: firstly, due to the perceived risk of imposing a quasi-colonial ontological and epistemological construct (i.e. the 'Healthy City') onto the Global South. Secondly, the application of healthy city analyses tends to discursively problematise the Global South. Broadly, these critiques often tend towards narratives of 'diseasogenic cities', whereby urban areas are the locus of polluted environments and drivers

L. Rice (✉)
World Health Organisation Collaborating Centre for Healthy Urban Environments, University of the West of England, Bristol, UK
e-mail: Louis.Rice@uwe.ac.uk

A. A. R. Ioris (ed.), *Environment and Development*,
https://doi.org/10.1007/978-3-030-55416-3_2

of ill health and disease. Thirdly, and more prosaically, the overwhelming data available on healthy cities address the well-being issues of the Global North, whereas the health requirements for Global South are frequently different and under-researched. These dominant agendas lead to theoretical and conceptual difficulties but also to practical problems for policy-makers, attempting to address the very real need to improve health and well-being in Global South cities.

As the majority of all future urbanisation is predicted to occur in the Global South, more research is urgently required to address the health inequalities in these urban contexts. There is mounting evidence of myriad health issues correlated to the design, construction, management and functioning of the built environment. The built environment is a significant determinant of human health; however, rather than cities acting to support health and well-being, the configuration of most cities is considered 'diseasogenic' and making health worse.

The chapter begins with a literature review that contributes to identifying the key dynamics in healthy city development in the Global North and reveals the key theories, mechanisms and practices involved in developing urbanism in the Global South Healthy City. The review requires an interdisciplinary lens through which to examine the complex interweaving of public health, development processes and urban growth in the Global South. The final section explores some narratives of nascent Southern healthy city strategies and practices. Primary fieldwork and secondary data sources from empirical case studies are critically examined to identify emerging praxis as well as tentatively reveal potential future scenarios for constructing, where appropriate and possible, Healthy Cities of the Global South.

This introductory section begins by contextualising the emergence of Healthy Cities in relation to *developmentalism* and the Global South. President Truman declared a 'fair deal' for the world in his 1949 inaugural speech, a key component of which was the ambition to solve the challenges of the 'underdeveloped world' (Escobar 1995; Sachs 1997). At the heart of subsequent transformation of the Global South lies Truman's theoretical framework for understanding development as the key to solve Global South problems. 'Developmentalism' became the byword for interventionism by the Global North, particularly the Euro-American nations, into the Global South, transforming many aspects including financial systems, agricultural practices, urban form, transport and health sectors. Arguably 'Healthy Cities' have become the latest iteration of developmentalism. Though terminology and processes have altered over time, developmentalism is taken as the general trend for the mechanisms of economic, social and spatial change to emerge in the Global North and be transferred to, or copied by, the Global South. One of the defining traits of developmentalism important to this debate is how the Global South attempts to emulate Western transformation as the 'best' model for development, and particularly that they follow through the developmental stages that the West went through (Domingues 2012; Grosfoguel 2000; Wallerstein 2005).

With the world's population urbanising at an unprecedented rate, the role of Southern cities and urban areas is increasingly important. Although they only take up 2 per cent of the world's surface, 'cities consume 78 per cent of the world's energy and produce more than 60 per cent of greenhouse gas emissions' (United Nations (UN) 2019). Half of the world's population live in cities and urban areas; by 2050, over five billion will live in low-middle-income countries and one billion in high-income countries (Dupont et al. 2015; Galea et al. 2019; UN Department of Economic and Social Affairs 2018), with 70 per cent of all urban inhabitants located in Asia and Africa (Cohen 2006; Roy 2005; Vlahov et al. 2005). It is predicted that the Global South will be the location of almost all new cities and the majority of all urban growth globally; the contribution of these Southern cities to global health will be significant. Truman envisaged a process of development lead by governments (especially the American government) and intergovernmental forces (e.g. the UN), which frame the overall context for development (Schuurman 2000).

However, in recent decades, non-governmental forces have played a more active role in development; there are complex alliances and synergies between governments and capitalist agencies, and in some instances, these global forces occur outside of the influence of governments (sometimes despite the influence) or of governmental intervention. Recent urban development theory relates towards 'market-oriented or self-determined solutions' to better contextualise the relationships between spatial factors, social issues and economic development mechanisms

(McFarlane 2008: 343). The tropes of globalisation and neoliberal economic forces are often used to position Southern urbanism (Brenner and Schmid 2015; Gugler 2004; Kanai 2014). These perspectives draw out the connections across and between cities in the processes of globalisation and how hegemonic development processes are played out across the globe. However, whilst the global flows of finance are transforming countries and regions, with the effect of homogenising urban conditions, there are also important less-visible local economic processes (particularly, but not limited to the informal economy) that remain more heterogeneous and under-reported. Inasmuch as the dominant global financial power emanates from the Global North, the Global South cities are often framed as copies (often suboptimal replicas) of Global North cities (Ferguson 2006).

A 'Health Turn' in Developmentalism

The Healthy City is sometimes portrayed as a Global North construct that is being played out across the Global South as a more recent form of developmentalism; "construed as a positive development process in terms of its implications for health and well-being: urbanisation=development =health" (Stephens 1996: 11). However, this is an oversimplification as the counterarguments and alternative trends emerging within healthy city discourse place greater importance on the contributions from emerging Southern urbanisms and healthy city ontologies of the Global South. Brenner and Schmid (2015: 151) ask of the Global South, through *"what categories, methods and cartographies should urban life be understood?"*. Whilst there has been a proliferation of much needed alternative theoretical and theorised perspectives, in terms of practical developmental forces, *health* has become one of the drivers for urban development globally. This is partly due to the increased importance of health placed on it by the United Nations (UN). The UN's Sustainable Development Goals have replaced the Millennium Development Goals with a greater focus on health—"*health has a central place as a major contributor to and beneficiary of sustainable development policies*" (World Health Organisation (WHO) 2015). In a rapidly urbanising world, it means that healthy city strategies are a big driver of development in the twenty-first century for cities (and indirectly for rural areas).

A large proportion of government's financial expenditure is targeted towards health (directly or indirectly) (Reeves et al. 2015). Governmental expenditure on health is often connected to Euro-American conceptions of medical practices and the healthcare industry, which de facto operate to drive a particular form of development in the Global South. Western health systems, from pills, drugs, pharmacies, hospitals, clinics, educational systems (to train doctors and nurses) through to sewage infrastructure, water supplies, air quality legislation, traffic safety regulations, and so on, are superimposed onto the Global South. The implications for a nation are profound, for example, in order to provide clean sanitation, a Western model of urban development is invariably required to achieve this—with centralised and municipal (thereby requiring a municipality) sanitation systems. This extends not just to urban areas (even though it is more often prevalent in discourse) but to food production and agribusiness. The contemporary use of pesticides, fertilisers, mechanisation, standardisation and regulation are required to ensure 'healthy' food standards for a nation. In the century of the urban, the emergence of 'healthy cities' as a developmental force is currently under-represented and under-theorised. There is a 'health turn' in development, emerging in the Global South, to which this chapter hopes to contribute to these debates.

Healthy City Qua Colonialism

This section explores the first theme through which healthy city strategies have been critiqued in relation to development in the Global South. Healthy city discourse and practice is critiqued as a form of colonial or ongoing postcolonial control of the Global South (by the Global North). In these narratives there exists a *"stereotype of the global south city as a 'pathological' space in need of salvation at the hands of Western experts"* (Kanna 2012: 360). Cities of the Global South become locations in need of help (invariably in the form of development) from the Global North, with 'health' being the latest battlefront in this pathologising trend. Through this lens, the move towards Healthy Cities is another form of developmentalism (wrapped up in a new guise). Escobar (1995: 27) posits how the "*discourses and practices of agricultural experts, health professionals, urban planners, and educators were deployed in the colonial context*" contribute to the conflation of "*colonial and developmentalist regimes*" in the Global South. Put simply, 'Healthy City' is *of* the Global North and alien to the Global South. The theoretical and epistemological premise for the Healthy City programme is firmly embedded in Western medicinal traditions. The WHO definition of a Healthy City is the most widely adopted: "*one that is continually creating and improving those physical and social environments and expanding those community resources which enable people to mutually support each other in performing all the functions of life and developing to their maximum potential.*" (WHO 1995). The WHO 'Healthy City' movement emerged as a Global North phenomenon "*starting in 1986, the first Healthy Cities programmes were launched in developed countries (i.e. Canada, USA, Australia, and many European nations)*" (WHO 2019). It was almost a decade later before Healthy Cities arrived in the Global South; "*around 1994, developing countries used the resources and implementation strategies of initial successes to begin their own programmes*" (ibid). It is clear that even when Healthy Cities arrived in the Global South, it was premised on the conceptions and practices developed in the Global North.

The tenets and canons of a 'Healthy City' are predicated on evidence, scientific methods, institutions and practices derived from the Global North and particularly medical research, epidemiology, public health professionals, pharmaceutical industry, administrative measures, architectural forms and scientific statements (Memmi 1992). This heterogeneous ensemble is the mechanism through which postcolonial control is produced and reproduced in the Global South. Although the end of colonial times in a political sense may have occurred, there is still an ongoing postcolonial presence through the use, adoption or enforcement of theories, standards and practices from the Global North (Latouche 1996). Postcolonial control continues to be exerted over the Global South through the implementation of Healthy Cities. Postcolonial control is experienced particularly in the elimination of the local endogenous communities' beliefs and knowledge through the imported Global North health epistemology (Quijano 2005). Postcolonial knowledge systems are invariably adopted by the Global South elite, sometimes perceived as more advanced or superior to local systems of medical knowledge (Alcadipani and Caldas 2012). This leads to the subalternisation and epistemicide of local knowledge systems within the Global South (Quijano 2008). Epistemicide concerns the destruction of systems of thinking, which de Sousa describes as the *murder of knowledge* (de Sousa Santos 2014: 92). In this context, epistemicide involves the extirpation and/or elimination of local and indigenous health knowledge systems whereby the dominance of Western medicine comes at the cost of alternative health beliefs (Mills 2014).

Across the Global South (and beyond), there are many different models and belief systems related to medicine, particularly notable in the context of this debate are parallel medical practices, including shamanism, magic, spirituality, paganism, necromancy, voodoo, occultism and witch doctors. Most of these exist outside of formal medical systems but are, nonetheless, still practised in many communities, often in parallel to formal healthcare (or in the absence of healthcare) (Bodeker et al. 2005). These alternative medicines compose a pluriverse of multiple and conflicting models of systems of thinking (and beliefs) about health (Escobar 2018; Ferguson 2007). Alternative Global South medical traditions operate parallel to Western medicine. Often as a form of 'private' healthcare, where individuals seek guidance for a health issue when Western medicine is too expensive, inaccessible or is not effective—or simply part of local cultural traditions (Micozzi 2018). Whilst most of these alternative medicinal practices are not currently recognised by the Western healthcare industry, they can provide a range of benefits. For local communities, they help maintain indigenous traditions and community structures, including the veneration of such knowledge held by older generations. Furthermore, Western medicine is looking into these alternative systems of medicine for inspiration and practice. Local potions and knowledge of plants and herbs are being harnessed to find ingredients for global pharmaceutical companies (Partridge 2006). These alternative beliefs provide pathways for mental health treatments, particularly looking for healing approaches that are 'practised' (such as meditation) rather than administered or swallowed.

That these alternative health epistemologies and ontologies continue to exist provides a critique of the failures of Western medicine (particularly for mental and social health). Western medicine has generally succeeded for many communicable diseases (particularly in terms of physical health); it has enabled life expectancy to improve hugely over the past few hundred years (Easterlin 1995). However, Western medical epistemology fails to cope with many non-communicable diseases, particularly those related to sedentary lives or mental health. For example, obesity is a risk factor for the most common health issues, diabetes, cancer and strokes; yet, there does not appear to be any Western solution on the horizon for a 'pill' to combat obesity. Mental health issues are rife in the cities and communities of the Global North, a situation that is getting worse (Thornicroft and Maingay 2002). Attempts by Western models of medicine, particularly the pharmaceutical industry, to intervene in this sector have arguably made the situation worse. The current scandal in the US over opioid addiction is a result of the growth in mental health issues combined with ineffective medical prescriptions (Meyer et al. 2014). The mental health crisis of the Global North provides succour to the view that the Global South should not look towards the West for inspiration for tackling mental and social health and resist 'medical imperialism' (Fernando and Moodley 2018). Indeed, there is a reverse process underway, with Western medicine looking to alternative medical cultures and practices for solutions (Partridge 2005). Alternative medical approaches are being derived from non-Western contexts, such as yoga from India and Buddhist meditation techniques from China, Japan and Thailand. A health pluriverse is emerging in which the multifarious ontologies of health and well-being are explored and exploited as part of a broadening of strategies to improve health, whilst simultaneously critiquing the flaws and failures of the Western model of medicine.

The design and development of cities to improve health as a form of spatial medicine is nothing new; for millennia, cities have implemented different strategies for improving public health (Frost 1997; Bankoff 2007; Douglas 2003). Over 2000 years ago, Vitruvious (1999) stated "*The architect should also have a knowledge of the study of medicine on account of the questions of climates, air, the healthiness and unhealthiness of sites, and the use of different waters. For without these considerations, the healthiness of a dwelling cannot be assured.*" Designers have controlled and manipulated the spaces of buildings and cities as a form of spatial medicine. One of the most common strategies for improving health spatially is through segregation; the removal of dirty, dangerous or unclean activities away from certain parts of a city. What is defined as dirty, unclean or dangerous varies throughout history, geography and culture, but there are often similarities. The spread of fires in cities is a common problem, destroying much of ancient London, Rome, Constantinople and Tokyo. Some cities prevented fire by moving certain industries (e.g. those that depend on the use of fires, furnaces and kilns) to the edge of cities. In Edo, the risk of fires leads to the self-segregation by the rich who built new palaces outside the city boundaries (Ichikawa 1994). Slaughterhouses and abattoirs are often located outside city walls (Fitzgerald 2010; Otter 2005); ill or sick populations such as lepers are separated into colonies (Poestges 2011). The dead are located in cemeteries, columbarium and necropolis (cities for the dead) to spatially segregate clean from dirty (Arnold 2008; Thomas 2015). Alongside such spatial segregation has been the widespread hygienisation of cities. Since the birth of the Industrial Revolution and the discovery of germs, city planning and urban development have been transformed with a view to control diseases (i.e. those spread by germs) (Buer 2013; Smith 2008; Szreter 2004). This move towards a hygienic city occurred through a plethora of changes to the fabric of city and society. McFarlane (2008: 10) describes how "*the aim of producing a civilised, clean, respectable, productive and healthy city necessarily drove the urban fabric and the city's moral condition into relation with one another*". This occurred throughout the Euro-American cities initially, particularly affecting the underclasses and urban poor, then exported to the colonies (Douglas 2003). In the colonies, segregation was taken a step further and extended to the physical exclusion of populations, according to ethnicity or perceived race (who were deemed dirty or unclean) (Christopher 1983; Njoh 2008). Anderson (1995) calls this process '*excremental colonialism*', and these practices of colonialism, particularly the spatial segregation of minority people, have a legacy on health and city design. These precedents highlight that health issues such as disease, inequalities and spatial injustice are prevalent throughout history, and that cities have been used as a control mechanism by those in power. The next subsection explores a positive example of how the Global South has embraced the pluriverse to pursue a new road map for more sustainable development.

Latin America's Healthy City Pluriverse

Latin America's healthy city pluriverse is leading the way in pursuing locally situated models for socio-economic reform to improve health and well-being in urban and rural settlements. Latin American nations, particularly Bolivia and Ecuador, are establishing innovative approaches to reorienting health ontologies, in order to decentre development. The *buen vivir* movement has emerged as a viable alternative to Western health and development models. *Buen vivir* emerged in Latin America; variations include *sumak kawsay* in Quechua, *suma kamaña* in Aymará, *ñande reko* in Guarani, *shiir waras* in Ashuar and *küme mongen* in Mapuche (Thomson 2011: 449) can be translated as 'well living' or 'well-being'. As conceived in Latin America, it is one of the first examples of the recognition of a nation embracing the pluriverse of coexisting multiple world views (with a similar process underway in Ecuador) (Escobar 1995). Alberto Acosta (2011: 189), President of Ecuador's Constituent Assembly, claims "*only by imagining other worlds will this one be changed*". The movement enshrines alternative ontological conceptions of well-being and healthy lifestyles into National Legislation.

The term 'well-being' is widely used in Euro-American nations and is rooted in a Western medicine ontology. The term *buen vivir* is embedded in a radically different ontological and cosmological understanding of reality (Caria and Domínguez 2016). Although there are differences across the myriad iterations of this concept, in many of the interpretations emerging in Latin America, the relationship between humans and the natural world is conceived quite differently to that of the West and, as such, place humans into a different relationship to their environments (Thomson 2011). Firstly, nature is paramount in these interpretations. Human health is situated within the natural world, as a constituent part of nature, rather than external to it or separated from it. Perhaps the most recognisable impact of these reforms is the changing attitudes towards the natural world. In rural areas, it is eschewing large agribusinesses, which, through their practices, pollute the environment, contaminate the soil and rivers, and damage the local communities. In urban areas, it is moving towards a greening of cities partly for mental health benefits but also using nature as green infrastructure to deal with natural disasters and the impacts of climate change.

Secondly, *buen vivir* often involves a greater emphasis on sociality, particularly being part of a community or cultural group (Gudynas 2011). Indigenous cultural traditions play much more of a role in determining legislative and policy reforms, which, in turn, require an alternative approach to development. *Buen vivir* provides a supporting framework for the sanctioning, encouragement and advocacy for existing indigenous health practices and beliefs. This has the effect of maintaining local community networks and cultural health practices. Thirdly, as a corollary of the previous two issues, *buen vivir* is often contextualised with, or aligned to anti-development movements, which support anti-growth, degrowth, slow development and anti-capitalism (Villalba 2013). *Buen vivir* revisits the question of development(alism) from a new and critical perspective. Rather than Healthy Cities becoming just another mode of developmentalism imposed on the Global South, *buen vivir* provides a road map for locally specific well-being-centred alternatives to development. Whilst these reforms are relatively new, they provide an important precedent for other nations to develop their own healthy city transitions based on indigenous cultures, practices and beliefs pertinent to each context.

Diseasogenic Cities

This section examines the second theme through which the development of the Global South and Healthy Cities has been critiqued, that of 'diseased' space. The reality is that most cities (in the Global North and Global South) are not very healthy places at all. Indeed, there is much discourse and evidence supporting the view that cities are *bad* for one's health (Marmot 2015), which reaffirms the need to embrace a pluriverse of alternative worlds for nurturing health. The term 'diseasogenic city' refers to contexts where poor health outcomes are a consequence of the built environment (Rice 2019a). Only a tiny proportion of all cities might be categorised as '*healthy*'—the greater majority would be categorised as *diseasogenic*. A diseasogenic city contributes to the greater prevalence of illness and disease through the creation and maintenance of social and physical environments. Many cities of the Global North and Global South can be described as diseasogenic; however, those descriptions vary according to context whether they be Global North or Global South. As an oversimplification of such narratives, the Global North cities are spaces for non-communicable diseases, whereas Global South cities are cradles for communicable diseases.

The WHO (2019) Healthy City definition*: "continually creating and improving…developing to their maximum potential*" does not set specific or absolute limits for environmental health. It is not the rate of, for example, car accidents or incidence of cholera that determines healthiness; there is no scale or threshold upon which to measure city-healthiness. Healthy City is thus conceived with a developmental relationality; it is measured in terms of how a city develops over time with respect to specific health indicators. This is perhaps counter-intuitive because there are many agreed-upon levels of environmental pollutants, which can be categorised as healthy or unhealthy. It is evident that there are positive relationships between spatial characteristics and good health: safe roads, good sanitation, drinkable water and so on. An illustrative example is the level of air pollution in urban areas; when hazardous particulate levels remain at a low level, human health is not adversely affected, but at higher thresholds, air pollution will damage health. There are alternative systems for assessing the liveability of cities or quality of urban life where such environmental standards are prescribed and must be achieved (Kaklauskas et al. 2018). However, whilst certain physical indicators might be measurable for city health, there are many other criteria, particularly for mental and cultural factors that are not so easy to limit or define. It is much harder to prescribe 'healthy operating limits' for the social determinants of health, such as the absence (or presence) of a supportive community, mental health stressors or the social fabric of urban contexts. The next section explores these narratives before critiquing and discussing the implications and repercussions of the development of diseasogenic cities.

The meaning of the term *diseasogenic city* varies considerably across Global North and Global South contexts. In the Global South, such cities are often characterised in a negative light, with portrayals of places as "*the most extreme instances of economic injustice, ecological unsustainability, and spatial apartheid ever confronted by humanity*" (Dawson and Hayes Edwards 2004: 6). These descriptions are widespread in literature and portray cities of the Global South as the unhealthiest environments on the planet in which to live (Davis 2007). Global South cities are locations of failing (or non-existent) infrastructure as the source of communicable diseases, overcrowding, poor environmental health, spatial injustice and conditions where urban violence can flourish (Gilbert 2007; Owusu et al. 2008) At the most extreme accounts, Global South cities are described as "*Dickensian conditions or worse, ravaged by emergent diseases and subject to a menu of megadisasters following in the wake of global warming*" (Davis 2004: 13). Whilst there is certainly evidence to support some of these claims, arguably, there is too much hyperbole in these narratives.

The cities of the Global South vary enormously in relation to all these criteria, and it is perhaps meaningless to attempt to categorise them all simultaneously. It is possible to find Global South nations suffering from extreme megadisasters (e.g. the Philippines), but, equally, others do not suffer much from natural disasters or climate change (e.g. South Africa); megadisasters are also experienced by cities in the Global North (California regularly suffers horrific urban fires). Similarly, some urban locations experience high levels of violence and crime (e.g. South Africa), but others are relatively safe (e.g. India). What is perhaps true is the prevalence of communicable diseases, such as cholera or malaria, which are higher in the cities of the Global South. The descriptions of overcrowding are often overstated; indeed, the Global North often suffers from the reverse with extreme loneliness partly because it has such low-density settlements that community relationships cannot be sustained. There is praise for the merits of Global South cities: "*the west has much to learn from societies and places which, while sometimes poorer in material terms are infinitely richer in the ways in which they live and organise themselves as communities*" (The Prince of Wales 2009). In particular, mental and social health is often greater in the Global South than in the Global North.

This subsection looks, in turn, at the diseasogenic cities of the Global North. In the Global North, most of the communicable diseases have been successfully controlled, and now non-communicable diseases are the biggest health challenges (Grant et al. 2017; Nieuwenhuijsen and Khreis 2018). Diseasogenic cities of the Global North are characterised by their relationship to non-communicable diseases. The redesign of cities was used as a mechanism to combat communicable diseases, which has led to many of the determinants of non-communicable diseases. The eradication of communicable diseases through the design of cities has literally built in the conditions for the proliferation of non-communicable diseases, which are now part of the deoxyribonucleic acid (DNA) of modern cities. Diseasogenic cities of the Global North, particularly American models of development, have some negative relationships between human health and urban environments. At their worst, diseasogenic cities lock people into unhealthy lifestyles and harmful behaviours (Barton and Grant 2006; Rice 2019b). Bryson (2019) describes the *design of modern cities as a form of suicide in slow motion* because of how cities modify aspects of human behaviour, which kill the inhabitants over a long time period. An illustration of how this has emerged is when, in order to combat urban air pollution, high-density slums were demolished and replaced with low-density settlements (Bruegman 2008). The lower-density spaces contribute to more sedentary lifestyles, which is associated with greater prevalence of obesity and related illnesses. They also contribute to far lower levels of social network and community spirit; this is particularly exacerbated for older populations that are no longer able to access cars for mobility. Low-density car-based settlements can lock people into sedentary lifestyles (Chaput et al. 2011; Marteau et al. 2011). Long before a person 'decides' to get behind the wheel, driving a car is a lifestyle behaviour that has been determined by government transport ministries, highway engineers, advertisers, car-manufacturer lobbyists and others (Sunstein and Thaler 2008). Furthermore, we are designing cars that kill millions of people annually, to developing materials that are hazardous, to polluting the air and natural environments with our waste. The damage done by these professions to an individual's well-being, the health of society and planetary health is considered a form of "*murder on a mass-production basis*" (Papanek 2009: iv).

Just as pollution, climate change, material waste and environmental degradation are deemed as 'externalities' of market-led developmental processes; human ill health is yet another externality of these same processes. For the Global North, non-communicable diseases are outcomes of the externalities of the design of our built environments. Diseasogenic environments encompass urban settlements, homes, factories, cars, roads, advertising as well as rural areas; the industrialisation of agriculture, and widespread use of pesticides and fertilisers; and mechanisation of almost all aspects of the economy. Similarly, the shift towards desk-based employment effectively traps large sectors of the population into low levels of activity for long periods. As the majority of ill health in the Global North is related to unhealthy lifestyles, the design and layout of cities play a complex but key role in enforcing these unhealthy behaviours (Rice 2018). Contemporary developmental forces generate the proliferation of junk-food outlets, mental health stressors and other contributors to urban ill health. Diseasogenic cities of the Global North comprise a complex interweaving of urban form, technological systems, social structures, legislative apparatus and economic forces that impinge on human behaviour and nudge it towards ill health (Callon 1991; Latour 2005).

While the previous sections explored how cities have become diseasogenic, differing in the Global North and Global South, this concluding part provides a brief summary and critique thereof. The diseasogenic cities of the Global South are frequently overstated and expounded in hyperbolic language. The depictions are an oversimplification of the realities in each context. Whilst there are numerous problems, these are not experienced in all Southern settings and nor are these challenges exclusive to the Global South. For example, in India, there are now more overweight people than underweight, thereby both diseases are evident simultaneously (i.e. communicable and non-communicable diseases); these are urban health problems characteristic of the Global North and South. The diseasogenic cities of the Global North suffer from less poetic and emotive narratives, but there are, nonetheless, many complex health problems built into the urban fabric. The manipulation of urban form to resolve health problems has merely shifted the types of diseases that impact on society. These lifestyle diseases are slower acting and less visible but, nonetheless, harmful to urban inhabitants. There are close associations between developmentalism, capitalism, industrialisation, climate change and the drivers of these urban environments, which are unsustainable in the long term. What is clear is the cities of the Global North do not provide a workable model for cities of the Global South to emulate. There is not (yet) a clear road map for how cities of the Global North or Global South to better support all aspects of human health; further research, innovation, experimentation and practice are required. The next section explores the issues raised previously, namely how developmentalism has led to inequalities in urban health outcomes, particularly how cities have become places of disease rather than wellness. Some of the problems are implicated in the legacy of colonial and postcolonial processes and the epistemicide of local health knowledge. Furthermore, contemporary developmental forces are continuing to exacerbate the health problems in the Global South. Whilst economic development and Western medical systems have produced some local socio-economic benefits and improvements to health (particularly communicable diseases), these accrued societal benefits mask severe structural health inequalities in the Global South.

Gaps in Healthy City Knowledge

This section explores the third theme through which Healthy Cities are critiqued—through the absence of knowledge on health issues for the Global South. There is an absence of data and research on urban health issues for the Global South. The overwhelming data available on healthy cities addresses the well-being issues of the Global North, whereas the health requirements for Global South are frequently different and under-researched. Increasingly, there are better data on health at a global level from agencies such as the UN, WHO, The Gates Foundation and others, but data on Healthy Cities of the South need further improvement. As previously mentioned, cities house large proportions of the world's population, and these urban conditions have specific environmental and social characteristics that impact on health. The urban communities of Global South cities are at higher risk of many illnesses, particularly a series of communicable diseases (WHO 2017). A challenge for filling the gaps in knowledge relates to the (flawed) model of developmentalism whereby

Global South cities proceed through the stages of development that the Global North have undertaken. This model of developmentalism suggests that there is no need for further research into solving the problems of urban conditions for the Global South because they have already been researched and solved in the Global North. However, this has been proven to be a flawed logic on many fronts as evidenced in the emergence of diseasogenic cities.

Some of the communicable diseases of the Global South are a legacy of developmentalism and colonial practices. "*Western powers were not merely actors in an already existing disease scenario; they also generated diseases in their colonies and territories*" (Farley 2003: 298). This occurred through changes to agribusiness, particularly the industrial-scale implementation of monoculture cropping leading to an imbalance in local ecological systems, the profusion of invasive parasites and a rise in communicable diseases (mostly affecting poorer communities).The emergence of 'tropical diseases' as a discipline emerged as a consequence of imperialism; the colonial power's incursion into tropical regions required a new branch of Western medicine to deal with these new non-Western illnesses.

Many determinants of health lie outside of the scope of Western health sectors meaning they remain under-reported and there is inaction to address them. For example, poverty is an important determinant of health and indirectly affects a range of related factors such as poor housing, unsafe transport and inadequate diets (Lu et al. 2016; Walker et al. 2007). Similarly, poor education affects the capacity of individuals and communities to improve their health through, for example, changing unhealthy behaviours, improving diet and vector control (Curtis 2018). Many other health determinants are not controlled by the medical and health sectors, yet have significant impacts on human health, including inadequate sanitation systems, poor housing, unsafe drinking water, dangerous living conditions, urban violence and air pollution; even seemingly minor factors, such as hunger in childhood or absenteeism from school, can profoundly impact on the health of the urban poor. At present, too many of the determinants of urban health fall into the gaps between different professions and governmental agencies. There is a need for greater interdisciplinarity to address these complex urban health challenges. The role of the health pluriverse is pertinent here, whereby different conceptual frameworks of health, society and nature might come out in differing structural, organisational and disciplinary networks in order to address and resolve health and lifestyle challenges in local contexts.

What is emerging, especially from a healthy city perspective, is that copying the Global North will not work. Global South cities need to develop technologies, practices, systems and mechanisms that are relevant to local cultural practices, climatic factors, financial resources, and geographical and urban conditions. The next section examines a significant consequence of the absence of health knowledge in the Global South—those neglected by developmental processes.

Neglected Diseases

There are neglected diseases that affect the urban poor of the Global South who "*tend to be marginalised by the health sector, as are many of the diseases that affect them*" (Ehrenberg and Ault 2005: 2). The most common communicable diseases, such as malaria and tuberculosis, are reported and accounted for in global and national health reports. However, many communicable diseases which affect smaller populations, particularly poorer populations, tend to go unreported or underreported, corresponding to the neglected diseases of neglected populations (Hotez 2017). Tropical diseases are less relevant to the Global North, particularly the former colonial powers (from more temperate planetary zones), and there are relatively lower levels of investment into medicines and treatments thereof. Diseases affecting small subpopulations with genetic predispositions to certain illnesses are also of low priority to health concerns to the Global North populations. The most vulnerable groups to these neglected risks are minority ethnic populations, indigenous people, the very old, the very young and certain groups such as those living in slums or itinerant migrant workers. Neglected diseases, neglected communities and neglected spaces remain part of the puzzle for Global South Healthy City strategies to resolve. There are a number of issues that need to be better addressed here—firstly, is an acceptance of more diverse and heterogeneous health conditions among the different Global South populations. Secondly, the problem is exacerbated as there is limited funding available for research on health issues affecting smaller subpopulations. Governmental funding is required to fill this gap as the large pharmaceutical companies will not focus on areas with limited opportunity for financial reward. Thirdly, there is a need for a more complex and nuanced understanding of the factors at play in improving human health in the Global South cities. Local and alternative models of medicine are mostly ignored by mainstream medical systems in the Global South. Cultural practices based on traditional medicines sometimes affect the uptake and/or effectiveness of Western medicines and treatments. These gaps in knowledge, neglected diseases and populations point to the need for more localised healthy Global South solutions and the need for multi-sectoral interventions.

This subsection explores this issue through an examination of one seemingly unimportant aspect of everyday health practices, that of going to the toilet. It reveals how this practice is a complex socio-spatio-technical network of relationships imbricated in the themes of health and developmentalism explored previously. Poor toilet provision is the cause of millions of unnecessary deaths and illnesses each year (Pruss et al. 2002; UNICEF-WHO 2009). For decades the 'solution' has been to attempt to emulate the toilets (and sanitation systems) of the Global North—without much success. Southern Cities have adopted the methods, mechanisms, technologies and practices of Northern sanitation. This subsection focuses on sewage to illustrate how poor health outcomes are enacted in contemporary Global South cities. The cause of so many deaths and severe illnesses in the Global South is a consequence of poor toilet provision, cultural practices and inadequate sanitation facilities (Jewitt 2011). Not to mention how decades of development has not led to any improvements for millions of people living in the Global South. No issue is so closely enmeshed with ideas of dirt/cleanliness as bodily excreta; it is a sensitive cultural concern (Swyngedouw 2004).

There are complex cultural issues to resolve, for example, eradicating open defecation in order to improve public health is more than a technical issue. Over one billion people practise open defecation globally; however, the practice of open defecation is deeply entrenched in cultural practices, beliefs as well as pragmatic issues (WHO 2012). In India, draconian measures of implementing the 'end open defecation' policy led to the lynching of two boys (BBC 2019). There are many challenges in play here; whilst part of the issue is related to a lack of finances, there are many structural and environmental factors that contribute. Even when sewage processing plants are built in the Global South, they are often simply replicas of Global North models (with little concern for differences in context, culture or climate) that do not work and, at times, are so ineffective that the sewage simply bypasses these facilities and further pollutes the environment (The Gates Foundation 2019). Most of the sewage systems are based on 'Western' models of sewage treatment, which, for example, in the US works for a large country with a relatively small population; these systems do not work for geographically smaller nations with larger relative populations (ibid). Scant research has been undertaken to develop sanitation systems that are appropriate for the Global South. However, this is beginning to change, particularly as the Gates Foundation has recently provided millions of dollars in research funding, particularly to researchers from the Global South, to explore systems of sanitation relevant to the needs of the Global South cities. Sanitation is just one aspect of radically different socio-techno-legal health ecosystems that are required as part of a healthy city strategy for Global South contexts.

Conclusions

This chapter has explored the role of Healthy City strategies, which have been criticised as being an iteration of developmentalism that fails to deliver necessary well-being outcomes in the urban Global South. The attainment of urban health involves a comprehensive developmental system leading towards cleaner cities, sanitised spaces, purified water, implementing hygiene standards, removal of dirt, spatial segregation of impurities, control of contamination and limits on pollution. This has often been achieved through the processes of developmentalism, industrialisation and globalisation. The conclusion has addressed how the critiques of healthy cities comprise three broad analytical themes concerning how healthy cities are related to colonialist legacies, problematisation of the Global South as diseasogenic and the absence of local medical knowledge.

Changes enacted during the colonial period, particularly changing practices in agriculture, industry and urban lifestyles, caused new illnesses and exacerbated some existing health issues. Developmentalism has further added to the postcolonial legacy of ill health in the Global South. One of the consequences of this history has been the loss of local health beliefs and practices replaced instead by a Western medicinal system. The most widely espoused Healthy City strategies are based on practices, knowledge and techniques emerging from the Global North, situated in a Western medical ontology and epistemology. One of the mechanisms through which this has been pursued is through the epistemicide of local systems of knowledge and the destruction of local health practices. From inception during colonial times to postcolonial and contemporary capitalist forces, a disequilibrium has been maintained that affords greater power to the Global North. This has led to divergent health outcomes in the Global South cities. The promise of better health has not been delivered to the entire population. There are now subpopulations that have benefitted from the westernisation of cities in terms of improved health outcomes, but there remain severe injustices in Southern cities, often associated to severe poverty and deprivation. The cities of the Global South suffer from relatively high levels of communicable diseases, many of which have already been eradicated in the Global North. Furthermore, as the cities of the Global South follow the developmental trends of the Global North, the urban patterns emerging are associated with many non-communicable diseases built into their DNA. There is evidence emerging that the cities of the Global South may experience the worst of both worlds, with populations struggling with high levels of communicable and non-communicable diseases simultaneously. This last challenge is further exacerbated by the lack of knowledge for many local health issues. There are significant gaps in knowledge related to urban health in the Global South, whereby there are neglected diseases affecting neglected subpopulations and neglected spaces. There is a paucity of funding to focus on these neglected health issues by the larger pharmaceutical and medical companies.

Whilst there may be valid criticisms of the current configuration of Healthy City strategies, there is also a clear, practical (and urgent) need for improving health and well-being in cities and urban areas. It is important not to throw the baby out with the bathwater; healthy Cities do deliver a range of positive health outcomes, particularly combatting communicable diseases. Western medicine and modern pharmaceutical companies have enabled a huge improvement to human health in the past few centuries and should be acknowledged accordingly. However, improvements are required, and, arguably, many of the answers lie outside of the remit of Western perspectives. The Global North does not currently provide a blueprint for how this might be achieved. Global North urbanisms perform poorly in terms of mental and social health, and non-communicable diseases. In terms of planetary health, the cities of the Global North contribute significantly to climate change and are environmentally unsustainable. This situation requires alternative approaches for how society is to sustainably and equitably inhabit urban areas. With the predicted levels of urbanisation, the Southern city is foregrounded as the battleground for combatting emerging health issues; there is a very real and urgent requirement for the Global South to develop a road map towards healthier places.

The way forward would appear to require a radical rethinking, beginning with the embrace of a pluriverse of alternative understandings of health, nature and humankind. (It is not to suggest that the benefits of Western medicine be rejected; instead, the worlds of mainstream and alternative health practices should be combined). The healthy pluriverse should facilitate the nurturing of different modes of living and myriad urbanisms that reflect local cultures, local customs and beliefs. Second, alternative approaches to socio-economic models, namely ones that do not irrevocably damage planetary health, are urgently required. Different models of healthy cities are required for different cultures, contexts and climates. There is evidence from locations such as Latin America that this is achievable and to be welcomed. There are some instances of patterns of urban living that are not so diseasogenic but rather support healthier lifestyles and behaviours. More research, particularly interdisciplinary collaboration, is needed to develop appropriate solutions for local populations and places. The narratives and practices emerging from the pluriverse provide a road map for more sustainable and endogenous models of Healthy Cities.

References

Acosta, A. (2011). *Sólo Imaginando otros Mundos, se Cambiará Éste: Reflexiones sobre el Buen Vivir*. www.plataformabuenvivir.com/wp-content/uploads/2012/07/AcostaReflexionesBuenVivir.pdf. Accessed 25 Oct 2019.

Alcadipani, R., & Caldas, M. (2012). Americanizing Brazilian Management. *Critical Perspectives on International Business, 8*(1), 37–55.

Anderson, W. (1995). Excremental Colonialism: Public Health and the Poetics of Pollution. *Critical Inquiry, 21*(3), 640–669.

Arnold, C. (2008). *Necropolis: London and its Dead*. London: Simon and Schuster.

Bankoff, G. (2007). Comparing Vulnerabilities: Toward Charting an Historical Trajectory of Disasters. *Historical Social Research, 32*(3), 103–114.

Barton, H., & Grant, M. (2006). A Health Map for the Local Human Habitat. *The Journal for the Royal Society for the Promotion of Health, 126*(6), 252–253.

BBC. (2019, September 26). *India: Two Held for Killing Children for 'Defecating in the Open'*. www.bbc.co.uk/news/world-asia-india-49835830. Accessed 2 Oct 2019.

Bodeker, G., Ong, C. K., Grundy, C., Burford, G., & Shein, K. (2005). *WHO Global Atlas of Traditional, Complementary and Alternative Medicine* (Vol. 1). Kobe: WHO.

Brenner, N., & Schmid, C. (2015). Towards a New Epistemology of the Urban? *City, 19*(2–3), 151–182.

Bruegman, R. (2008). *Urban Sprawl: A Compact History*. Chicago: University of Chicago Press.

Bryson, B. (2019). *The Body: A Guide for Occupants*. New York: Doubleday.

Buer, M. C. (2013). *Health, Wealth and Population in the Early Days of the Industrial Revolution*. London: Routledge.

Callon, M. (1991). Techno-economic Networks and Irreversibility. In J. Law (Ed.), *A Sociology of Monsters: Essays on Power, Technology, and Domination* (pp. 132–161). London: Routledge.

Caria, S., & Domínguez, R. (2016). Ecuador's Buen Vivir: A New Ideology for Development. *Latin American Perspectives, 43*(1), 18–33.

Chaput, J.-P., Klingenberg, L., Astrup, A., & Sjödin, A. M. (2011). Modern Sedentary Activities Promote Overconsumption of Food in our Current Obesogenic Environment. *Obesity Reviews, 12*(5), e12–e20.
Christopher, A. J. (1983). From Flint to Soweto: Reflections on the Colonial Origins of the Apartheid City. *Area, 15*(2), 145–149.
Cohen, B. (2006). Urbanization in Developing Countries: Current Trends, Future Projections, and Key Challenges for Sustainability. *Technology in Society, 28*(1–2), 63–80.
Curtis, B. (2018). Education as Poverty Reduction. In B. Curtis & S. Cosgrove (Eds.), *Understanding Global Poverty Causes, Capabilities and Human Development* (pp. 201–221). London: Routledge.
Davis, M. (2004). The Urbanization of Empire: Megacities and the Laws of Chaos. *Social Text, 22*(4), 9–15.
Davis, M. (2007). *Planet of Slums*. London: Verso.
Dawson, A., & Hayes Edwards, B. (2004). Introduction: Global cities of the South. In A. Dawson & B. Hayes Edwards (Eds.), *Social Text Collective* (pp. 1–7). Durham: Duke University Press.
de Sousa Santos, B. (2014). *Epistemologies of the South: Justice Against Epistemicide*. Oxon: Routledge.
Domingues, J. M. (2012). Development and Dependency, Developmentalism and Alternatives. In R. Boschi & C. Henrique (Eds.), *Development and Semi-periphery: Post-neoliberal Trajectories in South America and Central Eastern Europe* (pp. 83–101). London: Anthem Press.
Douglas, M. (2003). *Purity and Danger: An Analysis of Concepts of Pollution and Taboo*. London: Routledge.
Dupont, V., Jordhus-Lier, D., Sutherland, C., & Braathen, E. (Eds.). (2015). *The Politics of Slums in the Global South: Urban Informality in Brazil, India, South Africa and Peru*. London: Routledge.
Easterlin, R. A. (1995). Industrial Revolution and Mortality Revolution: Two of a Kind? *Journal of Evolutionary Economics, 5*(4), 393–408.
Ehrenberg, J. P., & Ault, S. K. (2005). Neglected Diseases of Neglected Populations: Thinking to Reshape the Determinants of Health in Latin America and the Caribbean. *BMC Public Health, 5*(1), 119.
Escobar, A. (1995). *Encountering Development: The Making and Unmaking of the Third World*. Princeton: Princeton University Press.
Escobar, A. (2018). *Designs for the Pluriverse*. Durham: Duke University Press.
Farley, J. (2003). *Bilharzia: A History of Imperial Tropical Medicine*. Cambridge: Cambridge University Press.

Ferguson, J. (2006). *Global Shadows: Africa in the Neoliberal World Order*. Durham: Duke University Press.
Ferguson, K. (2007). *William James: Politics in the Pluriverse*. New York: Rowman & Littlefield Publishers.
Fernando, S., & Moodley, R. (Eds.). (2018). *Global Psychologies: Mental Health and the Global South*. London: Springer.
Fitzgerald, A. J. (2010). A Social History of the Slaughterhouse: From Inception to Contemporary Implications. *Human Ecology Review, 17*(1), 58–69.
Frost, L. (1997). Coping in their Own Way: Asian Cities and the Problem of Fires. *Urban History, 24*(1), 5–16.
Galea, S., Ettman, C. K., & Vlahov, D. (Eds.). (2019). *Urban Health*. Oxford: Oxford University Press.
Gilbert, A. (2007). The Return of Slum: Does Language Matter? *International Journal of Urban and Regional Research, 31*(4), 697–713.
Grant, M., Brown, C., Waleska, T. C., Capon, A., Corburn, J., Coutts, C., & Crespo, C. J. (2017). Cities and Health: An Evolving Global Conversation. *Cities and Health, 1*(1), 1–9.
Grosfoguel, R. (2000). Developmentalism, Modernity, and Dependency Theory in Latin America. *Nepantla: Views from South, 1*(2), 347–374.
Gudynas, E. (2011). Buen Vivir: Today's Tomorrow. *Development, 54*(4), 441–447.
Gugler, J. (2004). *World Cities beyond the West: Globalization, Development and Inequality*. Cambridge: Cambridge University Press.
Hotez, P. J. (2017). Ten Failings in Global Neglected Tropical Diseases Control. *PLoS Neglected Tropical Diseases, 11*(12), e0005896. https://doi.org/10.1371/journal.pntd.0005896.
Ichikawa, H. (1994). The Evolutionary Process of Urban Form in Edo/Tokyo to 1900. *Town Planning Review, 65*(2), 179.
Jewitt, S. (2011). Geographies of Shit: Spatial and Temporal Variations in Attitudes towards Human Waste. *Progress in Human Geography, 35*(5), 608–626.
Kaklauskas, A., Zavadskas, E. K., Radzeviciene, A., Ubarte, I., Podviezko, A., Podvezko, V., Kuzminske, A., Banaitis, A., Binkyte, A., & Bucinskas, V. (2018). Quality of City Life Multiple Criteria Analysis. *Cities, 1*(72), 82–93.
Kanai, J. M. (2014). On the Peripheries of Planetary Urbanization: Globalizing Manaus and its Expanding Impact. *Environment and Planning D: Society and Space, 32*(6), 1071–1087.
Kanna, A. (2012). Urban Praxis and the Arab Spring. *City, 16*(3), 360–368.

Latouche, S. (1996). *The Westernization of the World: Significance, Scope and Limits of the Drive towards Global Uniformity*. Cambridge: Polity.
Latour, B. (2005). *Reassembling the Social: An Introduction to Actor-Network-Theory*. Oxford: Oxford University Press.
Lu, C., Black, M. M., & Richter, L. M. (2016). Risk of Poor Development in Young Children in Low-income and Middle-income Countries: An Estimation and Analysis at the Global, Regional, and Country Level. *The Lancet Global Health, 4*(12), e916–e922.
Marmot, M. (2015). *The Health Gap: The Challenge of an Unequal World*. London: Bloomsbury.
Marteau, T. M., Ogilvie, D., Roland, M., Suhrcke, M., & Kelly, M. P. (2011). Judging Nudging: Can Nudging Improve Population Health? *BMJ, 342*, d228.
McFarlane, C. (2008). Urban Shadows: Materiality, the 'Southern City' and Urban Theory. *Geography Compass, 2*(2), 340–358.
Memmi, A. (1992). *The Colonizer and the Colonized*. Boston: Beacon Press.
Meyer, R., Patel, A. M., Rattana, S. K., Quock, T. P., & Mody, S. H. (2014). Prescription Opioid Abuse: A Literature Review of the Clinical and Economic Burden in the United States. *Population Health Management, 17*(6), 372–387.
Micozzi, M. S. (2018). *Fundamentals of Complementary, Alternative, and Integrative Medicine*. St. Louis: Elsevier Health Sciences.
Mills, C. (2014). *Decolonizing Global Mental Health: The Psychiatrization of the Majority World*. London: Routledge.
Nieuwenhuijsen, M., & Khreis, H. (Eds.). (2018). *Integrating Human Health into Urban and Transport Planning: A Framework*. London: Springer.
Njoh, A. J. (2008). Colonial Philosophies, Urban Space, and Racial Segregation in British and French Colonial Africa. *Journal of Black Studies, 38*(4), 579–599.
Otter, C. (2005). Civilizing Slaughter: The Development of the British Public Abattoir, 1850–1910. *Food and History, 3*(2), 29–51.
Owusu, G., Agyei-Mensah, S., & Lund, R. (2008). Slums of Hope and Slums of Despair: Mobility and Livelihoods in Nima, Accra. *Norwegian Journal of Geography, 62*(3), 180–190.
Papanek, V. (2009). *Design for the Real World: Human Ecology and Social Change*. Chicago: Academy Chicago Publisher.
Partridge, C. (2005). *The Re-enchantment of the West, Vol 1: Alternative Spiritualities, Sacralization Popular Culture and Occulture*. London: T&T Clark.

Partridge, C. (2006). *The Re-enchantment of the West, Vol 2: Alternative Spiritualities, Sacralization Popular Culture and Occulture.* London: T&T Clark.
Poestges, H. (2011). Leprosy, the Key to Another Kingdom. *Leprosy Review, 82*(2), 155.
Pruss, A., Kay, D., Fewtrell, L., & Bartram, J. (2002). Estimating the Burden of Disease from Water, Sanitation, and Hygiene at a Global Level. *Environmental Health Perspectives, 110*(5), 537–542.
Quijano, A. (2005). Of Don Quixote and Windmills in Latin America. *Estudos Avançados, 19*(55), 9–31.
Quijano, A. (2008). Coloniality of Power, Eurocentrism, and Social Classification. In M. Morãna, E. Dussel, & C. Jáuregui (Eds.), *Coloniality at Large: Latin America and Postcolonial Debate* (pp. 540–567). Durham: Duke University Press.
Reeves, A., Gourtsoyannis, Y., Basu, S., McCoy, D., McKee, M., & Stuckler, D. (2015). Financing Universal Health Coverage - Effects of Alternative Tax Structures on Public Health Systems: Cross-national Modelling in 89 Low-income and Middle-income Countries. *The Lancet, 386*(9990), 274–280.
Rice, L. (2018). Nonhumans in Participatory Design. *CoDesign, 14*(3), 238–257.
Rice, L. (2019a). The Nature and Extent of Healthy Architecture: The Current State of Progress. *Archnet-IJAR: International Journal of Architectural Research, 13*(2), 244–259.
Rice, L. (2019b). A Health Map for Architecture: The Determinants of Health and Wellbeing in Buildings. In M. Jones, L. Rice, & F. Meraz (Eds.), *Designing for Health & Wellbeing: Home, City, Society* (pp. 155–184). Delaware: Vernon Books.
Roy, A. (2005). Urban Informality: Toward an Epistemology of Planning. *Journal American Planning Association, 71*(2), 147–158.
Sachs, W. (Ed.). (1997). *Development Dictionary: A Guide to Knowledge as Power.* New York: St Martin's Press.
Schuurman, F. J. (2000). Paradigms Lost, Paradigms Regained? Development Studies in the Twenty-first Century. *Third World Quarterly, 21*(1), 7–20.
Smith, V. S. (2008). *Clean: A History of Personal Hygiene and Purity.* Oxford: Oxford University Press.
Stephens, C. (1996). Healthy Cities or Unhealthy Islands? The Health and Social Implications of Urban Inequality. *Environment and Urbanization, 8*(2), 9–30.

Sunstein, C., & Thaler, R. (2008). *Nudge: Improving Decisions about Health, Wealth, and Happiness.* New Haven: Yale University Press.

Swyngedouw, E. (2004). *Social Power and the Urbanization of Water: Flows of Power.* Oxford: Oxford University Press.

Szreter, S. (2004). Industrialization and Health. *British Medical Bulletin, 69*(1), 75–86.

The Gates Foundation. (2019). *Water, Sanitation & Hygiene: Strategy Overview.* www.gatesfoundation.org/what-we-do/global-growth-and-opportunity/water-sanitation-and-hygiene. Accessed 30 July 2019.

The Prince of Wales. (2009, February 6). Charles Declares Mumbai Shanty Town Model for the World. *The Guardian.* www.theguardian.com/artanddesign/2009/feb/06/prince-charles-slum-comments. Accessed 26 June 2018.

Thomas, A. J. (2015). *Cholera: The Victorian Plague.* Barnsley: Pen and Sword.

Thomson, B. (2011). Pachakuti: Indigenous Perspectives, Buen Vivir, Sumaq Kawsay and Degrowth. *Development, 54*(4), 448–454.

Thornicroft, G., & Maingay, S. (2002). The Global Response to Mental Illness: An Enormous Health Burden is Increasingly Being Recognised. *BMJ, 325,* 608–609.

UN. (2019). *Cities and Pollution Contribute to Climate Change.* www.un.org/en/climatechange/cities-pollution.shtml. Accessed 16 Sept 2019.

UN Department of Economic and Social Affairs. (2018). *World Urbanization Prospects: The 2018 Revision.* New York: United Nations.

UNICEF-WHO. (2009). *Diarrhoea: Why Children Are Still Dying and What Can be Done.* Geneva: UNICEF.

Villalba, U. (2013). Buen Vivir vs Development: A Paradigm shift in the Andes? *Third World Quarterly, 34*(8), 1427–1442.

Vitruvious. (1999). *Ten Books on Architecture – Book 1.* Cambridge: Cambridge University Press.

Vlahov, D., Galea, S., & Freudenberg, N. (2005). The Urban Health 'Advantage'. *Journal of Urban Health, 82*(1), 1–4.

Walker, S. P., Wachs, T. D., Gardner, J. M., Lozoff, B., Wasserman, G. A., Pollitt, E., & Carter, J. A. (2007). Child Development: Risk Factors for Adverse Outcomes in Developing Countries. *The Lancet, 369*(9556), 145–157.

Wallerstein, I. (2005). After Developmentalism and Globalization, What? *Social Forces, 83*(3), 1263–1278.

WHO. (1995). *Building a Healthy City: A Practitioners Guide. A Step-by-Step Approach to Implementing Healthy City Projects in Low-income Countries.* Geneva: WHO.

WHO. (2012). *Water Sanitation Hygiene: Fast Facts*. www.who.int/water_sanitation_health/monitoring/jmp2012/fast_facts/en. Accessed 11 Dec 2019.
WHO. (2015). *Health in 2015: From MDGs, Millennium Development Goals to SGDs, Sustainable Development Goals*. Geneva: WHO.
WHO. (2017). *Integrating Neglected Tropical Diseases into Global Health and Development: Fourth WHO Report on Neglected Tropical Diseases*. Geneva: WHO.
WHO. (2019). *Types of Healthy Settings: Healthy Cities*. www.who.int/healthy_settinGlobalSouth/types/cities/en. Accessed 11 Dec 2019.

3

Regenerating the Socio-Ecological Quality of Urban Streams: The Potential of a Social Learning Approach

Daniele Tubino P. de Souza, Edson Grandisoli, Pedro Roberto Jacobi, and Arjen E. J. Wals

Introduction

Accelerated and unplanned urban growth and poor sanitation infrastructure, coupled with inequity and poverty, are factors directly associated with the degradation of urban streams (UN-Habitat 2016; WWAP 2015). Cities in the so-called Global South are particularly affected by this phenomenon (Capps et al. 2016; Wantzen 2018; Wantzen et al.

D. T. P. de Souza (✉) • P. R. Jacobi
Institute of Energy and the Environment, University of São Paulo (USP), São Paulo, Brazil
e-mail: danieletubino@usp.br; prjacobi@usp.br

E. Grandisoli
Institute of Advanced Studies, University of São Paulo (USP), São Paulo, Brazil
e-mail: edsongrandisoli@gmail.com

A. E. J. Wals
Wageningen University, Wageningen, The Netherlands
e-mail: arjen.wals@wur.nl

A. A. R. Ioris (ed.), *Environment and Development*,
https://doi.org/10.1007/978-3-030-55416-3_3

2019). The scale and severity of this problem mean that there is an urgent need to rethink approaches and processes in order to find appropriate solutions to regenerate urban streams and promote more sustainable urban environments. The regeneration of urban streams is a complex challenge that demands multi-stakeholder action in collaborative, multidisciplinary and learning-oriented platforms out of which joint solutions to socio-ecological challenges may emerge (Agramont et al. 2019; Silva-Sánchez and Jacobi 2016; Wantzen et al. 2019; Warner et al. 2008; Wehn et al. 2018).

This chapter seeks to identify potentialities and challenges in using the social learning approach as a framework for the multi-stakeholder interactions involved in initiatives for the regeneration of urban streams in contexts of socio-ecological vulnerability. The analysis is built on the case study of the Taquara Stream, located in the city of Porto Alegre, in south Brazil. This case study comprises a self-organised group of citizens—composed of members of the local community, the public sector and educational institutions—acting to re-establish the socio-ecological quality of the Taquara Stream and watershed, an area largely occupied by informal settlements. Firstly, we contextualise and problematise the issue of urban stream degradation, focusing on the Brazilian context. Secondly, we provide a brief literature review on social learning, and, finally, we describe and discuss the case of the Taquara Stream as an example of a social learning–oriented process to demonstrate the potential contribution and challenges this approach brings within processes for the regeneration of urban streams in vulnerable areas.

Urban Rivers and Streams in the Global South and Brazil

Urban rivers and streams and their banks are fundamental ecological structures for the sustainability of cities. They are essential areas for the protection and maintenance of biological diversity, connecting green areas within urban environments (e.g. through greenways), preservation of water quality, flooding reduction, provision of recreational and leisure uses that allow people to interact with nature, and environmental education activities (Ahern 1995; Flink and Searns 1993; Hough 2002; Newman and Jennings 2008; Spirn 1984). Increasing attention has been given in recent years to the need to restore rivers and streams in urban environments, as evidenced by multiple projects and initiatives around the world (Bernhardt et al. 2005; Tunstall et al. 2000; Wantzen et al. 2019). The concept of greenways, for example, has been extensively applied as a strategy for the requalification of urban streams and their banks (Fabos 1995; Searns 1995; Hellmund and Smith 2006). The rewilding of streambanks (Rhoads 1999) and the implementation of depollution processes based on ecological principles and ecosystems functioning (Lyle 1996) are seen as sustainability-oriented approaches to dealing with this challenge, and are increasingly widespread (Wantzen et al. 2016). At the same time, it has been also recognised that actions towards the regeneration of urban streams demand collaboration between different social actors in a multidisciplinary process that considers urban water bodies as socio-environmental systems characterised by the complex integration of factors such as ecology, water supply, landscape recovery, community memory, sanitation, housing etc. (Silva-Sánchez and Jacobi 2016).

Ecology-oriented regeneration strategies are well established in the so-called Global North, while, in the Global South, they are still relatively new (Wantzen et al. 2019). In the Global South factors such as rapid urbanisation, poverty, the growth of slum populations occupying riparian zones, the inability or unwillingness of local and national governments to deal with the issue of social housing for the poor, social conflicts around land and resources, inadequate waste-management policies and infrastructure, a lack of financial resources to invest in needed social and physical infrastructures, poor management of water resources, inefficiencies in environmental governance systems and lack of transboundary action are all significant factors limiting the implementation of these projects (Capps et al. 2016; Wantzen 2018, Wantzen et al. 2019; WWAP 2015). Furthermore, populations inhabiting riverbanks have little chance to participate in decision-making processes and tend to lack the political power and technical skills to develop urban stream areas in a sustainable way (Wantzen et al. 2019).

In Brazil, the degradation of urban rivers and streams is a pressing problem in many cities. This issue is usually related to conditions of social and environmental vulnerability, which are connected to poverty, the presence of informal settlements or slums—reported in 97.3% of Brazilian municipalities with more than 500,000 inhabitants (IBGE 2009)—precarious housing conditions, riparian forest suppression, huge deficits in sanitation infrastructure provision, the direct disposal of sewage and solid waste in water bodies, the exposure of populations to risks—such as landslides and floods—and waterborne diseases (ANA 2018; Maricato 2000; Moser 1998; Tucci 2005). The National Water Agency points out that water quality is very low along more than 110,000 kilometres of river due to an excess of organic load. Along 83,450 kilometres of river, it is no longer possible to capture water for public supply due to pollution, and along 27,040 kilometres, water can only be captured with the use of advanced treatment methods (ANA 2019). It is worth noting that the current Brazilian government intends to address the complex issue of sanitation in Brazil by privatising the sector; the text that establishes this legal framework was approved by the Chamber of Deputies in December 2019 (as noted in a Folha de São Paulo newspaper article on 12 December 2019). Privatising the sanitation systems points, however, to the implementation of top-down solutions and rising prices for the service—thus penalising the poorest segments of population—without guaranteeing the systems' improvement (Faria et al. 2005; de Sousa et al. 2017).

Vulnerable groups that inhabit urban river and streambanks in Brazil are, to a great extent, socially marginalised, excluded from effective political participation in decision-making processes (Acselrad 2015) and deprived of access to formal urban areas (Rolnik 1999). This situation is engendered by complex factors such as the striking social inequality that marks the Brazilian context, uncontrolled urban growth, lack of efficient housing policies and an absence of public power to limit the occupation of unsuitable areas (Balbim 2016; Maricato 2000, 2003). The great imbalance in income distribution in Brazil leads to marked socio-spatial inequalities, as segments of the population unable to acquire land eventually occupy illegal and lower market value areas—usually environmental preservation areas such as wetlands, steep hills and riverbanks—where there is no basic sanitation, and the implementation of such services is difficult (Rolnik 1999). In terms of political ecology, this situation reveals how the disharmonious relationship between social groups engenders environmental inequalities (Swyngedouw et al. 2002) related to the disputes over land access that arise in contexts of capitalist urbanisation (Denaldi and Ferrara 2018).

Concerted actions aimed at recovering urban rivers and streams in socio-ecologically vulnerable areas must therefore address specific economic, cultural, political and environmental features of these contexts so as to generate legitimate solutions that are technically appropriate, economically viable, socially acceptable and environmentally sound. Successful experiences of wastewater management and collection in contexts of financial resources constraint, as experienced by most countries in the Global South, point to the importance of expanding decentralised systems and low-cost, innovative place-based solutions that are technically easy to implement (e.g. local wastewater treatment, biogas production for reusing and recycling wastewater) (WWAP 2015). Furthermore, ensuring that communities directly affected by water-related problems can participate in decision-making processes is crucial for producing legitimate responses that fit local sociocultural contexts (Gallo et al. 2016). It is particularly important for the success of restoration projects in lower-income countries that actions not only offer feasible and effective short-term solutions to people who live in precarious situations along the banks of urban streams and rivers but also engage stakeholders in a broader debate that goes beyond purely technical approaches (Wantzen et al. 2019).

In this context, it is critical to create inclusive processes to facilitate the participation of marginalised social groups in collaborative action towards the transformation of their realities. In contexts of deep social disparities, collaborative efforts to improve the socio-ecological quality of urban rivers and their surroundings might create opportunities for dialogical interactions among diverse stakeholders and disadvantaged social groups. Such processes are likely to generate individual and collective critical reflections and knowledge exchange among participating actors as well as empowerment and greater autonomy of involved communities in local resource management (Jacobi 2013). From this standpoint, collaborative practices, dialogic learning processes and knowledge integration are central aspects in fostering transformations towards more sustainable practices and values to promote changes in local realities (Lotz-Sisitka et al. 2015) and to regenerate socio-ecological systems (Wahl 2016). In this context, social learning appears as an important theoretical framework that can inform processes with this aim.

Social Learning

The development of responses towards sustainability can benefit from a critical assessment of dominant collective mindsets that, often unwillingly, normalise or invite unsustainable behaviour and stand in the way of collective action on socio-environmental issues (Keen et al. 2005; Sterling 2011). From this perspective, the conception of sustainability encompasses both a subjective and a technical process. As a subjective process, it requires learning and critical reflection on values, views and concepts that underlie decision-making, and that are associated with unsustainable patterns of production and consumption, and unequal and exclusionary social structures (Abson et al. 2017; O'Sullivan and Taylor 2004; Röling 2002; Wahl and Baxter 2008). As a technical process, it demands the selection and explanation of specific knowledge and parameters that can guide the development of practical solutions that fit a specific social, ecological, cultural and political context (Macintyre et al. 2018; Šūmane et al. 2018). Social learning is a valuable approach for connecting the two as it allows us to think about how to engage groups in a critical assessment of knowledge and collective reflection on entrenched views, beliefs and values, and encourages meaningful conversations to produce innovative practices towards the construction of a sustainable future.

Wals (2011: 181) highlights four main aspects of social learning for sustainability: (1) it uses the diversity of views within a group as a way to foster co-creative processes; (2) it is based on processes aimed at creating opportunities for individual and collective reflection and reflexivity; (3) it unfolds through social cohesion and social capital to facilitate change in complex situations; and (4) it counts on the 'power of collaborative action that strengthens the (unique) qualities of each individual'. Social learning implies the promotion and intensification of learning processes that can lead to transformation, both individually and collectively, stimulated by dialogue and cooperation among a diverse group of people that share a common goal.

Diversity is a key element in social learning processes, as it is from the plurality of perspectives and the confrontation of divergent views in a conducive environment that critical learning processes can converge, and views, mindsets and values may be transformed (Wals and Heymann 2004). However, in complex social systems where people with different backgrounds and attitudes are brought together, social learning will not necessarily unfold automatically. It is based on a relational approach in the sense of establishing social connections, and depends on a favourable social environment, often informal, wherein dialogical interactions can unfold (Souza et al. 2019). Dialogue construction requires the willingness of all involved to communicate as equals in a process of a relatively open conversation that welcomes a diversity of views and conflict—when properly managed—as driving forces for collective learning (Wals and Bawden 2000). Diversity is, therefore, a potential force for learning, allowing participants to create new knowledge when the different perspectives, experiences, ideas and values of individuals are integrated in an enriching way through dialogical interactions (Loeber et al. 2009).

Integrating diversity into symmetrical dialogical practices is not, however, an easy task, as this depends on a climate of mutual trust and overcoming, or at least acknowledging and questioning, power imbalances among actors participating in the social learning process (Schusler et al. 2003). Hence, skilled facilitation is also considered an important factor in social learning processes as it can contribute to reducing power imbalances by creating equal opportunities for participation (Ernst 2019). Social learning also requires building trust among participants (Rist et al. 2009). In a group environment where there is mutual trust, dialogue can be deepened as participants feel more confident in expressing and questioning views (Rist et al. 2006).

Social Learning and Collaborative Projects

Social learning is directed towards problem-solving in specific contexts, where groups or communities are operating in new, uncertain and unpredictable circumstances (Wildemeersch 2009). It emphasises the learning that arises from practical experiences, recognising the reciprocal influence and transformation that result from dynamic relationships among individuals, and between individuals and the environment (Souza et al. 2019). The engagement of groups and communities in collaborative projects in the context of complex contemporary social and environmental challenges represents an opportunity to initiate social learning processes, especially when it is recognised that exclusively regulatory, authoritarian and technical approaches are inadequate in addressing challenges within complex systems (Mostert et al. 2007).

Projects focused on regaining the socio-ecological quality of urban streams in vulnerable areas call for comprehensive approaches that must take into account social and biophysical features as well as the technical knowledge needed to design practical responses. They require the construction of relationships among diverse stakeholders and the critical engagement of directly affected communities as legitimate groups in defining territorialised agendas (Gallo et al. 2016; Machado et al. 2019). The involvement of affected social groups as well as other stakeholders—such as representatives from the public sector and other civil society organisations—in collaborative action and learning practices is a key factor in promoting systemic and environmentally and culturally appropriate responses, as well as capacity building and greater autonomy of participants in local resource management (Hoff 2011).

When utilising social learning to help communities move towards sustainability, two kinds of social differences can be distinguished: healthy or desirable ones and unhealthy or undesirable ones. Healthy differences include variety in vantage points, values, ways of thinking and connecting that might increase the richness of the learning process and the chance of creative outcomes. These differences are only healthy when the people who are included in the learning process experience a certain level of trust, social cohesion and respect. Unhealthy differences, on the other hand, include disparities in agency and access to resources and power. Sustainability-oriented social learning needs to become transgressive in order to overcome such inequalities and to be able to include those who are marginalised. Here, transgressive refers to disrupting the structural mechanisms that keep inequality and associated differences in place, while empowering and equipping communities to take action (Bengtsson 2019). Social learning can help to build trust, commitment and social cohesion that can develop the capacities communities need to overcome structural inequality (Sol et al. 2013).

Another aspect that needs to be highlighted in a context of socio-ecological vulnerability marked by poverty and a lack of both basic sanitation infrastructures and the resources to solve problems locally is that the required solutions are unlikely to be achieved through exclusively community-based efforts. Complex problems such as those related to urban water system regeneration in vulnerable areas require a combination of top-down and bottom-up actions. Such contexts often demand complex technical interventions, which are generally top-down and not guided by reflexive interaction among the diverse actors involved (Smith 2008). Conversely, bottom-up initiatives emerge from community reactions to local challenges and have the potential to bring forth innovative solutions from locally available material, cultural and human resources, while promoting autonomy and increasing the ability of populations to deal with socio-ecological problems in their territories (Seyfang and Smith 2007). Nevertheless, the search for a balance between top-down and bottom-up approaches is key to concurrently meeting the ecological and social challenges involved in working towards sustainable justice, equanimity, community autonomy and regenerated ecosystems (Hopkins 2008).

Finally, it is also important to acknowledge that the ability of participants to direct the actions of a group in the search for sustainable responses is affected by the available resources and institutional arrangements, which might eventually restrict the possibilities of adequately responding to problems (Loeber et al. 2009). Contextual factors such as rules, governance systems, knowledge capacity and inter-stakeholder conflicts and tension will greatly influence social learning processes towards sustainability (Medema et al. 2014).

The Regeneration of an Urban Stream in Lomba do Pinheiro, Porto Alegre, Brazil

Lomba do Pinheiro is a neighbourhood located in the city of Porto Alegre, South Brazil (Fig. 3.1). It mixes densely populated nuclei and areas of prominent environmental value, such as native fields, woods, hilltops, watercourses and springs (PMPA 2013). The various springs found in the area compose the so-called Ring of Springs of Porto Alegre; most of them are located in the Saint-Hilaire Municipal Natural Park, a nature conservation unit adjacent to the neighbourhood (Menegat et al. 1999). Over the past few decades, Lomba do Pinheiro has been heavily occupied by informal settlements (PMPA 2013). Unplanned and informal urban growth has led to severe environmental degradation in some parts of the territory, mainly due to a lack of sanitation infrastructure and the occupation of streambanks (Oliveira and Oberrather 2010). A local initiative has been launched to recover a degraded urban stream, the Taquara Stream, involving representatives of the local community, the public sector and educational institutions. It is focused on the co-creation of solutions and the progressive implementation of improvements to the local situation using a participatory and learning-oriented approach.

This case was the subject of a participatory research project conducted by the lead author of this chapter. The research applied a range of methods, such as in-depth and semi-structured interviews, focus group conversations and participant observation. The participant observation was carried out during a period of approximately 18 months (between May

Fig. 3.1 Porto Alegre, Lomba do Pinheiro and the Taquara Stream watershed

2016 and December 2017). Individual and focus group interviews were conducted with a limited number of participants from the group leading the initiative—selected for their high attendance at meetings, relevant roles in organising actions in the community and ability to foster connections with other collaborators. This group was composed of 11 people (including residents, technicians and representatives of educational institutions). In the following sections, we analyse the case of the Taquara Stream, based on the main results of this research.

The Taquara Stream

The Taquara Stream, with an approximate length of 2800 metres, is located in the oldest and most urbanised part of Lomba do Pinheiro (see Fig. 3.1). Data from 2010 indicate a population of around 20,000 people, mainly with low incomes, living in the area encompassed by its watershed (IBGE 2010). Several interrelated issues have led to the degradation of the Taquara Stream over the past decades. Suppression of riparian forest due to occupation of the marginal strips has caused erosion and siltation of the stream bed. Inadequate or non-existent sanitation infrastructure—in spite of infrastructure work carried out by the city's water and sewage department over the past decade—means that wastewater is directly discharged into the stream (Fig. 3.2). Additionally, it has been verified that solid waste is being directly dumped into the watercourse and along its banks—despite the existence of regular household waste collection in the region. The risk of flooding is higher when accumulated waste descending the watercourse during heavy rain becomes trapped in certain channelled sections of the stream, and recurrent flooding has been observed. Moreover, the high level of pollution in the Taquara Stream has a direct impact on public health in the region, causing an elevated incidence of waterborne diseases (PMSB 2015).

The precarious socio-ecological conditions of the Taquara Stream are not an isolated case in Porto Alegre, where there are several watercourses in the same situation. The city's water and sewage department has been expanding sanitation infrastructures over the past few decades in order to improve water quality in urban streams. However, solutions are mostly

Fig. 3.2 Risk situations and discharge of effluents and solid waste in Taquara Stream. (Source: Souza 2019)

based on end-of-pipe centralised treatment facilities, which require large investments of capital (Ashley and Cashman 2006) and do not adequately meet the demand for sanitation in the poorest neighbourhoods of the city. Moreover, these responses are typically top-down and non-systemic—as they are mainly focused on solving a specific aspect of the problem rather than dealing with the complex interactive factors associated with it—and non-participatory, and thus unlikely to produce locally legitimised solutions (Smith 2008).

The first community mobilisations seeking solutions for the Taquara Stream took place in 1995, from an initiative created by a local community association. Between 1999 and 2009, Porto Alegre City Hall developed, through an intersectoral action in partnership with the Federal University of Rio Grande do Sul (UFRGS) and the community of Lomba do Pinheiro, the Integrated Project for the Sustainable Development of Lomba do Pinheiro, which included a linear park project as a strategy to recover the Taquara Stream and its watershed (PMPA 2003). However, this project was not completed due to a lack of political support and administrative discontinuities.

A Local Initiative for the Socio-Ecological Regeneration of the Taquara Stream

In 2015, a group of citizens—mainly connected to the Lomba do Pinheiro Community Garden, a key site for the articulation of actions related to public health issues in the neighbourhood—performed a diagnostic walk along the Taquara Stream. This mobilisation led to the creation of the Taquara Stream Watershed Working Group (WG) to seek solutions to the problem. The WG is a non-institutionalised group composed of community members (local leaders and residents), public sector technicians (with the most active participation from members of the health and water and sewage departments) and members of local educational institutions (local schools and UFRGS). The group also includes hybrid figures, that is, technicians with close connections to the community and representatives from educational institutions who live in the neighbourhood.

It is noteworthy that local leaders engaged in this initiative have a long history of acting in local social movements. Most technicians have been involved with issues affecting this community for many years and demonstrate personal commitment to the cause, and participants from educational institutions work with participatory research approaches at the interface of formal and informal education. These seem to indicate a preference of involved stakeholders for using participatory-oriented processes to deal with local problems. Additionally, the fact that most technicians are involved in tackling health issues and the participants' affinity with ecological perspectives have directly influenced the initiative's focus and approach. The WG members share the understanding that solutions to the complex socio-ecological problems found in the Taquara Stream and its surroundings depend on joint and collaborative efforts and integrated technical and learning-oriented actions. In this context, the WG has developed various actions in the territory since 2015, promoting a broader critical reflection on local issues, raising awareness and mobilising the local community and stakeholders, and improving local livelihoods and ecological conditions.

Results and Discussion

In this section, we address three key aspects of the Taquara Stream case that are relevant to our analysis. We first present the approach used in the initiative, introducing the normative knowledge adopted to guide problem-solving (e.g. rewilding streambanks and diverting streams represent two divergent technical strategies), and the social learning approach utilised to guide stakeholders' interactions. Then, we address the local community and stakeholders' engagement and interaction, focusing on the participatory and dialogical spaces and activities created to promote stakeholder interactions, as well as obstacles to cooperation that were identified. Lastly, the social learning practices verified in the leading group are analysed; at this point we address factors that enabled dialogical interactions and collective critical reflections among the participants of the initiative.

Approach Used in the Initiative

A general approach to the Taquara Stream initiative was defined by the WG members and is schematically illustrated by the chart in Fig. 3.3. This approach was set in order to guide group interactions, provide parameters to analyse local issues and design solutions, and evaluate outcomes of the process. The chart interconnects the initiative's guidelines, its general and specific goals, and short, medium and long-term coordinated actions performed by diverse collaborators. Social learning and ecological principles were adopted by the WG as the main guidelines for the movement.

The social learning approach is in line with the learning perspective advocated by the WG. It emphasises a symmetrical and inclusive way of interacting both among WG members and between these members and the local community (here we refer to the population inhabiting the Taquara streambanks). Social learning–oriented practices sought to create opportunities for the critical participation of social groups commonly excluded from the debate on issues concerning their territories in inclusive dialogical spaces where different world views could be appreciated

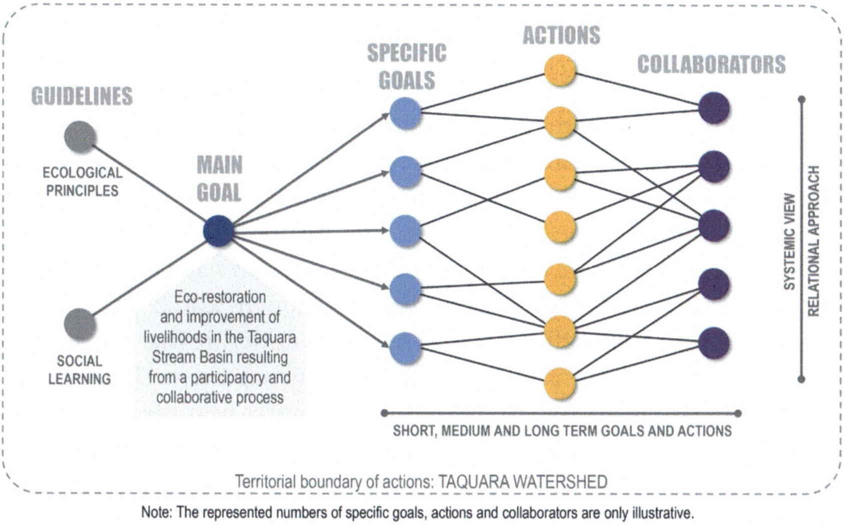

Fig. 3.3 WG approach to actions towards the regeneration of the Taquara Stream. (Adapted from Souza 2019)

and acknowledged (Balazs and Lubell 2014). The participation of disadvantaged communities in social learning processes tends to promote the critical reflection of these groups on the roots of their socio-ecological issues and to broaden their repertoire of solutions to local problems, thus enhancing their capacity building and autonomy in the management of their territories (Coudel et al. 2017). Such learning also enables the integration of knowledge brought by the various social actors involved (Murti et al. 2019). Furthermore, it can contribute to government representatives learning about community needs and perspectives on different water solutions, enabling them to identify the best way to address local issues and develop solutions tailored to local views (Balazs and Lubell 2014).

The ecological principles, on the other hand, highlight the need for life-supporting and life-enhancing practices for a balanced coexistence between human society and nature, as well as the importance of fostering a sense of co-responsibility among stakeholders (Capra 2002; Du Plessis and Brandon 2015). In line with innovative approaches to the regeneration of urban streams (Wantzen et al. 2019), these principles provide ecological parameters for the design of bio-inspired and low-cost solutions as well as a basis for the critical analysis of traditional responses to this type of problem. These principles provide normative knowledge and objective criteria to inform decision-making towards sustainability-focused solutions from the standpoint of ecosystems' supportability and minimisation of environmental impacts. For instance, sewage treatment solutions informed by these principles tend to differ from conventional centralised systems (Lyle 1996). They are usually smaller scale, implemented near effluent emissions and may be based on natural water purification systems such as bio-remediators (Hough 2002; Lyle 1996; Newman and Jennings 2008). Furthermore, considering that transportation greatly affects the costs of wastewater management, less energy-intensive decentralised systems that treat wastewater near the source, and the use of simple technologies based on recycling water and nutrients—which may also be associated with energy production, water reuse and local income generation—can lead to investment savings that are particularly important in economically constrained contexts (WWAP 2015).

Notwithstanding the benefits brought by small-scale alternative systems, the WG recognises that the pressing local demand for sanitation infrastructure compels the use of routine technical solutions—which are mostly large scale and centralised—since alternative practices are hampered by the lack of regulation. This points to a need to change and update norms, following a broader critical discussion on the sustainability of routine solutions within different institutional levels and government sectors (Medema et al. 2014), so that a wider range of options might be legitimised. The application of this principle could not, therefore, be observed in practice in the Taquara case, as technical actions implemented in response to the initiative were restricted to the emergency connection of some residences to the existing sewage system; the lack of regulation and resources meant that alternative solutions could not be tested.

Local Community and Stakeholders' Engagement and Interaction

Based on the approach presented earlier, the WG carried out several actions with different scopes and scales in the community of Lomba do Pinheiro. They aimed to create channels of dialogue with stakeholders and the local population, promote critical thinking on local issues, encourage residents' engagement in the movement and improve conditions at specific locations along the Taquara Stream. Several activities were carried out, such as collaborative map-making, diagnostic walks, lectures, and circle and one-to-one conversations with residents, among others. In social learning–oriented processes, the use of diverse participatory methodologies is key in order to sensitise and mobilise stakeholders in collaborative efforts towards sustainability (Brouwer and Woodhill 2016).

In cases involving marginalised populations, it is critical to understand the drivers that generate engagement and prioritisation of local environmental conservation. Marginalised groups are more likely to actively engage in sustainability-oriented social learning when they feel that the learning process recognises their understanding of their predicament and their own local circumstances, and when the learning, and those who support and facilitate it, open up spaces for participation, dialogue and building a common cause, while acknowledging the structural barriers to change. Monitoring and evaluating the changes that occur in the relationships and capacities within the community, as well as changes in socio-ecological terms, are critical for sustaining this engagement. Often, these changes are not felt or seen because such transformations can be subtle or slow, but on the occasions when they are noted and shared, they can be motivating and empowering (Wals et al. 2009). Additionally, re-establishing an empathic and emotional connection between the community and the local hydric ecosystem is key to promoting community engagement and fostering pro-environmental behaviour, and a culture of care towards the river (Ives et al. 2018; Wantzen et al. 2016).

The WG performed small-scale actions in specific areas of the watershed either to promote concrete improvements (such as the cleaning of

Fig. 3.4 Small-scale actions (A, B and C) and thematic events (D, E and F). (Source: Souza 2019)

specific sites through joint efforts), to establish conversations with residents or to diagnose socio-ecological conditions. These actions produced relevant outcomes—such as the regeneration of the Taquara Spring (Fig. 3.4)—and also revealed barriers to broader engagement of the most vulnerable population inhabiting areas along the Taquara streambanks. These barriers to dialogue construction included: (1) fear among these groups of being forced out of this territory due to illegal occupation of the riparian zone; (2) a lack of trust in actions involving the public sector due to perceived institutional negligence in meeting community's basic needs over time; and (3) the presence of drug trafficking in various locations, preventing communities living in areas controlled by traffickers from participating. These circumstances are common in vulnerable areas in Brazil and clearly expose the structural problems (Acselrad 2015) intrinsic to the harsh socio-environmental conditions around urban streams. These factors add complexity to the task of building the effective, inclusive approaches necessary to produce optimally legitimated solutions with the wide participation of local populations (Souza 2019).

The WG also promoted thematic events aimed specifically at expanding the local debate and creating opportunities for a wider knowledge exchange among the local community and stakeholders. The events included lectures and debates about research outcomes developed by supporting groups from local educational institutions and UFRGS on a variety of topics such as sanitation, water quality, waterborne diseases, land occupation patterns and the history of the area. These events stimulated the engagement of educational groups, as they created a means for integrating scientific information into the local process. The events were key moments in producing a boundary-crossing movement throughout a process that helped integrate local and technical knowledge (Engeström et al. 1995; Scholz and Steiner 2015), as requested by the projects for the socio-ecological regeneration of urban streams. Results from on-site research performed by supporting academic groups also led to the implementation of routine water monitoring by the health surveillance sector. This demonstrates the potential for connection between universities, schools, community and public sectors not just in terms of updating knowledge and constructing transdisciplinary approaches but also for broadening possibilities for generating new practices and routines (Peters and Wals 2013).

The outcomes of these actions were discussed in WG meetings, allowing for a better understanding of the local problems, as well as helping to identify new issues and to adjust strategies for future actions. The sequence of actions thus enabled 'learning by doing'. Agramont et al. (2019) call attention to the great epistemological differences among groups brought together in learning-oriented practices for water-related problem-solving in vulnerable areas and the importance of 'learning by doing' to overcome these disparities and integrate stakeholders in diversified ways. In this sense, a variety of interactive methods adapted to the local situation, developed by the participants themselves and guided by a social learning–oriented approach aimed at capacity building and empowerment of marginalised social groups, is essential in these cases (Phuong et al. 2018).

Social Learning Practices Within the Leading Group

The WG meetings included the planning of actions, conversations about the local problems and opportunities to critically evaluate activity outcomes (Fig. 3.5). Regular meetings (held biweekly or monthly) in a relatively small and diverse group (8–15 people), a friendly atmosphere and a balanced participation of group members were observed as facilitating conditions for the social learning process in the WG.

Informal interactions and a perceived egalitarian atmosphere between participants were also cited as favourable factors for the development of social learning–oriented practices in the group (Rist et al. 2006; Schusler et al. 2003). The egalitarian atmosphere, even in a group of people with very diverse backgrounds, was related to the perception of shared values amongst participants—which were also associated with perceived mutual trust and open communication among group members (Souza 2019). Also, the creation of opportunities for equal participation and the organisation of meeting dynamics benefited from the active facilitation of one participant. Her role was essential in organising practical aspects of the

Fig. 3.5 WG meetings at a resident's house. (Source: Souza 2019)

process and ensuring a balanced space for all participants to speak, which reveals the importance of facilitation in social learning–oriented practices (Ernst 2019), especially in contexts marked by power imbalances.

These aspects are fundamental to the search for joint and systemic solutions for the regeneration of urban streams, since the multiplicity of interests and perspectives underlying these processes can potentially generate excessively conflicting environments (Silva-Sánchez and Jacobi 2016) that might hinder collective learning. In the case of the Taquara Stream, divergence and diversity were drivers for learning since participants interacted in socially favourable conditions that were conducive to collective reflection and reflexivity (Wals 2009). The equalisation of diverse languages coexisting in the WG—popular, academic and technical—was also promoted by the group interactive space and facilitated knowledge exchange amongst participants. Community representatives in the WG also perceived that popular knowledge was respected, and this was cited as a stimulating and encouraging factor for their active participation in the movement. Furthermore, the interactive space provided by the WG also enhanced the relational capacities of group participants; this is particularly relevant in a context of deep social asymmetries, where a rebalancing of social forces is required to promote a more symmetrical communication between social actors.

Despite the reported positive factors, barriers to a more comprehensive collective learning process were also observed. A lack of engagement in the WG from specific public sector departments—such as the urban sanitation and housing departments—and the perceived absence of specific knowledge to deal with some local issues—such as violence in the area—appeared as barriers to a broader understanding of the problem. This calls attention to the need for institutional learning (Phuong et al. 2018), in order to raise technicians' awareness of the importance of co-constructing solutions along with communities and the need for transformations in institutional structures to facilitate participatory processes to tackle sanitation problems.

Lessons from the Taquara Stream Case

Table 3.1 recapitulates elements of the literature review and summarises lessons learned from the case study of the Taquara Stream, as presented earlier.

Table 3.1 Synthesis of the contents observed in the case study

Approach used in the initiative	Local community and stakeholders' engagement and interaction	Social learning practices within the leading group
Desirable characteristics (drawing from literature review)		
Socio-technical approach focused on sustainable solutions Technical and socially inclusive processes to support multi-stakeholder decision-making Sustainable parameters for decision-making and localised solutions	Promotion of inclusive and emancipatory participatory spaces where communities can engage in learning partnerships Embracing diversity and complexity Construction of transdisciplinary knowledge Fostering autonomy and co-responsibility among directly affected social groups Integration of subjectivities and cultural and environmental specificities in counter-hegemonic processes.	Promotion of reflexive processes for the transformation of views, values and mindsets Spaces for dialogical interaction Diversity as a driving force for collective learning Balanced participation of representatives from diverse social actors Trust building
Taquara Stream case		
Facilitating conditions		
Use of ecological principles to guide the design of sustainable solutions, low-impact and nature-integrated responses inspired by ecosystem functioning Social learning as a guiding principle to facilitate a relational approach and dialogical interaction between stakeholders	Use of diverse participatory methodologies for critical community engagement Actions at different scales involving different audiences, 'learning by doing' Thematic events to broaden the debate among stakeholders and the community Research carried out by schools and university on practical issues Research outcomes indicating issues to be monitored by the public sector	Regular meetings in small groups Egalitarian atmosphere based on perception of shared values among members Friendly atmosphere and informal interactions facilitating open communication and collective critical reflection Perception of trust among participants, facilitating dialogical interactions Facilitator in the group ensuring equal opportunities for participation

(*continued*)

Table 3.1 (continued)

Approach used in the initiative	Local community and stakeholders' engagement and interaction	Social learning practices within the leading group
Hindering aspects		
Lack of regulation on alternative systems for sustainable sewage treatment Urgent demands in a context of extreme need, requiring quick responses leading to replication of conventional technical solutions	Lack of community trust in actions involving the public sector Drug-related violence preventing communities living in areas controlled by traffickers from participating Community fear of being forced out from their land	Absence of specific knowledge to deal with local issues Lack of engagement from specific departments of the public sector in the WG

Conclusion

Actions for the regeneration of degraded urban streams in contexts of socio-ecological vulnerability call for inclusive problem-solving approaches to promote systemic and legitimate responses. The role of learning-oriented participatory processes is fundamental to generate dialogical spaces in order to enable individual and collective critical reflection on complex water-related issues, foster the co-responsibility of the social actors involved and arrive at collectively designed solutions. These processes can also lead to the capacity building and empowerment of local communities in the creation of territorialised agendas for the reversal of socio-ecological degradation of urban streams. Social learning can contribute in these contexts to the improvement of collaborative processes and coordinated actions, focusing on (1) the transformation of individual and collective views and practices; (2) bridging technical and socio-emancipatory approaches; and (3) the co-creation of local solutions aiming to improve the social and environmental conditions of urban streams, as well as social justice, equity and the well-being of communities inhabiting these areas.

The initiative on the Taquara Stream, led by the Taquara Stream Watershed Working Group, created conditions for the development of a social learning–oriented process that enabled a more balanced relationship between the local community, members of the public sector and educational institutions, facilitating the critical participation of these actors in the production of socially and environmentally sound actions and responses. The group also facilitated cross-sector articulations, which were made feasible by the participation of representatives from different public sector departments. Additionally, it is noteworthy that the approach advocated by the WG sought to transform prevailing top-down logic by promoting a dialogued process and educational activities to engage the local community and stakeholders in a critical debate in order to promote legitimate jointly designed solutions. This initiative demonstrated the potential of innovative and participatory learning environments to transform traditionally disadvantageous power relations and engage diverse stakeholders in social learning practices, thus making decision-making processes more democratic and inclusive.

This process did not occur, however, in an environment free of contradictions, and it was necessary to find a balance between the ideal responses from a sustainability point of view and those that were feasible, given contextual constraints and the urgent need for solutions in the face of extreme circumstances. Additionally, we emphasise the particularities of social learning processes, depending on the contextual conditions in which they unfold. Social asymmetries, specific circumstances of environmental degradation, economic constraints and political and institutional configurations, among others, influence and shape the learning processes according to existing possibilities and needs.

From a broader perspective, participatory initiatives towards the regeneration of urban streams are connected to water governance processes. Regenerative participatory water governance stimulates cooperation and collaboration between different sectors (e.g. the world of science, governance, civil society and of work). In this context, social learning furthers communicative practices that encourage cooperative and non-directive engagement of the various actors involved, stimulating what we might call a reflexive collaboration. Initiatives that seek to increase dialogue on the water crisis, vulnerability, ecological degradation of hydric ecosystems and the uncertainties inherent in the unsustainable model of society we are building bring to the foreground the weaknesses of the current governance processes used to tackle these problems and the potential for self-organised citizen actions. In terms of water justice, such reflexive movements offer the chance to expose and mediate power imbalances, which are conditional for establishing ecologically sustainable urban rivers and streams and more equitable access to water resources.

References

Abson, D. J., Fischer, J., Leventon, J., et al. (2017). Leverage Points for Sustainability Transformation. *Ambio, 46*, 30–39. https://doi.org/10.1007/s13280-016-0800-y.

Acselrad, H. (2015, Jan–June). Vulnerabilidade Social, Conflitos Ambientais e Regulação Urbana. *O Social em Questão - Revista do Departamento de Serviço Social da PUC-Rio, 33*

Agramont, A., Craps, M., Balderrama, M., & Huysmans, M. (2019). Transdisciplinary Learning Communities to Involve Vulnerable Social Groups in Solving Complex Water-Related Problems in Bolivia. *Water, 11*, 385. https://doi.org/10.3390/w11020385.

Ahern, J. (1995). Greenways as a Planning Strategy. *Landscape and Urban Planning, 33*, 131–155. https://doi.org/10.1016/0169-2046(95)02039-V.

ANA. (2018). *Conjuntura dos Recursos Hídricos no Brasil 2018: Informe Anual.* Brasília: Agência Nacional de Águas.

ANA. (2019). *Atlas Esgotos: Despoluição das Bacias Hidrográficas.* Brasília: Agência Nacional de Águas.

Ashley, R., & Cashman, A. (2006). The Impacts of Change on the Long-term Future Demand for Water Sector Infrastructure. In *Infrastructure to 2030: Telecom, Land Transport, Water and Electricity*. Paris: OECD Publishing. https://doi.org/10.1787/9789264023994-6-en.

Balazs, C. L., & Lubell, M. (2014). Social Learning in an Environmental Justice Context: A Case Study of Integrated Regional Water Management. *Water Policy, 16*, 97–120. https://doi.org/10.2166/wp.2014.101.

Balbim, R. (2016). *Geopolítica das Cidades: Velhos Desafios, Novos Problemas.* http://www.ipea.gov.br/portal/index.php?option=com_content&view=article&id=28688. Accessed 3 Jan 2020.

Bengtsson, S. (2019). Engaging with the Beyond – Diffracting Conceptions of T-Learning. *Sustainability, 11*, 3430. https://doi.org/10.3390/su11123430.

Bernhardt, E. S., Palmer, M. A., Allan, J. D., et al. (2005). Ecology. Synthesizing U.S. River Restoration Efforts. *Science, 308*, 636–637. https://doi.org/10.1126/science.1109769.

Brouwer, H., & Woodhill, J. (2016). *The MSP Guide: How to Design and Facilitate Multi-stakeholder Partnerships*. Wageningen: Wageningen University.

Capps, K. A., Bentsen, C. N., & Ramírez, A. (2016). Poverty, Urbanization, and Environmental Degradation: Urban Streams in the Developing World. *Freshwater Science, 35*, 429–435. https://doi.org/10.1086/684945.

Capra, F. (2002). *The Hidden Connections: Integrating the Hidden connections among the Biological, Cognitive, and Social Dimensions of Life*. New York: Doubleday.

Coudel, É. J. P., Tonneau, F., Bousquet, M., et al. (2017). Social Learning for Territorial Development. In T. Wassenaar, P. Caron, V. Papazian, et al. (Eds.), *Living Territories to Transform the World* (pp. 157–162). Paris: Quae.

Oliveira, C. B. H., & Oberrather, A. (2010). A Experiência Integrada da Lomba do Pinheiro: O Diálogo Territorial em Porto Alegre para Além do Orçamento Participativo. *Revista Crítica de Ciências Sociais, 91*, 255–274. https://doi.org/10.4000/rccs.4475.

de Sousa, A. C. A., Barrocas, P. R. G., de Sousa, A. C. A., & Barrocas, P. R. G. (2017). To Privatize or Not to Privatize: That Is the Question. But the Only Question? Reediting the Liberal Agenda for Drinking Water Supply and Sanitation in Brazil. *Cadernos de Saúde Pública, 33*. https://doi.org/10.1590/0102-311x00048917.

Denaldi, R., & Ferrara, L. N. (2018). A Dimensão Ambiental da Urbanização em Favelas. *Ambiente & Sociedade, 21*, e01950.

Du Plessis, C., & Brandon, P. (2015). An Ecological Worldview as Basis for a Regenerative Sustainability Paradigm for the Built Environment. *Journal of Cleaner Production, 109*, 53–61. https://doi.org/10.1016/j.jclepro.2014.09.098.

Engeström, Y., Engeström, R., & Kärkkäinen, M. (1995). Polycontextuality and Boundary Crossing in Expert Cognition: Learning and Problem Solving in Complex Work Activities. *Learning and Instruction, 5*, 319–336. https://doi.org/10.1016/0959-4752(95)00021-6.

Ernst, A. (2019). Review of Factors Influencing Social Learning within Participatory Environmental Governance. *Ecology and Society, 24*, 3. https://doi.org/10.5751/ES-10599-240103.

Fabos, J. G. (1995). Introduction and Overview: The Greenway Movement, Uses and Potentials of Greenways. *Landscape and Urban Planning, 33*(1–3), 1–13.

Faria, R. C., Faria, S. A., & Moreira, T. B. S. (2005). A Privatização no Setor de Saneamento tem Melhorado a Performance dos Serviços? *Planejamento e Políticas Públicas* 28. www.ipea.gov.br/ppp/index.php/PPP/article/view/46/45

Flink, C. A., & Searns, R. M. (1993). Greenways: A Guide to Planning. In *Design and Development*. Washington: Island Press.

Gallo, E., Setti, A. F. F., Ruprecht, T., et al. (2016). Territorial Solutions, Governance and Climate Change: Ecological Sanitation at Praia do Sono, Paraty, Rio de Janeiro, Brazil. In W. Leal Filho, U. M. Azeiteiro, & F. Alves (Eds.), *Climate Change and Health: Improving Resilience and Reducing Risks* (pp. 515–532). Cham: Springer International Publishing.

Hellmund, P. C., & Smith, D. (2006). *Designing Greenways: Sustainable Landscapes for Nature and People* (2nd ed.). Washington: Island Press.

Hoff, H. (2011, November 16–18). *Understanding the Nexus*. Background Paper Presented at the Bonn 2011 Nexus Conference, Bonn.

Hopkins, R. (2008). *The Transition Handbook: From Oil Dependency to Local Resilience*. Totnes: Green Books.

Hough, M. (2002). *Cities and Natural Process*. Abington: Routledge.

IBGE (Instituto Brasileiro de Geografia e Estatística). (2009). *Perfil dos Municípios Brasileiros*. ww2.ibge.gov.br/home/estatistica/economia/perfilmunic/2009/default.shtm. Accessed 15 Dec 2019.

IBGE (Instituto Brasileiro de Geografia e Estatística). (2010). *Primeiros Dados do Censo 2010*. ww2.ibge.gov.br/home/estatistica/populacao/censo2010/default.shtm. Accessed 7 Dec 2018

Ives, C. D., Abson, D. J., von Wehrden, H., et al. (2018). Reconnecting with Nature for Sustainability. *Sustainability Science, 13*, 1389–1397. https://doi.org/10.1007/s11625-018-0542-9.

Jacobi, P. R. (2013). Governança Ambiental e Práticas Participativas. In P. R. Jacobi (Ed.), *Aprendizagem Social e Unidades de Conservação: Aprender Juntos para Cuidar dos Recursos Naturais* (pp. 11–17). São Paulo: IEE/PROCAM.

Keen, M., Brown, V. A., & Dyball, R. (Eds.). (2005). *Social Learning in Environmental Management: Towards a Sustainable Future*. New York: Earthscan.

Loeber, A., van Mierlo, B., Grin, J., & Leeuwis, C. (2009). The Practical Value of Theory: Conceptualising Learning in the Pursuit of a Sustainable Development. In A. E. J. Wals (Ed.), *Social Learning towards a Sustainable World* (pp. 83–98). Wageningen: Wageningen Academic Publishers.

Lotz-Sisitka, H., Wals, A. E., Kronlid, D., & McGarry, D. (2015). Transformative, Transgressive Social Learning: Rethinking Higher Education Pedagogy in Times of Systemic Global Dysfunction. *Current Opinion in Environmental Sustainability, 16*, 73–80. https://doi.org/10.1016/j.cosust.2015.07.018.

Lyle, J. T. (1996). *Regenerative Design for Sustainable Development*. New York: John Wiley & Sons.

Machado, G. C. X. M. P., Maciel, T. M. F. B., & Thiollent, M. (2019). *Uma Abordagem Integral para Saneamento Ecológico em Comunidades Tradicionais e Rurais*. www.cienciaesaudecoletiva.com.br/artigos/uma-abordagem-integral-para-saneamento-ecologico-em-comunidades-tradicionais-e-rurais/17291?id=17291. Accessed 3 Jan 2020.

Macintyre, T., Lotz-Sisitka, H., Wals, A., et al. (2018). Towards Transformative Social Learning on the Path to 1.5 Degrees. *Current Opinion in Environmental Sustainability, 31*, 80–87. https://doi.org/10.1016/j.cosust.2017.12.003.

Maricato, E. (2000). As Ideais Fora do Lugar e o Lugar fora das Ideias. In O. B. F. Arantes, C. Vainer, & E. Maricato (Eds.), *A Cidade do Pensamento Único: Desmanchando Consensos* (pp. 121–192). Petrópolis: Vozes.

Maricato, E. (2003). Metrópole, Legislação e Desigualdade. *Estudos Avançados, 17*, 151–166. https://doi.org/10.1590/S0103-40142003000200013.

Medema, W., Wals, A., & Adamowski, J. (2014). Multi-Loop Social Learning for Sustainable Land and Water Governance: Towards a Research Agenda on the Potential of Virtual Learning Platforms. *NJAS - Wageningen Journal of Life Sciences, 69*, 23–38. https://doi.org/10.1016/j.njas.2014.03.003.

Menegat, R., et al. (1999). *Atlas Ambiental de Porto Alegre*. Porto Alegre: Universidade Federal do Rio Grande do Sul.

Moser, C. O. N. (1998). The Asset Vulnerability Framework: Reassessing Urban Poverty Reduction Strategies. *World Development, 26*, 1–19. https://doi.org/10.1016/S0305-750X(97)10015-8.

Mostert, E., Pahl-Wostl, C., Rees, Y., et al. (2007). Social Learning in European River-Basin Management: Barriers and Fostering Mechanisms from 10 River Basins. *Ecology and Society, 12*. https://doi.org/10.5751/ES-01960-120119.

Murti, R., Mathez-Stiefel, S.-L., & Rist, S. (2019). A Methodological Orientation for Social Learning Based Adaptation Planning: Lessons from Pilot Interventions in Rural Communities of Burkina Faso, Chile and

Senegal. *Systemic Practice and Action Research*. (in press). https://doi.org/10.1007/s11213-019-09495-8.

Newman, P., & Jennings, I. (2008). *Cities as Sustainable Ecosystems: Principles and Practices*. Washington: Island Press.

O'Sullivan, E., & Taylor, M. M. (Eds.). (2004). *Learning Toward an Ecological Consciousness: Selected Transformative Practices*. New York: Palgrave Macmillan.

Peters, S., & Wals, A. E. J. (2013). Learning and Knowing in Pursuit of Sustainability: Concepts and Tools for Trans-Disciplinary Environmental Research. In M. Krasny & J. Dillon (Eds.), *Trading Zones in Environmental Education: Creating Transdisciplinary Dialogue* (pp. 79–104). New York: Peter Lang.

Phuong, L. T. H., Wals, A. E. J., Sen, L. T. H., et al. (2018). Using a Social Learning Configuration to Increase Vietnamese Smallholder Farmers' Adaptive Capacity to Respond to Climate Change. *Local Environment, 23*, 879–897. https://doi.org/10.1080/13549839.2018.1482859.

PMPA (Prefeitura Municipal de Porto Alegre). (2003). *Plano Integrado para o Desenvolvimento Sustentável da Lomba do Pinheiro: construindo a Lomba do Futuro*. www2.portoalegre.rs.gov.br/spm/default.php?p_secao=288. Accessed 12 July 2015.

PMPA (Prefeitura Municipal de Porto Alegre). (2013). *Fundo Nacional do Meio Ambiente: Plano de Recuperação das Matas Ciliares do Arroio Dilúvio*. Convênio 025/2007–Adequação Ambiental das Propriedades Beneficiadas. Porto Alegre: PMPA.

PMSB (Plano Municipal de Saneamento Básico). (2015). *Diagnóstico*. Porto Alegre: PMPA.

Rhoads, B. L. (1999). Interaction Between Scientists and Nonscientists in Community-Based Watershed Management: Emergence of the Concept of Stream Naturalization. *Environmental Management, 24*, 297–308. https://doi.org/10.1007/s002679900234.

Rist, S., Chiddambaranathan, M., Escobar, C., & Wiesmann, U. (2006). "It was Hard to Come to Mutual Understanding ..."—The Multidimensionality of Social Learning Processes Concerned with Sustainable Natural Resource Use in India, Africa and Latin America. *Systemic Practice and Action Research, 19*, 219–237. https://doi.org/10.1007/s11213-006-9014-8.

Rist, S., Delgado, F., & Wiesmann, U. (2009). Social Learning Processes and Sustainable Development: The Emergence and Transformation of an Indigenous Land Use System in the Andes of Bolivia. In A. E. J. Wals (Ed.),

Social Learning Towards a Sustainable World (pp. 229–244). Wageningen: Wageningen Academic Publishers.

Röling, N. (2002). Beyond the Aggregation of Individual Preferences: Moving from Multiple to Distributed Cognition in Resource Dilemmas. In C. Leeuwis & R. Pyburn (Eds.), *Wheelbarrows Full of Frogs: Social Learning in Rural Resource Management* (pp. 25–47). van Gorcum: Koninklijke.

Rolnik, R. (1999). Exclusão Territorial e Violência. *São Paulo em Perspectiva, 13*, 100–111. https://doi.org/10.1590/S0102-88391999000400011.

Scholz, R. W., & Steiner, G. (2015). The Real Type and Ideal Type of Transdisciplinary Processes: Part I – Theoretical Foundations. *Sustainability Science, 10*, 527–544. https://doi.org/10.1007/s11625-015-0326-4.

Schusler, T. M., Decker, D. J., & Pfeffer, M. J. (2003). Social Learning for Collaborative Natural Resource Management. *Society & Natural Resources, 15*, 309–326. https://doi.org/10.1080/08941920390178874.

Searns, R. M. (1995). The Evolution of Greenways as an Adaptive Urban Landscape Form. *Landscape and Urban Planning, 33*, 65–80. https://doi.org/10.1016/0169-2046(94)02014-7.

Seyfang, G., & Smith, A. (2007). Grassroots Innovations for Sustainable Development: Towards a New Research and Policy Agenda. *Environmental Politics, 16*, 584–603. https://doi.org/10.1080/09644010701419121.

Silva-Sánchez, S., & Jacobi, P. R. (2016). Implementation of Riverside Parks in the City of São Paulo: Progress and Constraints. *Local Environment, 21*, 65–84. https://doi.org/10.1080/13549839.2014.922060.

Smith, J. L. (2008). A Critical Appreciation of the "Bottom-up" Approach to Sustainable Water Management: Embracing Complexity Rather than Desirability. *Local Environment, 13*, 353–366. https://doi.org/10.1080/13549830701803323.

Sol, J., Beers, P. J., & Wals, A. E. J. (2013). Social Learning in Regional Innovation Networks: Trust, Commitment and Reframing as Emergent Properties of Interaction. *Journal of Cleaner Production, 49*, 35–43. https://doi.org/10.1016/j.jclepro.2012.07.041.

Souza, D. T. P. (2019). *Caminhos para a Transformação de uma realidade local: uma experiência de Aprendizagem Social para a Sustentabilidade na Comunidade da Lomba do Pinheiro – O Caso do Arroio Taquara*. PhD thesis, University of São Paulo (USP).

Souza, D. T. P., Wals, A. E. J., & Jacobi, P. R. (2019). Learning-based Transformations Towards Sustainability: A Relational Approach Based on

Humberto Maturana and Paulo Freire. *Environmental Education Research, 25*(11), 1605–1619. https://doi.org/10.1080/13504622.2019.1641183.
Spirn, A. W. (1984). *The Granite Garden: Urban Nature and Human Design*. New York: Basic Books.
Sterling, S. (2011). Transformative Learning and Sustainability: Sketching the Conceptual Ground. *Learning and Teaching in Higher Education, 5*, 17–33.
Šūmane, S., Kunda, I., Knickel, K., et al. (2018). Local and Farmers' Knowledge Matters! How Integrating Informal and Formal Knowledge Enhances Sustainable and Resilient Agriculture. *Journal of Rural Studies, 59*, 232–241. https://doi.org/10.1016/j.jrurstud.2017.01.020.
Swyngedouw, E., Kaïka, M., & Castro, E. (2002). Urban Water: A Political-Ecology Perspective. *Built Environment, 28*, 124–137.
Tucci, C. E. (2005). *Gestão de Águas Pluviais Urbanas: Saneamento para Todos*. Brasília: Ministério das Cidades.
Tunstall, S. M., Penning-Rowsell, E. C., Tapsell, S. M., & Eden, S. E. (2000). River Restoration: Public Attitudes and Expectations. *Water and Environment Journal, 14*, 363–370. https://doi.org/10.1111/j.1747-6593.2000.tb00274.x.
UN-Habitat. (2016). *World Cities Report 2016. Urbanization and Development: Emerging Futures–Key Findings and Messages*. Nairobi: United Nations Human Settlements Programme.
Wahl, D. C. (2016). *Designing Regenerative Cultures*. Axminister: Triarchy Press.
Wahl, D. C., & Baxter, S. (2008). The Designer's Role in Facilitating Sustainable Solutions. *Design Issues, 24*, 72–83. https://doi.org/10.1162/desi.2008.24.2.72.
Wals, A. E. J. (Ed.). (2009). *Social Learning Towards a Sustainable World: Principals, Perspectives and Praxis*. Wageningen: Wageningen Academic Publishers.
Wals, A. E. J. (2011). Learning Our Way to Sustainability. *Journal of Education for Sustainable Development, 5*, 177–186. https://doi.org/10.1177/0973408211100500208.
Wals, A. E. J., & Bawden, R. (2000). *Integrating Sustainability into Agricultural Education: Dealing with Complexity, Uncertainty and Diverging Worldviews*. Ghent: ICA.
Wals, A. E. J., & Heymann, F. (2004). Learning on the Edge: Exploring the Change Potential of Conflict in Social Learning for Sustainable Living. In A. Wenden (Ed.), *Educating for a Culture of Social and Ecological Peace* (pp. 123–144). Albany: State University of New York Press.

Wals, A. E. J., Van der Hoeven, N., & Blanken, H. (2009). *The Acoustics of Social Learning: Designing Learning Processes that Contribute to a More Sustainable World.* Wageningen: Wageningen Academic Publishers.

Wantzen, K. M. (2018, June 4–8). *Urban River Restoration in the Global South–Problem Analysis and Suggestions by the UNESCO Chair for River Culture/Fleuves et Patrimoine.* Paper Presented at the 3rd International Conference on Integrative Sciences and Sustainable Development of Rivers, Lyon.

Wantzen, K. M., Ballouche, A., Longuet, I., et al. (2016). River Culture: An Eco-social Approach to Mitigate the Biological and Cultural Diversity Crisis in Riverscapes. *Ecohydrology & Hydrobiology, 16*, 7–18. https://doi.org/10.1016/j.ecohyd.2015.12.003.

Wantzen, K. M., Alves, C. B. M., Badiane, S. D., et al. (2019). Urban Stream and Wetland Restoration in the Global South—A DPSIR Analysis. *Sustainability, 11*, 4975. https://doi.org/10.3390/su11184975.

Warner, J., Wester, P., & Bolding, A. (2008). Going with the Flow: River Basins as the Natural Units for Water Management? *Water Policy, 10*, 121–138. https://doi.org/10.2166/wp.2008.210.

Wehn, U., Collins, K., Anema, K., et al. (2018). Stakeholder Engagement in Water Governance as Social Learning: Lessons from Practice. *Water International, 43*, 34–59. https://doi.org/10.1080/02508060.2018.1403083.

Wildemeersch, D. (2009). Social Learning Revisited: Lessons Learned from North and South. In A. E. J. Wals (Ed.), *Social Learning Towards a Sustainable World: Principles, Perspectives, and Praxis* (pp. 99–116). Wageningen: Wageningen Academic Publishers.

WWAP (United Nations World Water Assessment Programme). (2015). *The United Nations World Water Development Report 2015: Water for a Sustainable World.* https://unesdoc.unesco.org/ark:/48223/pf0000231823. Accessed 3 Jan 2020.

4

A Systems Analysis Approach to Addressing Contemporary Water Challenges: Management Improvements in Brazil and Beyond

Nazli Koseoglu

Introduction

Development of water resources and socio-economic development are interlinked and entangled in a vicious circle. Reduced economic development makes financing water and sanitation services and ensuring a reliable water supply a challenge, while a lack of water-related infrastructure is a barrier to a sustainable socio-economic progress (Goswami and Bisht 2017). The effects of this synergetic relationship between socio-economic development and development of water resources are particularly pronounced in developing regions. Environmental issues started to appear more prominently in the economics debate worldwide in the 1950s. Water pollution and its harmful effects, including rising costs of water treatment and public health expenses, increased as a consequence of the rapid demographic and economic boom after the Second World War

N. Koseoglu (✉)
Social, Economic and Geographical Sciences, The James Hutton Institute, Aberdeen, UK
e-mail: Nazli.Koseoglu@hutton.ac.uk

A. A. R. Ioris (ed.), *Environment and Development*,
https://doi.org/10.1007/978-3-030-55416-3_4

(Du Pisani 2007). In addition to pollution, the allocation of water also became a challenge, as additional industrial demands and increased household consumption led to competition for finite resources. Each policy decision involves a different set of trade-offs, with winners and losers. The isolated economic approach to resource planning and environmental management at the time began to fall short in its management of human-environment systems.

Management of water resources is inseparable from politics, culture and economy. Effective coordination of water resources goes beyond human interventions to hydrological processes. It considers political, socio-economic and ecological dynamics in complex interactions between stakeholders, with the aim of preventing future water conflicts that might occur due to environmental change and increasing demand. Water resources cannot be considered as hydrological systems in isolation from socio-economic and sociopolitical dynamics (Ioris 2001, 2011; Linton and Budds 2014). In the management of water resources, questions of power, access, struggle and representation go hand in hand with material aspects of environmental change, approaches to which are shaped not only by technical and scientific representations of nature but also scientific research cultures and subjective interests of the scientists (Demeritt 2001). Finding a balanced and intersectional account of nature-society interactions between the critical discourse of political economy and the positivist perspective of environmental science discourse in water resources management is often challenging, if not impossible (Budds 2009).

In theory, principles such as equity, efficiency, productivity and maintaining basic domestic and ecological demands are incorporated into water policy. However, it is almost impossible to allocate water in a way that simultaneously enforces all these principles. As a result, compromises and priorities are made under asymmetric political power (Wang et al. 2017). Therefore, water conflicts are closely linked to the asymmetrical distribution of political power, and thus, the outcomes of water policies usually reflect the existing distribution of social costs and benefits in a society. The development of water resources, like any resource development, is theoretically beneficial to the overall progress and welfare of society. The resulting socio-economic gains can either be an opportunity for redistribution and improvement of the lives of the many or can replicate the existing distribution of power and wealth in society, benefitting the few. Distributional justice is not limited to the distribution of benefits. The development of water resources, both in the form of physical interventions such as the building of dams, irrigation canals and multipurpose reservoirs and in the form of technocratic measures such as zoning and water allocation plans that reform water-related rights (Wang et al. 2017), usually brings about significant environmental change and creates externalities that affect certain communities and individuals negatively.

In this chapter, I explore the contribution of systems models to water management and their evolution over time in relation to socio-economic and political systems. I focus on the last four decades of the SA modelling tradition applied to water resources management in parallel with the important historical trends and themes of each period. The timeline is divided into two main periods (1980–2000 and 2000–2020), and the major models that have addressed critical water management issues in different parts of the world in each era are introduced. I also discuss relevant and contemporary water challenges in the Global South that could be addressed by examples of systems thinking and modelling developed in Brazil.

A Brief History of Water Resources Development and Systems Models

Environmental economics was established as a separate field in the 1950s, borrowing concepts such as public goods, cost-benefit analysis and externalities from the field of neoclassical welfare economics (Sandmo 2015). However, economics or natural resource allocation purely based on monetary outcomes alone were not sufficient for making informed decisions that might result in long-term and unintended legacies such as soil contamination or the prioritisation of one sector over another. Economic models lacked holistic consideration of the natural environment where water was sourced and returned (Cosgrove and Loucks 2015), and multi-objective analysis was not yet an established concept. Typically, the sole objective of water resources planning was to maximise net economic benefits without internalising environmental costs or considering the equitable distribution of benefits (Lund et al. 2006). This narrative dominated early water resource development discourse. However, soon, the critical view began to form in scientific circles that water as a national resource had to be developed in a way that would serve and maximise the national interest, and the national interest went beyond maximising net economic output (Brown et al. 2015; Maass et al. 1962).

Systems Analysis (SA) initially emerged from innovations during the Second World War and transformed into an academic field stemming from cybernetics and operational research in the late 1940s (Rindzevičiūtė 2016). It was proposed that SA could address the shortcomings of monodisciplinary natural resource management discourses, which had previously dominated water management and other policy-relevant environmental resource issues, by combining qualitative and quantitative modelling practices (Dolado 1992; Hodgson 1992), bringing together different types of evaluation methods for analysis based on different criteria. Water systems are dynamic and influenced by a number of factors such as human activity, different stakeholder interests, climatic conditions, natural and anthropogenic processes, and physical constraints, such as water supply shortages caused by limited capacity, which have a high societal cost (Wolstenholme 2004). As a result of many dimensions and objectives interacting vertically and horizontally, there is a certain degree of complexity in water systems, like other human-environment systems involving feedbacks and discontinuous non-linear relations (Sahin et al. 2016). Analysing such systems required an interdisciplinary approach and simultaneous acknowledgement of conflicting and complementary outcomes and constraints arising from spatial and temporal aspects, such as the water rights framework of a given legislature (Beard 1971).

An interdisciplinary understanding of a system's structure and complexity makes its ex-ante projection by SA models more reliable, and SA models are found to be more robust in addressing uncertainties (Sterman et al. 1994). The SA approach breaks down systems that are composed of smaller constituent parts and uses these parts and the relationships among them to explain how the system as a whole works, with the aim of understanding how complex systems evolve and behave over time (Sahin et al. 2016). Initial system conceptualisation in SA is based on determining the positive and negative relationships between different factors as well as the feedback loops, systematic behaviour patterns called system archetypes, and delays in the system (Wolstenholme 2004). Feedback loops are the basic structural elements of dynamic systems that reflect a chain of causal relations among the interacting components of a system (Kirkwood 1998). They are often linked together and display non-linear relationships that often cause counterintuitive behaviour (Forrester 1971). SA models enable a more comprehensive representation of these relationships, making it possible to simulate, forecast and optimise the complex human-environment systems in decision-making processes (Homer 1996; Sterman 2000; Winz et al. 2009).

The Early Era of Systems Analysis Models in Water Management

The themes of complexity, informational and network control have been discussed among policy scientists since the 1960s. In the early 1960s, political scientist Arthur Maass, in collaboration with engineers and economists, launched an interdisciplinary and applied field of inquiry known as water resources systems analysis (Brown et al. 2015), using the technical principles of systems analysis research. The early SA studies on water resources share the zeitgeist of the 1960s, known as the Decade of Development (UN 2017b), and were largely formulated in line with real-life issues, in order to provide solutions for policymakers (Rindzevičiūtė 2016). They focused on developing water resources to meet the multiple objectives of water supply provision required for economic development, that is, power generation, low-flow regulation and flood control, and informed associated infrastructure planning decisions (Maass et al. 1962). While water quality and water basin management under uncertainity have remained recurring themes, in each decade there were specific water management challanges that attracted researchers' attention.

The International Institute for Applied Systems Analysis (IIASA), established in 1972, was remarkable not only for being the first international think tank but also because it was established by opposing Cold War superpowers, the Soviet Union and the USA, to facilitate collaboration among the transnational community of systems scholars on both sides of the Iron Curtain. The establishment of the IIASA consolidated existing SA research networks, promoting the epistemological agenda of systems analysis and contributing to mainstreaming its applications (Rindzevičiūtė 2016). A detailed overview of SA-based models developed during this era and applied to water management can be found in Fleming (1975). In the 1980s, the focus of systems analysis research shifted from planning and development of water resources to developing analytic approaches for operational efficiency, for example, storage (Orlovsky et al. 1982) and pollution control. The negative externalities of rapid and uncontrolled development in the previous decades became observable and began to significantly impact areas of public policy such as health and water quality (Osborne 2008). Regional and central authorities became more concerned with sustaining the natural environment in response to these impacts. Thus, the role of water management in regional policies, economic development and sectorial control of water quality and land use gained further importance and visibility in the 1980s.

This specific policy interest led to the application of multi-objective models, which examine the trade-offs between different scenarios to determine 'optimal' management of water resources at various geographic and administrative scales from the perspective of policymakers. Several new themes developed during the 1980s, such as transboundary water management across different legislatures; the relationship between different types of land use; water resources and regional development; introduction of the concepts of reliability, vulnerability and resilience; minimising uncertainty in future forecasts; and the frequent use of multipurpose and sectorial water allocations in SA models. While policy-relevant research started as early as the 1960s, evidence-based policymaking did not become mainstream until the end of the 1990s (Rindzevičiūtė 2016). During the 1990s the capability of SA applications increased, together with developments in computer science and technology, in general, and more user-friendly system dynamics software with interactive features and real-time simulation (Winz et al. 2009). The recognition of the multiple and complex links between people and natural systems, the hydrological system in particular, and of humans' dependence on nature beyond resource extraction led to the coining of the term 'coupled human-environment systems' in the 1990s. This term characterises the dynamic two-way interactions between humans (e.g. economic, social) and natural (e.g. hydrologic, atmospheric, biological, geological) systems (Carter et al. 2014) and implies that the human and environmental systems can no longer be treated separately from each other (Sheppard and McMaster 2004). This has also led to recognising the significant role of human activities on climate change.

Until the mid-1980s, industrialised countries in the European Community and elsewhere generally had separate sets of policies for agriculture and for the environment. While agricultural policies encouraged intensification without considering environmental impacts, environmental policies did not address the consequences of intensified agricultural practices on water resources. Therefore, agriculture in the late 1970s and the early 1980s became a major source of diffuse pollution of water resources, including drinking water, which affected public health (Scheierling 1995). Before significant policy changes took place in the 1990s, the first significant response to this came from the scientific community, which developed SA models linking regional land use with water quality. Chemicals, Runoff, and Erosion from Agricultural Management Systems (CREAMS), was a land-use-based model that linked agricultural land use with water pollution (Knisel 1980). It was developed to identify the impact of erosion on surface and subsurface flows via sub-models of erosion and chemical changes on water quality in the Švihov Reservoir in the Sedlicky catchment in then Czechoslovakia (Holy et al. 1981). Later, another model, Groundwater Loading Effects of Agricultural Management Systems (GLEAMS), was developed as a modification of CREAMS, with the addition of vertical pesticides flux, to estimate diffuse pollution affecting groundwater recharge areas in agricultural regions (Leonard et al. 1987).

Beyond agricultural land uses, SA models in this period looked at wider regional land use and planning, and assessed the policy for regional development and how different options competed for and impacted on available water resources (Kindler and Loucks 1989). Seo and Sakawa (1989) looked into the role of industrial land-use programmes in water quality control in Osaka, Japan. They analysed the trade-offs between the conflicting objectives of achieving better water quality and maximising profit from economic activities by using a dyanmic multiple citeria decision analysis model designed for multi-level systems. The results informed the evaluation of industrial development options for the region. In addition to being more scientific, models were also perceived as being less biased tools of policy assessment and were particularly used not only in sectoral water allocation but also in transboundary conflicts around water. Developing effective means of managing water quality using a SA approach became a necessity in operational water management (Beck 1981) and in regional planning, especially for intensively used transboundary rivers and lakes, such as the Vistula and Danube river systems in the Central and the Eastern Europe. Models were employed as a neutral, transparent and collaborative process for conflict resolution among different regional and national interests. The models adressing agricultural pollution–related problems in the Southern Peel, Canada (Orlovsky and van Walsum 1984), and the models addressing problems relating to the groundwater quantity and quality management in the mining region of Lusatia, between Germany and Poland (Kaden 1985), are important examples. These analytical models, were coupled with interactive computer graphics and were used for estimating the additional cost of resolving water-use conflicts.

During this period, regional planning models designed to aid decision-making and policy became known as 'decision support tools'. These used systems analysis tools to define regions, using impact diagrams, hierarchies of interdependencies providing decision-makers with evidence for regional socio-economic structure, schemes of regional problems, scenario analyses, parameters for uncertainty, and deterministic formulation (Orlovsky et al. 1986). Decision Support Systems (DSS) were applied to resource management and environmental protection in the form of an interactive computer programme in two further case studies: the Danube basin in central Europe and the Zambezi river basin in Western Africa. This approach was novel as it required at least two clients with conflicting interest to collaborate for joint solutions rather than modelling the system from the perspective of a single stakeholder (Kindler and Loucks 1989). Meanwhile water quality management remained a significant topic in the systems analysis research in the 1990s. Somlyody (1995) reviewed trends and developments in water quality management since the 1970s in terms of wastewater treatment, monitoring, environmental impact assessment and legislation, and developed the Use Attainability Analysis (UAA) method. UAA is an integrated water quality management and planning method based on the SA approach. The UAA consisted of three interacting sub-components: a water body assessment to identify the degree and the source of water quality problems in each receiving water body; the total maximum daily load process, which accesses the feasibility of water quality targets and the future abatement limits for point and non-point resources; and a socio-economic analysis, which assesses the availability of funds for water quality management (Novotny 1996).

Another important milestone in water management that took place in the 1990s was the formalisation of the concept of Integrated Water Resources Management (IWRM). This new concept aimed to reflect the multidimensional nature of water management and extended the collaborative watershed management concept that had emerged in response to the failure of the centralised command and control methods in preventing water pollution and over-allocation. The concept updated long-established integrated approaches with an emphasis on sustainable development and the inclusion of environmental protection, participation, efficiency and equity concepts. Since the early 1990s, IWRM has been the dominant water management paradigm in both rich and poor countries. Over the following two decades, 80% of the countries around the world adopted IWRM principles in their water law or policies, and 60% developed national IWRM plans (Mehta et al. 2017). The rise of IWRM in the 1990s highlighted the need for standardised interdisciplinary frameworks and technical tools capable of accommodating the new multi-criteria analysis and optimisation paradigm shift informing water resources development decisions worldwide.

The Recent Era of Systems Analysis Models in Water Management

By the 2000s, the drivers of global change were becoming better understood, as were the strategies for reducing the associated risks and uncertainties. Models developed to understand and reduce the uncertainties around the climate change became a major part of the SA research. The water-energy-land-food nexus approach emerged in an attempt to recognise the interdependencies and the interlinkages between water, agriculture and energy production (Ringler et al. 2013). The Global Biosphere Management Model (GLOBIOM) is a bottom-up model that represents the supply side of agricultural, forestry and bioenergy production via a spatially explicit description of land resources. This was one of the first global models to analyse the competition for land between the land-based production sectors, e.g. agriculture, forestry and bioenergy, using future scenarios (Havlik et al. 2011) and to evaluate the rational allocation of land and water between the production of food, wood fibre and biofuels at the global scale. Sauer et al. (2010) further integrated alternative irrigation systems into GLOBIOM to estimate the outcomes of regional adaptations in agricultural water use in different development scenarios.

A consensus had been reached by the 2000s about climate change and its game-changing influence on the water cycle and the rest of the environment (Oreskes 2005). The debate on climate change and its already observable effects made the topic of resilient and adaptive water management more of a priority and increased the importance of groundwater and of developing other mitigation options for safeguarding water supply in the event of extreme weather events. The nexus approach exploring the connections between land-energy-water also became part of the debate around regional and sectorial resource allocation for economic growth and human development. In the higher-income countries where SA was born and flourished, in terms of both model development and implementation, environmental policies like the Water Framework Directive of the European Commission (2000) were introduced to safeguard water quality with a more holistic approach and to guide the allocation of scarce water resources. A growing number of countries in Europe started using economic instruments mandated by the Dublin Principles (ICWE 1992) to reform their inefficient water allocation regimes (Rey et al. 2019). This was also addressed in SA models at the time. Riegels et al. (2013) applied the SA approach to designing and comparing different economic instruments for the efficient management of ground and surface water resources. The resulting model was used to identify the optimal prices for groundwater, controlling for the energy prices within the Water Framework Directive's ecological flow and groundwater sustainability targets.

The role of groundwater in water security and resilience became better understood with the changing availability of precipitation and surface waters due to climate change (Margat and van der Gun 2013). Issues related to the alarming rates of depletion and decline in groundwater became increasingly evident worldwide in the late 1990s and 2000s (FAO 2011). Excessive withdrawal, waterlogging and salinisation, insufficient coordination between the use of surface- and groundwater resources, and diffuse pollution, mainly as a result of agriculture, were the underlying reasons for this situation. This was particularly common in regions with high population density, dynamic tube-well irrigated agriculture and insufficient surface water resources (Gorelick and Zheng 2015; Shah et al. 2000). Groundwater models progressively aimed to investigate spatial outcomes (Ahmad and Simonovic 2004) and operational planning. Besides agriculture, the uncontrolled growth of cities had already led to over-exploitation and contamination of groundwater resources in water-stressed urban locations. Carrera-Hernández and Gaskin (2007a, b, 2008) developed a regional database for borehole, run-off and climate data and an open-source groundwater modelling tool for Geographic Resources Analysis Support System a regional database of the boreholes, run-off and climate to optimise groundwater recharge using reclaimed wastewater in the Mexico City Basin.

Rapid economic growth and accelerated development and urbanisation in many developing countries worldwide affected the availability and the quality of water resources in these regions. China emerged as the world's largest economy[1] by 2014 (Woetzel et al. 2019). However, this unprecedented scale of growth had significant effects on the water

[1] When estimated by gross domestic production at purchasing power parity.

resources and overall environment. Water scarcity and pollution becoming a threat to the public health and the continuity of economic growth of the country, led researchers to focus on these issues. Zhang et al. (2010) used an ecological network utility matrix structure and ecological relationships in an urban water metabolic system in Beijing using the 'urban disease' concept to optimise urban water use. They did this by addressing the relationships and mutualism among the distinct flows (local environment, rainwater collection, industry, agriculture, domestic sector, and wastewater recycling) and classifying each as producer, consumer and reducer of trophic levels. Yao et al. (2015) looked into the sectorial water uses and their socio-economic drivers in the Pearl River Delta using homogenised annual water use data from 2000 to 2010. Their model built on the global water use model WaterGAP2 (Alcamo et al. 2003) to identify the relationship between development and water use intensity in different cities in the region. Xu et al. (2015) simulated the spatially detailed analysis of expansion in coal and agricultural production in the Shanxi region against national, sub-national and regional trends established by global models (e.g. MESSAGE, GLOBIOM). The model calculated shadow prices based on each location's water shortage risk and set investment priorities to improve water saving, retention and transfer technologies, with the possibility of trading water between locations. It also projected aggregate water demands in terms of local resource constraints such as land availability, environmental and social circumstances.

Another major challenge in recent decades has been the inclusion of environmental water demand in the portfolio of competing water uses. The concept of Environmental Flow Requirement (EFR) has arisen in an attempt to define the right of nature to a certain amount and quality of water, in order to sustain healthy ecosystems. Several SA models have looked into environmental flow quantification, bringing flow, ecological, geomorphological and social elements together in an interdisciplinary way. This has provided more holistic outcomes in environmental flow quantification than conventional methodologies based only on hydrological, hydraulic or habitat simulations (Pastor et al. 2014). Van Vliet et al. (2012, 2013a, b) examined global stream flow and the thermal habitats of freshwater fish under climate change, and found that increasing water temperatures had far-reaching impacts on ecosystems, such as an increased need for cooling water in the energy sector, which, in turn, affects electricity supply and the availability of environmental flows.

The development of the world's water and land resources led to steady improvements in quality of life for a growing world population throughout the 1980s, 1990s and 2000s. However, in many cases, development planning was short sighted and came at a cost paid by the general public or certain communities. Predicted economic and social benefits were not fully realised or equitably distributed. Unanticipated externalities of development often surfaced in a relatively short period of time, causing production loss and adverse effects on health and the environment, as well as socio-economic problems (Ghassemi et al. 1995). Although novel at the time of its introduction, it gradually became clear that the IWRM concept was unable to fully address dynamic water management issues in a context of uncertainty caused by climate change and general global change. The centralised and normative framework of IWRM has attracted considerable controversy and criticism for limiting alternative thinking on pragmatic solutions and preventing water-related and other policy reforms (Giordano and Shah 2014). In the 2010s, more adaptation-focused perspectives such as Nexus Approach, Problem-Driven Iterative Adaptation and Adaptive Water Management began to take over the hegemony of the IWRM brand in water management and extended the concept from the original river basin scale of IWRM to multiple geographical scales (Butterworth 2014).

With the shift from IWRM to the Nexus-focused perspectives, elements of governance and participation have also evolved. It has been recognised that for stakeholder participation to be meaningful, it has to go beyond passive involvement. Increasingly, the objective became active inter-sectorial collaboration through bottom-up initiatives and institutions operating in a multi-tiered and decentralised manner (Benson et al. 2015).

Incorporation of Stakeholder Participation in SA Models

Despite the constant evolution of policies, modelling methods and technological tools to predict, monitor and manage water systems, some chronic water management issues have not been fully resolved for decades, such as water quality and water allocation issues. Gradually, both researchers and policymakers have acknowledged that successful water management goes beyond physical interventions to hydrological processes. While scientific models, evidence-based policy and technical innovations offered great potential for improvement without involving stakeholders in decision-making, it was not possible to achieve adaptive and resilient human-environment systems. Designing and implementing policies that resonate with the physical, cultural and socio-economic realities of societies enable higher compliance. Planning and development activities, such as major infrastructure plans and policies, that fail to acknowledge the various interests and values involved have received adverse reactions from the public. These reactions have ranged from protests to non-compliance to legal cases delaying or obstructing construction. Undertaking effective participatory processes early on could have prevented some of the dissatisfaction with policy decisions.

Participatory and stakeholder models introduced public consultation in research and development (Palmer et al. 1993). Such SA models became mainstream in the 2000s due to the need to understand the human agency involved in water use and management beyond the assumed rationality of users, a concept borrowed from neoclassical economics. The models took account of and adapted to societal, economic and environmental conditions characterised by uncertainty, variability and considered the interdependence, competition and coordination of the stakeholders involved in the system (Wang et al. 2017). Besides facilitating collection of qualitative input for the construction and calibration of the actual models, engagement events also provided an opportunity for shared and social learning. The consultations offered the opportunity to develop a better common understanding between scientists and a common space to meditate different stakeholder interests.

Stakeholder consultations benefited from different engagement and communication techniques and tools. The Shared Vision Model (SVM) was at the centre of the Shared Vision Approach, in which stakeholders were involved in the process of constructing the model. The model, which used a Participatory Integrated Assessment process (Connor et al. 2004), was produced using a user-friendly graphic interface software platform that facilitated non-technical stakeholder cooperation in preventing water-related conflicts in the Rappahannock River Basin in the USA. Cohen et al. (2006) also used a Participatory Integrated Assessment process centred on an SA model to investigate the participatory assessment of integrated water management and climate change in the Okanagan Basin in Canada. The purpose of the activity was to assist the local community in evaluating their water resources in a systems context and to help them cooperate to achieve more climate change-resilient planning and policy development. The 'concept maps' technique (Cañas et al. 2005; Novak and Cañas 2008) was used to represent knowledge in an organised and simplistic way in the development of the Water Futures and Solutions conceptual framework. The concept maps technique was adopted in engaging practitioners with different backgrounds involved in the stakeholder consultation and to explain complex concepts and the relationships between them (Magnuszewski et al. 2015).

The use of participatory models became common in the 2000s, especially in the field of flood management in relation to the themes of regime shifts, resilience and transformability (Sendzimir et al. 2008). The role of stakeholder involvement in adapting to and mitigating the impacts of climate change had become evident and models designed in this period responded accordingly. Development on the transboundary Tisza River valley in Central Europe at the end of the twentieth century led to social and environmental consequences in the region. Sendzimir et al. (2008) explored the different world views of stakeholder subgroups in the Hungarian reaches of the Tisza River Basin, where the impact and frequency of floods had increased in the previous two decades (Koncsos and Balogh 2007; Sendzimir et al. 2007). Tools such as causal loop diagrams commonly used in SA models for informal capacity building and an adaptive learning process (Stave and Sterman 2002) were adopted in the study (Sendzimir et al. 2008). While flood management issues were the focus of adaptive management approaches in some countries, management against droughts was the objective in others. In Iran, Yazdanpanah et al. (2014a, b) used a systems analytic perspective to study stakeholders' collective action and memory in water governance, the interrelated risks associated with droughts and behavioural patterns of water use in the agricultural sector.

Games were also used as a method of improving stakeholder engagement in this period, especially to mimic democratic processes in policy-making and implementation. Games provided scientists with an opportunity to examine the decision-making processes of different stakeholder groups with interests in the floodplains. Stefanska et al. (2011) developed a system dynamics model of integrated floodplain agriculture, the Floodplain Management Game, based on an interactive game about agriculture, landscape and river management. The game was applied to the Hungarian part of the transboundary Tisza River. The games were instrumental in combining technical and relational perspectives into a single tool that allowed farmers, local authorities and water authorities to experience the challenges of making policies and decisions to manage rivers in agricultural floodplains (Stefanska et al. 2011). Viglione et al. (2014) analysed the role of interactions between hydrology and society in the context of flood risk looking into collective memory, risk-taking attitudes and trust in their dynamic socio-hydrology modelling. The study found that the stakeholders' perception of risk was more important than the actual risk of flooding, and that a combination of overestimation of risk and risk-averse behaviour led to economic loss in floodplains.

Despite the multiple benefits of stakeholder participation, achieving meaningful stakeholder engagement that lives up to the expectations of all those involved is a challenge, particularly in relation to different spatial scales and power imbalances among stakeholders. There is a strong interdependence between different scales (Minang et al. 2015), for example, watershed-level decisions have a considerable impact on river basin–level water quantity and quality, and vice versa. This requires a more integrated consideration of activities and processes. SA models built on the principle of understanding entire systems and the interaction between their parts factor in the interconnectedness of different spatial contexts and natural systems. However, the ecological scales at which ecosystems services, including water supply and quality provision, are generated rarely coincide with the institutional scales, that is, international, national, state and municipal administrations, where natural resources are valued and managed. Stakeholders involved in water resource management often act across a range of ecological and institutional scales (Hein et al. 2006). While participatory processes can bridge discontinuity between the ecological, geographical and jurisdictional boundaries of water resources management (Priscoli 2004), the different positions and conflicts of interest between the stakeholders at each institutional scale need to be mediated (Hein et al. 2006). Failure to recognise power imbalances and overall politics reproduces the invisibility of marginalised stakeholders in decision-making processes (Brandt et al. 2018). Stakeholders' perceptions of unequal power relations often influence the composition of stakeholder platforms (Brouwer et al. 2013). In such settings, both powerful and less powerful stakeholders might not see the point in engaging, even though acting collaboratively would be beneficial to all.

When resources are insufficient, it becomes difficult to properly address the power dynamics between the local stakeholders that lead to inequalities (Barnaud and van Paassen 2013) and to identify solutions for any positive transformation or power redistribution (Brandt et al. 2018). In long-term governance of water systems, engagement might place excessive demands on the time and resources of stakeholders, and costs associated with preparation, communication and deliberation might be prohibitive to the reseachers organising the process.

All in all, stakeholder engagement is a means of converting the public from passive recipients to active contributors in water resource–related decision processes (Priscoli 2004). The earlier top-down experiences of water resources development have shown that complex and dynamic environments benefit from flexible and transparent decision-making and the incorporation of diverse expertise and perspectives, such as local and indigenous knowledge and informal governance networks (Pahl-Wost et al. 2007). Participatory SA models have been groundbreaking in their inclusion of the stakeholder views previously missing from systematic decision-making processes. They have facilitated consideration of different human perspectives and values in the design and calibration of systematic decision-making tools without placing an excessive burden on stakeholders by combining their input with various different types of data and using mechanisms like feedback loops and system archetypes.

Application of SA Models in the Global South

While extreme events like floods and droughts are experienced indiscriminately worldwide and expected to increase in frequency and severity with climate change (IPCC 2018), water issues are different in industrialised and non-industrialised countries. While in most parts of the world there are more pressing priorities such as availability and access to adequate water and wastewater services, in countries where these problems have already been dealt with, the economic efficiency of allocating water to its competing uses and the resilience of supply aganist climate change and other distruptors have become the priority of water management operations. The focus in the rest of this section is the complex water problems faced by the Global South and how they could be addressed with the application of systems thinking, using examples from Brazil.

Brazil, as a recently industrialised country (Boddin 2016), suffers simultaneously from water challenges that are common in the Global North and the Global South due to its patchy trajectory of water and wastewater service development and its economy heavily dependent on the assumption that water is an abundant resource. Despite the abundance of water resources in Brazil (Ussami and Guilhoto 2018), these resources are not distributed across the country in the same pattern as the population (Maddocks et al. 2014; Nobre et al. 2016), making heavily urbanised regions like the State of São Paulo particularly vulnerable to droughts (Ussami and Guilhoto 2018). Like many other countries, water resource management in Brazil has historically relied upon heavy investment in hydraulic infrastructure, making hydropower generation the focus of water resource development from the early 1900s onwards. The large share of hydropower in the energy mix is beneficial in terms of low-carbon emissions. However, it simultaneously makes the electricity supply highly dependent on the hydrological cycle and causes volatility in electricity prices during periods of water scarcity (Corrêa et al. 2016; EPE 2017).

The resilience of water resources is central to the United Nation's (UN's) strategy for achieving a more equitable and inclusive society as well as to promoting sustainable growth (UN 2018). Brazil has deep inequalities in access to sanitation and safe water supply among its large population of 209 million people. These inequalities become more marked between geographical regions, and between rural and urban communities and households. According to the World Health Organization (WHO) and the United Nations International Children's Emergency Fund—UNICEF (2015), there are six million people in Brazil without access to safe water and 24 million without adequate sanitation who suffer from water supply downtime, disruptions in service and deficiencies in drinking water systems. The sectors that contribute most to the national economy are also those that are most dependent on water, such as extractive industries, agriculture and biofuel production. Irrigation alone consumes 72% of Brazil's available water supply (World Bank 2016). Additionally, 62% of Brazil's energy and 92% of its renewable energy (OECD 2015) are generated through hydropower plants, so extreme water shortages also mean power shortages or even blackouts.

Some of the current issues are due to practical and financial constraints like the cost of improving aging and insufficient water and wastewater infrastructure. Others are linked to the way water and the overall natural environment has been managed. The uncontrolled expansion of economic activities into the Amazon and the subsequent deforestation and illegal occupation of cleared land for agriculture have contributed to disturbances in the hydrological cycle (Marengo et al. 2018). The development of aquifer recharge areas, intense pumping of groundwater for irrigation, rapid urbanisation and direct dumping of waste and discharge of sewage into rivers have been further elements of the mismanagement (Ribeiro 2019; Slater 2019).

Conclusions: How Could Systems Analysis Contribute Further to Water Management?

In the past decade, future scenarios of water allocation have gained importance under the risk of supply-demand imbalances. Analysis of future scenarios and extensive literature reviews (Cosgrove 2013; Cosgrove and Loucks 2015) have identified the main issues in the future of water management worldwide as freshwater stress due to finite resources against exponentially expanding demands; globalisation of trade and export of water-intensive goods from arid regions; volatility of supply due to regime shifts and high variability in the hydrological cycle; and disruption in environmental flows due to increased demand. These drivers interact with each other and have synergetic effects at different scales and in different parts of the world (Gallopín 2012). The accelerating rate of climate change and discontinuity in its driving forces increase the uncertainty faced by decision-makers in water management (Cosgrove and Loucks 2015).

To address these multiple challenges, there is a need for a more holistic approach that can link wider production with water demand, opportunities and risks, and the total beneficiaries and losers in each allocation decision not only of water but also of other natural resources such as land, and financial resources, such as government incentives for sectorial or geographical development. Complex and non-linear human-environment systems like water resources are composed of parts which might be well understood individually but will interact to produce unexpected large-scale phenomena. Traditionally, there has not been a recognition of the synergetic effect of separate parts and processes in the water supply network systems (i.e. abstraction, purification and supply of potable water; the sewer networks; wastewater treatment; the receiving water body) on water quality and the hydrological cycle.

Systems science methods have been successfully applied to understanding complex challenges in a variety of sectors such as engineering, defence, transport, healthcare, economics, ecology and business since the mid-1950s. The SA approach is particularly well suited to multidisciplinary and multi-actor problems in the management of water resources.

It is instrumental in improving understanding of how water systems function in parts and as a whole by considering different social, economic and physical aspects of water systems in combination (Barlas and Carpenter 1990). Its interdisciplinarity and versatile structure also makes SA a resource-intensive discipline that requires trans-disciplinary know-how and teamwork. The main issues that have affected the practical applicability of SA in this sense are the requirement of significant effort and diverse technical expertise in syncing various simulations, optimising and assessing sub-models and analysing the resulting large-scale integrated models (Huang and Xia 2001) and interpreting their possible policy implications. Such interdisciplinary scientific production demands certain capacity in research funding and cooperation. Most of the SA models developed prior to the 2000s focused primarily on industrialised countries due to these resource constraints. However, most of the world's population and water-related problems are concentrated in developing countries (UN Water 2015).

Water security issues are expected to be exacerbated by climate change, which will also increase the frequency and scale of extreme weather conditions in Brazil as in the rest of the world (UN 2017a; Barrett 2014), making it more difficult to predict future water supply availability using only statistical models and historical data (Venkateswaran et al. 2018). Brazil is a large country composed of regions with diverse ecosystems, climates, population densities and levels of development. Its different regions experience water problems that vary in kind, frequency and scale. The application of SA to different Brazilian case studies will provide lessons applicable to different parts of the world. An internationally relevant topic in Brazil is the functioning and performance of multi-level water governance. The mismatch between ecological scales of production and administrative scales of natural resource management is magnified in large and highly decentralised countries that have to simultaneously manage state and federal water policies and priorities (OECD 2015). Multi-level governance efforts can be expanded by building on existing transboundary systems analyses developed for the European Union (EU) under the Water Framework Directive and applied to member countries of different sizes, interests and national legislatures, and with different levels of water availability (Voulvoulis et al. 2017). Other useful models may include those built for single countries that cover large territories and have federal adminstrations, such as Australia, Russia, the USA, India and Canada.

There are two water management challenges in particular where insights from Brazil can provide benefits internationally. The first is the governance of water supply systems in intensely populated urban areas. These are complex systems characterised by relentless increases in demand and failing infrastructure, where the conflicting interests of stakeholders make communication, cooperation and governance difficult (dos Santos et al. 2019). While the São Paulo drought between 2013 and 2015 is a typical case, such crises are not unique to Brazil. Many megacities, mostly located in the Global South, are particularly vulnerable to episodic mismatches between water supply and demand that climate change is expected to exacerbate. The second is the competition between sectorial water uses and the environment. Conflicts of interest around water resources often take place not only between different human interests, such as irrigation versus hydropower but also between human withdrawal or pollution of water and qualitative and quantitative water demands for ecosystem services. There are already examples (Nepstad et al. 2008; Lapola et al. 2009; Leonardo and Freitas 2011; Dias et al. 2018; Marengo et al. 2018; Mercure et al. 2019) of SA studies looking into interlinked water issues in Brazil, mostly concentrating on the water-energy nexus. However, more research in the synergies and trade-offs between sectorial water demands and ecosystem services is essential to inform water policy in a way that will conserve unique ecosystems and maintain economic development in the future. Having the necessary scientific capacity, resources and national interest, Brazil can lead the quest for SA research with internationally beneficial applications.

While economic growth and globalisation enhance the risks associated with economic and physical water scarcity by increasing water demand and the value of exposed assets (Cosgrove and Loucks 2015), they can also increase the availability of monetary and human capital, as well as interest in mitigating water-related risks specific to different regions, sectors and communities (Sadoff et al. 2015). In an era when critical scientific and environmental policies at different scales are disregarded or even attacked by populist political circles, the way scientific evidence is presented and communicated is as important as the quality of the research (Muradian and Pascual 2020) in attracting the resources required to prevent global water crises.

In this regard, communicating results to audiences of varying scientific literacy and world views and engaging them at some level are essential in supporting decisions relating to human-environment systems. Hard scientific evidence will not directly lead to change in actual behaviour because it is neither easily communicable to non-expert audiences nor directly operational by policymakers. Participatory processes are found to initiate a higher degree of ownership and motivation towards behavioural change in the participants involved (Maskrey et al. 2016) due to the co-creation of the research and its outputs. The element of co-creation and end-user engagement could be particularly instrumental in connecting with the end-users and influencing the demand side, especially in heavily bureaucratic water governance systems (Rodrigues et al. 2015). Once the SA research at IIASA led to collaboration between two conflicting sides at the height of the Cold War to achieve the common goal of better water management in the transboundary catchments they shared in Europe. Similarly, different political interests of today might be brought to reconcile their distinct priorities of continuous economic growth and environmental protection through consideration of the multiple capitals[2] involved in planning decisions.

References

Ahmad, S., & Simonovic, S. (2004). Spatial System Dynamics: New Approach for Simulation of Water Resources Systems. *Journal of Computing in Civil Engineering, 18*(4), 331–340.

Alcamo, J., Döll, P., Henrichs, T., Kaspar, F., Lehner, B., Rösch, T., & Siebert, S. (2003). Development and Testing of the WaterGAP 2 Global Model of Water Use and Availability. *Hydrological Sciences Journal, 48*(3), 317–337.

Barlas, Y., & Carpenter, S. (1990). Philosophical Roots of Model Validation: Two paradigms. *System Dynamics Review, 6*(2), 148–166.

[2] Financial, human, natural, manufactured, social and intellectual capitals are those included in the six capital frameworks (Romolini et al. 2017).

Barnaud, C., & van Paassen, A. (2013). Equity, Power Games, and Legitimacy: Dilemmas of Participatory Natural Resource Management. *Ecology and Society, 18*(2), 1–12.

Barrett, O. L. (2014). *Brazil: Water Woes Climate Change and Security*. https://climateandsecurity.org/2014/09/03/brazil-water-woes-climate-change-and-security. Accessed 10 June 2019.

Beard, R. L. (1971). *Status of Water Resource Systems Analysis, Technical Report Paper*. US Army Corps of Engineers.

Beck, M. B. (1981). *Water Quality Management: Beyond Planning and Design*. Laxenburg: International Institute for Applied Systems Analysis.

Benson, D., Gain, A. K., & Rouillard, J. J. (2015). Water Governance in a Comparative Perspective: From IWRM to a "Nexus" Approach? *Water Alternatives, 8*(1), 756–773.

Boddin, D. (2016). *The Role of Newly Industrialized Economies in Global Value Chains*. Germany: Kiel: International Monetary Fund.

Brandt, F., Josefsson, J., & Spierenburg, M. (2018). Power and Politics in Stakeholder Engagement: Farm Dweller (In)visibility and Conversions to Game Farming in South Africa. *Ecology and Society, 23*(3).

Brouwer, H., Hiemstra, W., Vugt, S. V., & Walters, H. (2013). Analysing Stakeholder Power Dynamics in Multi-stakeholder Processes: Insights of Practice from Africa and Asia. *Knowledge Management for Development Journal, 9*(3), 11–31.

Brown, C. M., et al. (2015). The future of water resources systems analysis: Toward a scientific framework for sustainable water management, *Water Resour. Res., 51*, 6110–6124, https://doi.org/10.1002/2015WR017114.

Budds, J. (2009). Contested H2O: Science, Policy and Politics in Water Resources Management in Chile. *Geoforum, 40*(3), 418–430.

Butterworth, J. (2014, August, 1–9). Is There Mileage Left in the IWRM Concept? Or Is It Time to Move On? *IRC Wash*. https://www.ircwash.org/blog/is-there-mileage-left-in-iwrm

Cañas, A. J., et al. (2005). Concept Maps: Integrating Knowledge and Information Visualization. In S. O. Tergan & T. Keller (Eds.), *Knowledge and Information Visualization* (Lecture Notes in Computer Science) (Vol. 3426, pp. 205–219). Heidelberg: Springer.

Carrera-Hernández, J. J., & Gaskin, S. J. (2007a). Spatial Dynamics of Aquifer Recharge in the Basin of Mexico. *32nd IAHR Conference: Harmonizing the Demands of Art and Nature in Hydraulics, 1–6 July 2007*. Venice, June 2007.

Carrera-Hernández, J. J., & Gaskin, S. J. (2007b). The Basin of Mexico Aquifer System: Regional Groundwater Level Dynamics and Database Development. *Hydrogeology Journal, 15*(8), 1577–1590.

Carrera-Hernández, J. J., & Gaskin, S. J. (2008). Spatio- temporal Analysis of Potential Aquifer Recharge: Application to the Basin of Mexico. *Journal of Hydrology, 353*(3–4), 228–246.

Carter, N. H., et al. (2014). Coupled Human and Natural Systems Approach to Wildlife Research and Conservation. *Ecology and Society, 19*(3), 43.

Cohen, S., et al. (2006). Learning with Local Help: Expanding the Dialogue on Climate Change and Water Management in the Okanagan Region, British Columbia, Canada. *Climatic Change, 75*(3), 331–358.

Connor, J., Cartwright, L., & Stephenson, K. (2004). *Collaborative Water Supply Planning: A Shared Vision Approach for the Rappahannock Basin in Virginia. World Water and Environmental Resources Congress* (pp. 2700–2708). USA: Salt Lake.

Corrêa, R., Neto, I. D. M., & Seifert, S. S. (2016). Electricity Supply Security and the Future Role of Renewable Energy Sources in Brazil. In *Renewable and Sustainable Energy Reviews* (Vol. 59, pp. 328–341). Oxford: Elsevier.

Cosgrove, W. (2013). Water Futures: The Evolution of Water Scenarios. *Current Opinion in Environmental Sustainability, 5*(6), 559–565.

Cosgrove, W. J., & Loucks, D. P. (2015). Water Resources Research. *Water Resources Research, 51*, 4823–4839.

Demeritt, D. (2001). The Construction of Global Warming and the Politics of Science. *Annals of the Association of American Geographers, 91*(2), 307–337.

Dias, V. D. S., et al. (2018). An Overview of Hydropower Reservoirs in Brazil: Current Situation, Future Perspectives and Impacts of Climate Change. *Water, 10*, 592.

Dolado, J. (1992). Qualitative Simulation and Systems Dynamics. *System Dynamics Review, 8*(1), 55–81.

dos Santos, J. R., et al. (2019). Water Used to Be Infinite: A Brazilian Tale of Climate Change. *Kybernetes, 48*(1), 143–162.

Du Pisani, J. (2007). Sustainable Development: Historical Roots of the Concept. *Environmental Sciences, 3430*(3), 83–96.

EC (European Commission). (2000). *Water Framework Directive, Official Journal (OJ L 327) on 22 December 2000*. Brussels: European Commission.

EPE. (2017). *Balanço Energético Nacional Relatório Síntese – Ano base 2016*. Brasília: EPE.

FAO. (2011). *The State of The Worlds Land and Water Resources for Food and Agriculture: Managing Systems at Risk*. Rome: Earthscan.
Fleming, G. (1975). *Computer Simulation Techniques in Hydrology*. New York: Elsevier.
Forrester, J. W. (1971). World dynamics, Cambridge, Mass., USA: Wright-Allen Press.
Gallopín, G. C. (2012). *Global Water Futures 2050: Five Stylized Scenarios*. Paris: United Nations World Water Assesment Programme.
Ghassemi, F., Jakeman, A. J., & Nix, H. A. (1995). Avoiding Disasters: The Role of Systems Analysis and an Integrated Approach in Water Resources Development. *GeoJournal, 35*(1), 49–51.
Giordano, M., & Shah, T. (2014). From IWRM back to Integrated Water Resources Management. *International Journal of Water Resources Development, 30*(3), 364–376.
Gorelick, S. M., & Zheng, C. (2015). Global Change and the Groundwater Management Challenge: Groundwater Management Challenge. *Water Resources Research, 51*.
Goswami, K. B., & Bisht, P. (2017). The Role of Water Resources in Socio-Economic Development. *International Journal for Research in Applied Science and Engineering Technologyechnology, 5*.(XII, 1669–1674.
Havlik, P., et al. (2011). Global Land-use Implications of First and Second Generation Biofuel Targets. *Energy Policy, 39*(10), 5690–5702.
Hein, L., et al. (2006). Spatial Scales, Stakeholders and the Valuation of Ecosystem Services. *Ecological Economics, 57*(2), 209–228.
Hodgson, A. M. (1992). Hexagons for Systems Thinking. *European Journal of Operational Research, 59*, 220–230.
Holy, M., Handova, Z., & Kos, Z. (1981). *Erosion and Water Quality as Modeled by CREAMS: A Case Study of the Sedlicky Catchment, Collaborative Paper*. Laxenburg: IIASA.
Homer, J. B. (1996). Why We Iterate: Scientific Modelling in Theory and Practice. *Systems Dynamics Review, 12*(1), 1–19.
Huang, G. H., & Xia, J. (2001). Barriers to Sustainable Water-quality Management. *Journal of Environmental Management, 61*(1), 1–23.
ICWE. (1992). *The Dublin Statement on Water and Sustainable Development*. Dublin: International Conference on Water and the Environment.
Ioris, A. A. R. (2001). Water Resources Development in the São Francisco River Basin (Brazil): Conflicts and Management Perspectives. *Water International, 26*(1), 24–39.

Ioris, A. A. R. (2011). Values, Meanings, and Positionalities: The Controversial Valuation of Water in Rio de Janeiro. *Environment and Planning C: Government and Policy, 29*(5), 872–888.

IPCC. (2018). *Special Report on the Impacts of Global Warming of 1.5°C*. Geneve: IPCC.

Kaden, S. (1985). Decision Support Model Systems for Regional Water Policies in Open-pit Lignite Mining Areas. *International Journal of Mine Water, 4*(1), 1–16.

Kindler, J., & Loucks, D. (1989). Water Resources Research at IIASA: 1973–1988. *Water Resource Management, 3*(3), 169–190.

Kirkwood, C. W. (1998). System Dynamics Methods: A Quick Introduction [Online]. Available: http://www.public.asu.edu/~kirkwood/sysdyn/SDIntro/SDIntro.htm

Knisel, W. G. (1980). *CREAMS: A Fieldscale Model for Chemical, Runoff, and Erosion from Agricultural Management Systems, Conservation Report*. Washington, DC: USDA, Science and Education Administration.

Koncsos, L., & Balogh, E. (2007). Flood Damage Calculation Supported by Inundation Model in the Tisza Valley. *32nd Congress of the International Association of Hydrolic Engineering and Research*. Venice, June 2007.

Lapola, D. M., et al. (2009). Indirect Land-use Changes can Overcome Carbon Savings from Biofuels in Brazil. *PNAS, 107*(8), 1–6.

Leonard, R. A., Knisel, W. G., & Still, D. A. (1987). GLEAMS: Groundwater Loading Effects of Agricultural Management Systems. *Transactions of ASAE, 30*, 1403–1418.

Leonardo, J. A., & Freitas, V. M. (2011). Amazon and the Expansion of Hydropower in Brazil: Vulnerability, Impacts and Possibilities for Adaptation to Global Climate Change. *Renewable and Sustainable Energy Reviews, 15*(6), 3165–3177.

Linton, J., & Budds, J. (2014). The Hydrosocial cycle: Defining and Mobilizing a Relational-dialectical Approach to Water. *Geoforum, 57*, 170–180. https://doi.org/10.1016/j.geoforum.2013.10.008.

Lund, J. R., Cai, X., & Characklis, G. W. (2006). Economic Engineering of Environmental and Water Resource Systems. *Journal of Water Resources Planning and Management, 132*(6), 399–402. 10.1061/(ASCE)0733-9496(2006)132:6(399).

Maass, A., et al. (1962). *Design of Water-Resources Systems*. Cambridge, MA: Harvard University Press.

Maddocks, A., Shiao, T., & Biderman, R. (2014, November). Three Maps Explain São Paulo, Brazil's Water Crisis. *GreenBiz,* 1–8.

Magnuszewski, P., et al. (2015). *Interim Report Conceptual Framework for Scenarios Development in the Water Futures and Solutions Project.* Vienna: IIASA.

Marengo, J. A., et al. (2018). Changes in Climate and Land Use Over the Amazon Region: Current and Future Variability and Trends. *Frontiers in Earth Science, 6*, 228.

Margat, J., & Van der Gun, J. (2013). *Groundwater Around the World: A Geographic Synopsis.* Boca Raton: CRC Press.

Maskrey, S. A., et al. (2016). Environmental Modelling and Software Participatory Modelling for Stakeholder Involvement in the Development of Flood Risk Management Intervention Options. *Environmental Modelling and Software, 82*, 275–294.

Mehta, L., et al. (2017). Introduction – Flows and Practices: The Politics of Integrated Water Resources Management (IWRM) in Eastern and Southern Africa. In L. Mehta, B. Derman, & E. Manzungu (Eds.), *Flows and Practices: The Politics of Integrated Water Resources Management in Eastern and Southern Africa* (pp. 1–29). Harare: Weaver Press.

Mercure, J., et al. (2019). System Complexity and Policy Integration Challenges: The Brazilian Energy- Water-Food Nexus. *Renewable and Sustainable Energy Reviews, 105*, 230–243. https://doi.org/10.1016/j.rser.2019.01.045.

Minang, P. A., et al. (2015). Scale Considerations in Landscape Approaches. In P. A. Minang et al. (Eds.), *Climate-Smart Landscapes: Multifunctionality in Practice* (p. 300). Nairobi: World Agroforestry Centre.

Muradian, R., & Pascual, U. (2020). Ecological Economics in the Age of Fear. *Ecological Economics, 169*, 106498. https://doi.org/10.1016/j.ecolecon.2019.106498.

Nepstad, D. C., et al. (2008). Interactions Among Amazon Land Use, Forests and Climate: Prospects for a Near-term Forest Tipping Point. *Philosophical Transactions of the Royal Society: Biological Sciences, 363*, 1737–1746.

Nobre, C. A., et al. (2016). Some Characteristics and Impacts of the Drought and Water Crisis in Southeastern Brazil During 2014 and 2015. *Journal of Water Resource and Protection, 8*, 252–262.

Novak, J. D., & Cañas, A. J. (2008). *The Theory Underlying Concept Maps and How to Construct Them.* Pensacola: Florida Institute for Human and Machine Cognition.

Novotny, V. (1996). Integrated Water Quality Management. *Water Science and Technology, 33*(4/5), 1–7.

OECD. (2015). *Brazil Policy Brief: Water*. Paris. www.oecd.org/policy-briefs. Accessed 10 Nov 2019.

Oreskes, N. (2005). The Scientific Consensus on Climate Change. *Essay beyond the Ivory Tower, 306*, 1684–1686.

Orlovsky, S., & van Walsum, P. E. V. (1984). *Water Policies: Regions with Intense Agriculture, IIASA Working Paper*. Austria: IIASA.

Orlovsky, S., Rinaldi, S., & Soncini-Sessa, R. (1982). *A Min-Max Approach to Reservoir Management, IIASA Collaborative Paper*. Laxenburg: IIASA.

Orlovsky, S., Kaden, S., & van Walsum, P. E. V. (1986). *Decision Support Systems for the Analysis of Regional Water Policies: Final Report of the Collaborative Regional Water Policies Project (1984–1985)*. Laxenburg: IIASA.

Osborne, T. (2008). *Eco-problems of the 80s Return to Haunt Us*. https://www.newscientist.com/article/dn16189-eco-problems-of-the-80s-return-to-haunt-us. Accessed 10 Nov 2019.

Pahl-Wost, et al. (2007). Managing change toward adaptive water management through social learning. *Ecology and Society*, 12(2), 30.

Palmer, R. N., Keyes, A. M., & Fisher, S. M. (1993). Empowering Stakeholders Through Simulation in Water Resources Planning. In *Water Management in the 90s. 20th Anniversary Water Resource Planning and Management Conference*. Washington, DC.

Pastor, A. V., et al. (2014). Accounting for Environmental Flow Requirements in Global Water Assessments. *Hydrology and Earth System Sciences, 18*(12), 5041–5059.

Priscoli, J. D. (2004). What is Public Participation in Water Resources Management and Why Is It Important? *Water International, 29*(2), 221–227.

Rey, D., et al. (2019). Role of Economic Instruments in Water Allocation Reform: Lessons from Europe. *International Journal of Water Resources Development, 627*, 1–34.

Ribeiro, M. A. (2019). *Emerging Water Issues In Brazil*. https://www.radicalecologicaldemocracy.org/emerging-water-issues-in-Brazil. Accessed 15 Nov 2019.

Riegels, N., et al. (2013). Systems Analysis Approach to the Design of Efficient Water Pricing Policies under the EU Water Framework Directive. *Journal of Water Resources Planning and Management, 139*(5). https://ascelibrary.org/doi/abs/10.1061/%28ASCE%29WR.1943-5452.0000284.

Rindzevičiūtė, E. (2016). *The Power of Systems How Policy Sciences Opened Up the Cold War World*. Ithaca: Cornell University Press.

Ringler, C., Bhaduri, A., & Lawford, R. (2013). The Nexus Across Water, Energy, Land and Food (WELF): Potential for Improved Resource use Efficiency? *Current Opinion in Environmental Sustainability, 5*(6), 617–624. https://doi.org/10.1016/j.cosust.2013.11.002.

Rodrigues, D. B. B., et al. (2015). Contrasting American and Brazilian Systems for Water Allocation and Transfers. *Journal of Water Resources Planning and Management, 141*(7). https://doi.org/10.1061/(ASCE)WR.1943-5452.0000483.

Romolini, A., Fissi, S., & Gori, E. (2017). Exploring Integrated Reporting Research: Results and Perspectives. *International Journal of Accounting and Financial Reporting, 7*(1).

Sadoff, C. W., et al. (2015). *Securing Water, Sustaining Growth*. Oxford: University of Oxford/Global Water Partnership/OECD Task Force on Water Security and Sustainable Growth.

Sahin, O., et al. (2016). Paradigm Shift to Enhanced Water Supply Planning through Augmented Grids, Scarcity Pricing and Adaptive Factory Water: A System Dynamics Approach. *Environmental Modelling and Software, 75*, 348–361.

Sandmo, A. (2015). The Early History of Environmental Economics. *Review of Environmental Economics and Policy, 9*(1), 43–63.

Sauer, T., et al. (2010). Agriculture and Resource Availability in a Changing World: The Role of Irrigation. *Water Resources Research, 46*(6), 1–12.

Scheierling, S. M. (1995). *Overcoming Agricultural Pollution of Water: The Challenge of Integrating Agricultural and Environmental Policies in the European Union, Finance and Development*. Washington, DC.

Sendzimir, J., et al. (2007). Anticipatory Modeling of Biocomplexity in the Tisza River Basin: First Steps to Establish a Participatory Adaptive Framework. *Environmental Modelling and Software, 22*(5), 599–609.

Sendzimir, J., et al. (2008). Assessing the Resilience of a River Management Regime: Informal Learning in a Shadow Network in the Tisza River Basin. *Ecology and Society, 13*(1), 11.

Seo, F., & Sakawa, M. (1989). *Multiple Criteria Decision Analysis in Regional Planning, Concepts, Methods and Applications*. Dordrecht: Springer.

Shah, T., et al. (2000). *The Global Groundwater Situation: Overview of Opportunities and Challenges*. Colombo: International Water Management Institute.

Sheppard, E., & McMaster, R. B. (Eds.). (2004). *Scale and Geographic Inquiry: Nature, Society, and Method*. Oxford: Blackwell Publishing.

Slater, D. (2019). *Water Scarcity in Brazil: A Case Study*. www.usmcu.edu/Outreach/Publishing/Marine-Corps-University-Press/Expeditions-with-MCUP-digital-journal/Water-Scarcity-in-Brazil. Accessed 19 Sept 2019.

Somlyody, L. (1995). Water Quality Management: Can We Improve Integration to Face Future Problems? *Water Science and Technology, 31*(8), 249–259.

Stave, K. A., & Sterman, J. D. (2002). Using System Dynamics to Improve Public Participation in Environmental Decisions. *System Dynamics Review, 18*(2), 139–167.

Stefanska, J., et al. (2011). A Gaming Exercise to Explore Problem-solving versus Relational Activities for River Floodplain Management. *Environmental Policy and Governance, 21*(6), 454–471.

Sterman, J. D. (2000). Learning In and About Complex Systems. *Reflections: The SoL Journal, 1*(3), 24–51.

Sterman, J. D., et al. (1994). The Meaning of Models. *Science, 264*(5157), 329–331.

UN. (2017a). *Water and Climate Change: Main Impacts in the Southeast Region of Brazil and Possible Ways to Address this Challenge Agenda*. United Nations Global Compact.

UN. (2017b). *World Economic and Social Survey: Post-war Reconstruction and Development in the Golden Age of Capitalism*. New York: UN-Development Policy and Analysis Department.

UN. (2018). *Sustainable Development Goal 6 Synthesis Report on Water and Sanitation 2018*. New York: United Nations.

UN Water. (2015). *The United Nations World Water Development Report*. Zurich: United Nations.

Ussami, K. A., & Guilhoto, J. J. M. (2018). Economic and Water Dependence Among Regions : The Case of Alto Tiete, Sao Paulo State, Brazil. *Economia, 19*(3), 350–376.

van Vliet, M. T. H., et al. (2012). Coupled Daily Streamflow and Water Temperature Modelling in Large River Basins. *Hydrology and Earth System Sciences, 16*(11), 4303–4321.

van Vliet, M. T. H., Ludwig, F., & Kabat, P. (2013a). Global Streamflow and Thermal Habitats of Freshwater Fishes Under Climate change. *Climatic Change, 121*(4), 739–754.

van Vliet, M. T. H., et al. (2013b). Global River Discharge and Water Temperature Under Climate Change. *Global Environmental Change, 23*(2), 450–464.

Venkateswaran, K., Macclune, K., & Tincani, L. (2018). *Using Climate Information for Climate-Resilient Water Management: Moving from Science to Action, Action on Climate Today*. Oxford Policy Management.

Viglione, A., et al. (2014). Insights from Socio-hydrology Modelling on Dealing with Flood Risk – Roles of Collective Memory, Risk-Taking Attitude and Trust. *Journal of Hydrology, 518*, 71–82.

Voulvoulis, N., Arpon, K. D., & Giakoumis, T. (2017). The EU Water Framework Directive: From Great Expectations to Problems with Implementation. *Science of the Total Environment, 575*, 358–366.

Wang, R. Y., et al. (2017). Unpacking Water Conflicts: A Reinterpretation of Coordination Problems in China's Water-governance System. *International Journal of Water Resources Development, 33*(4), 553–569. https://doi.org/10.1080/07900627.2016.1197824.

WHO and UNICEF. (2015). *25 year Progress on Sanitation and Drinking Water: 2015 Update and MDG Assessment*. Geneva: WHO.

Winz, I., Brierley, G., & Trowsdale, S. (2009). The Use of System Dynamics Simulation in Water Resources Management. *Water Resource Management Manage, 23*(7), 1301–1323.

Woetzel, J., et al. (2019). *China and the World: Inside the Dynamics of a Changing Relationship*. Shangai: McKinsey Global Institute.

Wolstenholme, E. (2004). Using Generic System Archetypes to Support Thinking and Modelling. *System Dynamics Review, 20*(4), 341–356. https://doi.org/10.1002/sdr.302.

World Bank. (2016). *Brazil May Be the Owner of 20% of the World's Water Supply but It Is Very Thristy. World Bank News Letter*. www.worldbank.org/en/news/feature/2016/07/27/how-brazil-managing-water-resources-new-report-scd. Accessed 21 Oct 2019.

Xu, X., et al. (2015). *Modeling Water-energy-food Nexus for Planning Energy and Agriculture Developments: Case Study of Coal Mining Industry in Shanxi Province, China, Interim Report*. Laxenburg: IIASA.

Yao, M., et al. (2015). Sectorial Water Use Trends in the Urbanizing Pearl River Delta, China. *PLoS One, 10*(2).

Yazdanpanah, M., Hayati, D., Hochrainer-Stigler, S., et al. (2014a). Understanding Farmers' Intention and Behavior Regarding Water Conservation in the Middle- East and North Africa: A Case Study in Iran. *Journal of Environmental Management, 135*, 63–72.

Yazdanpanah, M., Hayati, D., Thompson, M., et al. (2014b). Policy and Plural Responsiveness: Taking Constructive Account of the Ways in Which Iranian Farmers Think About and Behave in Relation to Water. *Journal of Hydrology, 514*, 347.

Zhang, Y., Yang, Z., & Fath, B. D. (2010). Ecological Network Analysis of an Urban Water Metabolic System: Model Development, and a Case Study for Beijing. *Science of the Total Environment, 408*(20), 4702–4711.

5

Doce River Large-Scale Environmental Catastrophe: Decision and Policy-Making Outcomes

Ana T. Lima, Felipe A. Bastos, Fernando Jakes Teubner Junior, Renato Rodrigues Neto, Helena I. Gomes, and Gilberto F. Barroso

Abbreviations and Acronyms

FUNAI	National Indian Foundation (Fundação Nacional do Índio)
Samarco	Samarco Mineração S.A., mining industry co-owned by VALE and BHP Billiton
BHP Billiton	BHP Billiton Brasil Ltda.; Samarco's co-share participant

A. T. Lima • R. R. Neto • G. F. Barroso
Department of Oceanography and Ecology, Federal University of Espírito Santo (UFES), Vitória, Brazil
e-mail: ana.t.lima@ufes.br

F. A. Bastos
State Institute of Environment and Water Resources of Espírito Santo (IEMA), Vitória, Brazil

A. A. R. Ioris (ed.), *Environment and Development*,
https://doi.org/10.1007/978-3-030-55416-3_5

VALE	Samarco's co-share participant
MMA	Ministry of the Environment (Ministério do Meio Ambiente)
MME	Mining and Energy Ministry (Ministério de Minas e Energia)
IBAMA	Brazilian Institute for the Environment and Renewable Natural Resources (Instituto Brasileiro do Meio Ambiente e dos Recursos Naturais renováveis)
ICMbio	Biodiversity Conservation Chico Mendes Institute (Instituto Chico Mendes de Conservação e Biodiversidade)
DNPM	National Department of Mineral Research (Departamento Nacional de Pesquisas Minerais)
ANA	Water National Agency (Agência Nacional de Águas)
SEAMA	Espírito Santo State Secretary for the Environment (Secretaria de Meio Ambiente para o Estado de Espírito Santo)
SEAG	Espírito Santo State Secretary for Agriculture and Fisheries (Secretaria de Agricultura e Pesca do Estado do Espírito Santo)
IEMA	Institute of Environmental and Water Resources of Espírito Santo (Instituto Estadual do Meio Ambiente e Recursos Hídricos)

F. J. Teubner Junior
Brazilian Institute of Geography and Statistics (IBGE), Vitória, Brazil

H. I. Gomes (✉)
Food, Water, Waste Research Group, Faculty of Engineering, University of Nottingham, Nottingham, UK
e-mail: Helena.Gomes1@nottingham.ac.uk

IDAF	Espírito Santo Agriculture, Animal Husbandry and Forestry Institute (Instituto de Defesa Agropecuária e Florestal do Espírito Santo)
AGERH	Espírito Santo State Agency of Water Resources (Agência Estadual de Recursos Hídricos do Espírito Santo)
SEMAD	Minas Gerais State Secretary for the Environment and Sustainable Development (Secretaria de Estado de Meio Ambiente e Desenvolvimento Sustentável de Minas Gerais)
FEAM	State Environmental Agency of Minas Gerais (Fundação Estadual do Meio Ambiente – Minas Gerais)
IGAM	Minas Gerais Water State Institute (Instituto Mineiro de Gestão das Águas)
IEF	Minas Gerais Forestry State Institute (Instituto Estadual de Florestas – Minas Gerais)
CPRM	Mineral Resources Research Company (Companhia de Pesquisa de Recursos Minerais)
CIF	Inter-Federative Committee (Comitê Inter-Federativo)
RENOVA Foundation	Foundation managing the new Framework Agreement
MPF	Federal Prosecutors' Office (Ministério Público Federal)

Introduction

The mining industry has experienced several significant impoundment failures over the past 30 years (Davies 2002; Rico et al. 2008) (Table 5.1). The after-effects of the continuing pressure on resource exploitation resulted in tailings dam failures accounting for roughly 75% of mining-related environmental disasters (MMSD 2002). On 5 November 2015, a tailings dam collapsed upstream the Doce River in the state of Minas Gerais, Brazil, in one of the worst large-scale mining disasters in terms of

Table 5.1 Environmental effects of mine tailings and industrial wastes impoundments failures

Impoundment location and year of failure	Main ore/waste materials released	Volume of tailings/wastes released M (m^3)	Active (A)/inactive (I) and cause of failure	Affected water bodies	Environmental effects	Population affected	Reference
Omai River (Guyana) 1994	Cyanide-laden	2.9	Piping failure	F Essequibo River	346 dead fish	No measurable effects	Vick (1997)
The Merriespruit (South Africa) 1994	Gold tailings	0.6	Moisture/static liquefaction build-up in the tailings due to rainfall	F Sand River	Bird sanctuary destruction	17 killed	Fourie et al. (2000), Van Niekerk and Viljoen (2005)
Ingá, Sepetiba Bay (RJ, Brazil) 1996	Wastes of Zn ingots production for export	Unknown	Dam collapse after intensive rainfall	F, C Sepetiba Bay	Bay and mangrove pollution with metals mainly Zn and Cd Impairment of coastal fisheries		Freitas and Rodrigues (2014)
Los Frailes (Spain) 1998	Zinc, lead, copper and manganese-rich pyrite deposits	5	Static liquefaction	F, C Guadiamar River and estuary	Affected a wide surface area, 4634 acres/over 30,000 kg of dead fish were collected	Nine municipalities	Pain et al. (2003)
The Baia Mare (Romania) 2000	Cyanide from former gold and silver extraction	100,000 containing 50–100 tons of cyanide (CN)	Design, operation and surveillance failure	F, C Lapus river, Somes, Tisza, Danube and Black Sea	1200 tons of fish killed; 2000 km of the Danube catchment area were affected	Interruption in the water supply in 24 localities; interdiction to use the river water for consumption, domestic needs, and animals drinking	UNEP/OCHA (2000)

Cataguases (MG, Brazil) 2003	Caustic soda, and Al, Si, and Na wastes of pulp mill processing plant	1.4	A dam collapse after intensive rainfall	F, C Paraíba do Sul River, North Rio de Janeiro and South Espírito Santo coasts	River and coastal waters pollution with caustic effluents, extensive fish kill, collapse of water supply, and impairment of coastal fisheries		Costa (2001)
Imperial Metals, Mount Polley (BC, Canada) 2014	Au and Cu ore tailings	18.6	An impoundment wall fail	F, L Hazeltine Creek, Polley Lake, and Quesnel Lake	Erosion of channel and the floodplain 136 ha impacted		MPMC (2015), Petticrew et al. (2015)
Gold King Mine, Silverton (CO, USA) 2015	Wastewater spill with Cd, Pb, As, Be, Zn, Fe, and Cu	Unknown	A Accident destroying the plug of groundwater	F Cement Creek and Animas River			Bourcy and Weeks (2000)
Kolontar plant (Hungary) 2010	Al and alkaline wastes	6.5	A Unknow	F Torna, Marcal, Rába, and Danube	All aquatic life was destroyed; rivers and soil with high alkaline pH level;	10 people killed 400 evacuated 6 municipalities were affected	The Kolontar report Herard (2010)
Doce river, (MG-ES, Brazil) 2015	Iron ore tailings	56.4	Fundão tailings dam collapse. Foundation failure/poor maintenance	F, L, R, and C Doce River	River and coastal water pollution; and collapse of water supply. Irrigation and coastal fisheries impairment	19 fatalities 700,000 people without drinking water; 179 indigenous and 12 municipalities impacted	Miranda and Marques (2016), ANA (2016)
Brumadinho, Belo Horizonte 2019	Iron ore tailings	12	I Tailings dam N°1 failure (stress-induced liquefaction)	F Paraopeba River and its tributaries	River water pollution (250 km extension)	300 fatalities 3485 people	Cambridge and Shaw (2019), Freitas et al. (2019)

F fluvial, *L* lake, *R* reservoir, *C* coastal

tailings volume and affected watershed, travelling up to the Atlantic Ocean. The Fundão tailings dam released 34 Mm^3 of tailings that contaminated the Doce River watershed. This disrupted the entire fluvial-marine continuum, including impacts on local population (circa 700,000 inhabitants), potable water supply, and irrigation. On 21 Nov 2015, the tailings reached the Espírito Santo coast, leaving behind 19 people dead and 14 tons of dead fish (IEMA 2017). Recent studies indicate ecosystem service losses of over US$521 million per year (Garcia et al. 2017) in the Doce River watershed.

The mining industry is one of the major economic drivers of the Brazilian economy. In 2015, one of the leading mining companies in Brazil—Samarco—had made a profit of approximately US$2 billion, that is, 0.3% of the gross domestic product (GDP) of Brazil, according to the World Bank (2017). Resource economy–based countries, like Brazil, support mining activities as a key commodity. In some cases, a well-structured environmental governance framework is established to minimise environmental disturbance and prevent large-scale accidents, for example, Australia and Canada (Schoenberger 2016). However, this does not ensure the absence of large-scale disasters (Table 5.1). Mining is the primary driver of landscape changes, providing enormous flows of target minerals and associated wastes; altering drainage systems, land uses, and vegetation structure, promoting soil loss and erosion; and introducing both new structures and new meanings to the landscape (Bridge 2004). Located in river basins, mineral ore tillage impoundments represent a considerable threat to water security and ecological integrity of the drainage systems.

Environmental risks have been associated with dam stability and rupture, surface and groundwater contamination, acid mine drainage, and precipitation of secondary minerals (Grangeia et al. 2011; Kossoff et al. 2014). Previous large-scale environmental disasters show that post-disaster recovery time can take decades and will likely never return to the original state (Foley et al. 2005; Lima et al. 2016). In fact, the mining industry has experienced several significant dam failures in the recent history: the Merriespruit (South Africa) gold tailings spill resulted in 17 fatalities in 1994 (Fourie et al. 200; Van Niekerk and Viljoen 2005); also in 1994, there was a dam failure and release of cyanide-laden water into the Omai River—a tributary to the Essequibo River, Guyana. In Los Frailes (Spain), a zinc mine collapsed in 1998, releasing an estimated 5 Mm^3 of acidic waste rich in toxic metals from pyrite ore processing (Pain et al. 2003); 20,000 tons of residue was flushed into the Vișeu River in 2000; and 0.1 Mm^3 of liquid and suspended waste of gold- and silver-producing plant was released into the Lapus/Somes/Tisza/Danube river catchment system at Baia Mare (Romania) (UNEP/OCHA 2000). Environmental disasters of impoundment failures have massive effects, not only regarding the quantity of tillage released but also due to the transported slurry, dissolved metals and fine reactive metal particles, and huge loads of suspended sediments, which may reach coastal waters. Widespread effects are related to biophysical, social, and economic impacts that occur in an acute mode, with long-standing consequences of overwhelming local, regional, and even transboundary economies and human well-being.

In the Doce River, the ruptured dam is part of a complex system created to contain Fe-ore mining waste slurry. Iron ore mining is spread worldwide, with around 60% of the mining carried out in developing countries. While some research has been directed towards the tailings toxicity within the riverine environment (Segura et al. 2016), the massive total suspended solids were one of the leading causes in the mass destruction of the riverine fauna (CPRM 2015). Long-term quality and sediment safety are still being assessed, although it has been speculated that the tailings are laced with high amounts of metals, including arsenic, copper, and mercury (Escobar 2015). Tailings have been previously studied by Pires et al. (2003) at the Germano dam (Fig. 5.1), and they concluded that sediments were high in Cr (600 ppm). Nevertheless, tailings are classified as non-hazardous by the Brazilian national standard Norma Brasileira Regulamentadora (NBR) 10.004 (Pires et al. 2003). Despite toxicity, over 2.33 billion metric tons of Fe-ore are produced annually, with Brazil representing 18% of this total production (Tuck 2015). Iron ore mining is mainly located in the Upper Doce River basin in the State of Minas Gerais, where approximately 18% of the global Fe-ore is produced.

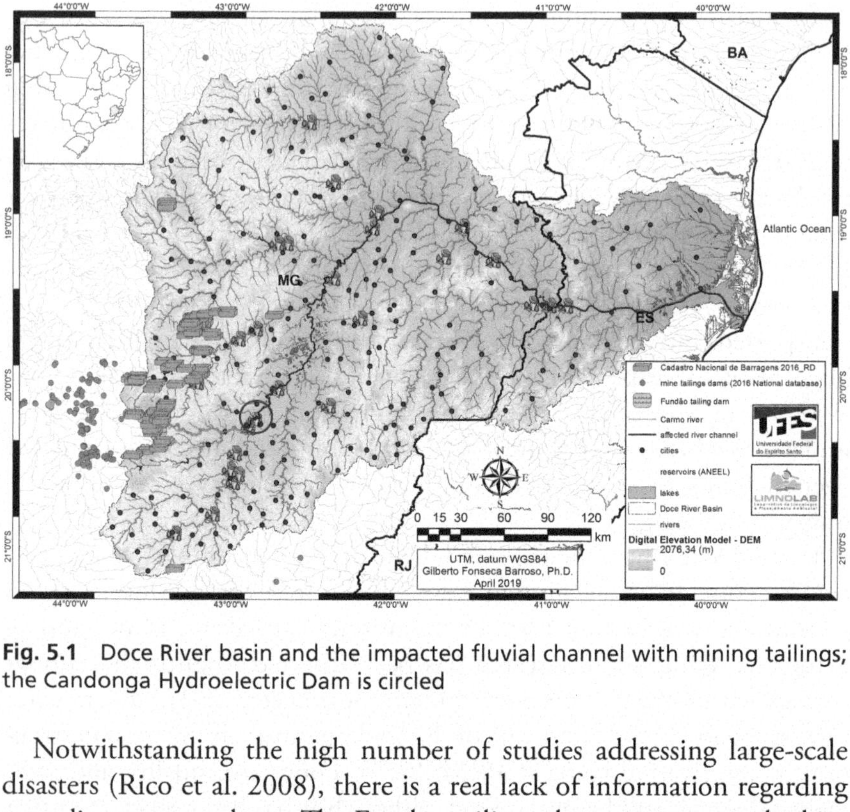

Fig. 5.1 Doce River basin and the impacted fluvial channel with mining tailings; the Candonga Hydroelectric Dam is circled

Notwithstanding the high number of studies addressing large-scale disasters (Rico et al. 2008), there is a real lack of information regarding post-disaster procedures. The Fundão tailings dam rupture was the biggest environmental disaster in Brazilian history, but also one of the most notorious in the world in terms of volume (Table 5.1). Immediate opinions diverged regarding post-disaster procedures: while some defended that fines and prosecutions would mostly finance ecosystem restoration and preventive protection of riverbanks and coast (Meira et al. 2016), others disputed that biodiversity was threatened by weak official policies and inadequate monitoring, management, and legislation (Nazareno and Vitule 2016). To rehabilitate the Doce River watershed, we need to establish a reference time point to reclaim the ecosystem to its previous condition (Lima et al. 2016; Rooney et al. 2012). However, the Doce River was already a degraded watershed (Ribeiro et al. 2011), with deficiencies in both water flow quantity and quality. These watershed management issues can be linked to governance systems, mainly because they reflect the result of mismanagement and unregulated resource exploitation (Biermann et al. 2016).

In a post-disaster period, a series of actions were taken to manage it and provide environmental rehabilitation. Of late, Brazilian authorities have created a Framework Agreement, together with the mining industry (Samarco), to combine and coordinate social and environmental recovery programmes. Effective environmental and resource governance is essential to prevent and remediate environmental impacts of the mining industry, particularly tailings dam failures. In this chapter, we discuss governmental actions after the Fundão tailings dam rupture and its environmental impact on the Doce River watershed. Remediation and integrated watershed management have not been considered in the perspective of the hydrological continuum from basin headwaters to the adjacent coastal ocean. This chapter addresses Brazilian environmental governance in the decision-making process during and after the large-scale Doce River disaster and its implications for watershed rehabilitation.

The Doce River

Pre-disaster

The Doce River is located in the southeast (SE) of Brazil, and it spreads over two states—Minas Gerais and Espírito Santo (Fig. 5.1). Due to its transboundary status, the Doce administration is done by the Federal Water Agency (ANA— Agência Nacional de Águas) with regional watershed management committees. With over 83,400 km^2, the Doce River watershed spreads over the states of Minas Gerais (70%) and Espírito Santo (30%), making it one of the largest Southeastern Atlantic watersheds. The overall land use in 2014 consisted of 72% farmland, 0.9% urban area, 6.6% husbandry, 19.2% natural area, and 1.3% the Doce River itself (IBGE 2016, 2017). As a tropical or subtropical region, it has two distinct seasons: wet summer from September to March and dry winter from April to August. The Doce River is among the most important in the east Brazilian coast, after the São Francisco River, in terms of water and suspended sediment loads to the ocean (de Oliveira et al. 2012). The

Doce River watershed hosts a population of circa 3.5 million inhabitants. From the native population, we highlight two of the original local indigenous communities—the Krenak and the Pataxó. These two groups amount to 179 individuals and are under the National Indian Foundation (Fundação Nacional do Índio—FUNAI) tutelage. Historically, the region of Minas Gerais in southeast Brazil is well known for mining activities. The city of Ouro Preto was one of the most important towns in the country, with most of its history associated with gold mining, beginning in the twentieth century.

The ruptured tailings dam is located at the Iron Quadrangle (Quadrilátero Ferrífero), Minas Gerais state, and is considered the world's largest open-pit mining industry (Santolin et al. 2015). Samarco, one of the mining industries exploring the area and owner of the ruptured dam, has an annual production capacity of more than 25 million tons of Fe-ore pellets and one million tons of Fe-concentrate. Samarco has a revenue of US$2.6 billion per year in Espírito Santo (Samarco 2014). Part of the Brazilian economic growth can be linked to the financial success of the mining industry and its export of mineral commodities in the past few years (from 1.6% in 2000 to 4.0% in 2014 of GDP). Samarco's sales revenue is equivalent to .4% of Espírito Santo GDP and 1.6% of the Minas Gerais GDP (Samarco 2014). Vale S.A. and BHP Billiton Brasil Ltda' are national companies that focus on mining, transportation, and production of ore. The two companies share ownership over Samarco (50% each). The elemental steps of the productive chain of mining industry can be seen as exploration, extraction, processing, and tailings production (Boger 2013). In extraction, ore is mined from rocks. Fine-grained rocks are then washed and, depending on the mineral, some chemicals are added to improve processing and separating minerals. After the extraction of minerals, tailings are discarded as slurries in open impoundments known as tailings dams.

The rupture of tailings dams is not new at the Iron Quadrangle region. This specific tailings dam endured a series of incidents before the rupture in November 2015. The Fundão tailings dam (Fig. 5.1), with 60 Mm^3 capacity (500 m in length and 90 m in height), started operations in

2008 (Samarco 2008). The first incident occurred in 2009 due to base drainage defects, shortly after the start of dam operations. In 2011, a second incident occurred, with tailings and wastewater released to one affluent of Doce River. Therefore, in 2012, the tailings dam was restructured and upgraded before resuming operations (IBAMA 2016a). Reports show that some misinformation was transmitted by Samarco and Vale to the National Department of Mineral Research (DNPM—Departamento Nacional de Pesquisas Minerais; governing body overseeing dam operations) about the overuse of the dam (4 $Mm^{3,}$ exceeding its capacity) (IBAMA 2016b). Several changes were made to the original project design to increase dam stability. These changes were untested, and the dam resumed operations unscathed (IBAMA 2016b). No industrial action was registered or taken up after this; no alert was given. Tailings dams are not like water retention dams. They are built in stages, as mining and waste production progresses, and they are built usually of mine wastes rather than concrete (Schoenberger 2016). The dam was used for mine wastes until the rupture on 5 Nov 2015. Ultimately, there was no contingency plan in place for the Fundão tailings dam, nor for the Doce River watershed.

The Doce River and its tributaries have an extensive dam system, reservoirs, and aqueducts with multiple purposes, for example, hydropower (27 reservoirs and 113 smaller ones [< 3.0 km^2]). The Doce River has undergone a remarkable surface water flow reduction in the past 70 years, from 1260 m^3/s in the 1940s to 648 m^3/s in 2010s (ANA 2015). The mean annual discharge at the most downstream gauge at Colatina, about 140 km from the river mouth, is 800 m^3/s. In 2015, the Doce River surface water flow had a record minimum of 114 m^3/s, just before the tailings dam rupture. The amount of river discharge is related to watershed rainfall intensity and variability. Intra-annual rainfall variability is remarkably high in the centre and southeastern Brazil. According to Zhang et al. (2018), around 80% of the annual rainfall occurs in the SE wet season (November–March), with 68% of annual river discharges produced between December and April. Annual rainfall variability has been 0.6–1.4 times the climatological mean, with 0.3–2.0 times the river discharge mean. For Colatina, 140 km upstream of the river mouth in the Atlantic Ocean, the Doce River discharge varied from 111 m^3/s to 9000 m^3/s from 2011 to 2014. Such variability brings many challenges to water resources management, including hydropower production. In the period 2014–2015, SE Brazil faced an intense drought that threatened water and energy security. Lack of rainfall and scorching summers generated the most pronounced water crisis in the past 50 years (Nobre et al. 2016).

The mining industry, together with hydroelectric dams, increasing forestry exploitation, and unplanned urban sprawl have turned the Doce River into a heavily impacted watershed. Before the tailings dam rupture, the watershed had recurrent flooding and the population was continuously exposed to waterborne diseases (Guedes et al. 2015). Contamination by faecal coliforms, the presence of resistant yeast strains in surface waters (Medeiros et al. 2008) and high cyanobacteria counts (ANA 2015) have been reported due to the lack of sanitation and wastewater treatment. It is estimated that 70% of domestic wastewater is discharged untreated to the basin fluvial network (ANA 2017). Land-use changes, such as unplanned disordered occupation, mining, cattle breeding, and *Eucalyptus* spp. forestry, are leading causes of Doce River watershed environmental degradation (Medeiros et al. 2008; Segura et al. 2016). Regarding water and sediment quality, previous studies report Fe and metal enrichment in the sediments (Oliveira and Quaresma 2017; Rodrigues et al. 2014), most of them exceeding international quality guidelines (Santolin et al. 2015), including high levels of arsenic in soils surrounding Samarco mines (Alves and Rietzler 2015).

The Disaster

The Fundão tailings dam, owned and managed by Samarco, ruptured on 5 Nov 2015. A total of 34 Mm^3 of mining ore tailings were released to the Doce River watershed (ANA 2015). Failure of the Fundão tailings dam affected more than 600 km of the river channel and the adjacent coastal area. From the total length of Doce River (881.4 km), 67.8% (598.3 km) was affected by the incident. If we add the most affected tributaries Gualaxo do Norte and Carmo River, it sums up to a total of

685.8 km of river channel that is impacted (Fig. 5.1). The most severely impacted section of the fluvial system was the 87.5 km of Gualaxo do Norte (small river tributary to Carmo River).

The fast-flash dam rupture increased the surface flow from 114 m^3/s to 810 m^3/s (CPRM 2015). Immediately after the incident, the tailings induced a hydraulic pressure, created a wave combining with the downstream volume of the Santarem Dam, flooding the towns of Bento Rodrigues and Paracatu de Baixo and killing 19 people. A significant part of sediment loadings was intercepted by the first reservoir for the generation of hydroelectricity. Candonga reservoir (circled in Fig. 5.1) has a volume of 54.4 Mm^3, but it was already silted by natural sediments, so only 10.5 Mm^3 of the tailings were retained there. The hydroelectric dam is undergoing dredging to restore its capacity for hydropower generation.

The wave progressed through the Carmo River and along the Doce River, killing 14 tons of freshwater fish. It took 16 days to travel approximately 660 km and four hydropower dams before reaching the Atlantic Ocean on 21 Nov 2015. At this inlet, the local population led by the Ministério Público Federal—MPF (Federal Prosecutors' Office) began to collect live fish and safely guarded them in the nearby ponds and lakes. Researchers and the population took suspended sediment samples, in general, and locals and nationals made numerous unsupervised visits. The lack of a contingency plan either supervised by the industry or governmental agencies made decision-making and actions difficult to coordinate in the immediate disaster aftermath (Fig. 5.2). According to Instituto Brasileiro do Meio Ambiente e dos Recursos Naturais renováveis—IBAMA (Brazilian Institute for the Environment and Renewable Natural Resources), the rupture of the Fundão tailings dam destroyed 1469 ha, along 77 km of watercourses, including protected areas and degraded local indigenous lowlands (IBAMA 2016b). Soil erosion increased drastically, adding to the total suspended solids during the flood. Turbidity was high (Table 5.2), translating into a suspended sediment loading of up to 33,000 $mg L^{-1}$ (Hatje et al. 2017). Sediments reached the highest enrichment factors for Hg (4234), Co (133), Fe (43), and Ni (16), whereas highest EFs were observed for suspended particulate matter (SPM) in the case of As (55), Ba (64), Cr (16), Cu (17), Mn (41), Pb (38) and Zn (82) (Hatje et al. 2017). Gomes et al. (2017) concluded that the tailings' plume increased trace metal concentrations of up

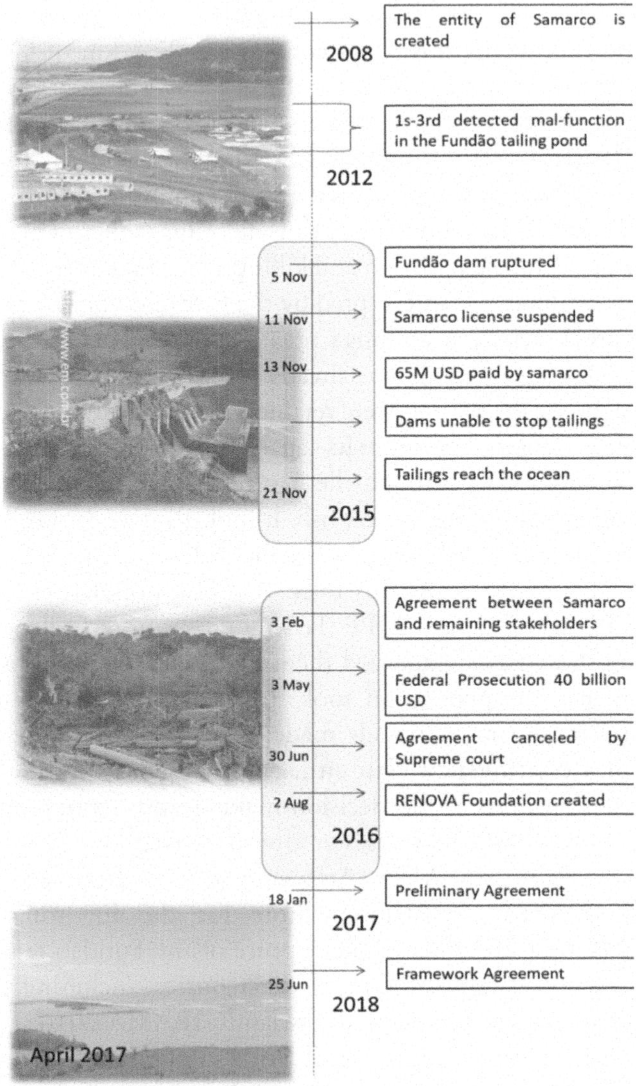

Fig. 5.2 Timeline from the creation of Samarco, the start of Fe-ore exploitation to the latest events regarding the ruptured tailings dam

Table 5.2 Doce River list of impacts in the post-disaster and loss of environmental services

Sector	Compartment	Description of impact	Quantification
Environment	Land (83,400 km^2)	Disturbance of riverine margins	1469 hectares; 77 km of watercourses
		Loss of riverbanks and soil along the river	Unknown/n.d.
		Alteration of geomorphology	Changed the overall natural character of the river
	River	Resuspension of airborne particulate matter from dry sediment at riverbank	Unknown/n.d.
		River bed silting	56.6 m^3 released
		Water quality decline[a]	As, B, Cr, Ni, Mn, Pb, V, and Zn exceed water quality regulations
		Sediment quality decline[a]	As, Cr, and Ni exceeded the norm of Conama 454 for sediment quality
		Temporary perturbation of the food web	Unknown/n.d.
		Biodiversity losses	Unknown/n.d. –14 t of dead fish, total of 29,292 collected specimens
		Temporary water turbidity	800,000 NTU
		Habitat alterations	Unknown/n.d.
		Endemic species extinction	Unknown/n.d.
	Ocean (1500 km^2)	Impacts on aquatic habitat	Turtle-nesting area (4000 births in 2015–2016)
		Beach erosion	400 m still trying to calculate this area
		Biodiversity losses	Unknown/n.d.
		Water and sediment quality decline[a]	
	Lakes	Water and sediment quality decline[a]	

(*continued*)

Table 5.2 (continued)

Sector	Compartment	Description of impact	Quantification
Social	Local communities	Flooding and destruction of villages	19 people dead
	Fisheries	Interruption of fishery activities	Forbidden at the coast and until 25 m depth at Doce River mouth
	Tourism	Temporary suspension of touristic activities	
	Water supply	Suspension of water supply	12 municipalities
Economic	Industries	Interruption of industrial activities	At least 16 huge companies
	Power plants	Interruption of power generation	Downstream hydroelectric power plants ceased activities to retain the tailings. Candonga is still closed.
	Irrigation and cattle breeding	High turbidity causing damage to the pumping systems, distribution networks, and water spray equipment	Water turbidity of 800,000 NTU

Data retrieved from IEMA (2015)
n.d. non determined
[a]in Hatje et al. (2017)

to five times to the estuary. Lastly, riverine riverbanks underwent massive rupture, with waste depositions in the margins.

Post-Disaster Actions

Brazilian Federal Police performed an investigation to assess the Fundão tailings dam status and indict responsibility for the accident. A parallel investigation by the mining company and stakeholders was carried out by an international law office to determine the immediate cause of the rupture. This study concluded that a series of incidents during the dam construction, together with its operation, led to the conditions that allowed the rupture (Chiaretti 2017). According to Morgenstern et al. (2016), these included damage to the original dam due to increased saturation, slime deposition, and existing structural problems. Therefore, the adjustment in the dam outline and the saturated conditions made the structure vulnerable to liquefaction. In 2013, the National Department of Mineral Production (DNPM) (MPF 2016a) already informed the company that the drainage system was insufficient, and there was a lack of monitoring instruments. The Brazilian Federal Police argued that the mining company took a risk for profit and prosecuted eight Samarco chief executives and awarded prison sentences.

Parallel to responsibility attribution, the country was left with a devastated watershed. A monitoring campaign led by Instituto Estadual do Meio Ambiente e Recursos Hídricos (IEMA) comprised an overall environmental impact assessment in the impacted Espírito Santo area (IEMA 2017). A summarised global overview is presented in Table 5.2. Main impacts consisted of:

- a total of 19 fatalities;
- a total of 14 tons of macro-fauna killed by asphyxia, mainly fishes;
- destruction of riparian vegetation: 1469 ha, along 77 km of watercourses; and
- impact over 660 km of the Doce River, the estuary, and its coastal region

Regarding surface water quality, besides the suspension of water supply in the affected municipalities, the presence of metals and changes on other parameters indicate the need for continuously monitoring the affected environment, as well as the remediation or recovery. Immediate actions consisted of the following:

- all marine fishery activities were interrupted at the coast (sea area of 1500 km^2) for unlimited time due to federal mandate;
- freshwater fisheries were interrupted in the middle and upper sections of Doce River, as requested by the state of Minas Gerais' attorneys. Some places have officially resumed fisheries (Rodrigues 2017);
- water supply was suspended;
- carry out risk assessment to other tailings dams within the Iron Quadrangle (Morgenstern et al. 2016); and
- Samarco committed to removing 1.3 Mm^3 of 10.5 Mm^3 tailings retained at Candonga's hydroelectric dam by February 2018 (Morgenstern et al. 2016).

Although a major economic driver and lucrative business, mineral resource exploitation has an environmental cost that impacts humans directly. It disrupts ecosystems that provide food and water (provisioning services), regulation of disturbances (regulating services), habitat for wildlife (supporting services), and aesthetics (cultural services). However, incorporation of ecosystem services into ecosystem management policy and decision-making remains inadequate (Antón et al. 2011; Costanza et al. 1997). Four years after the disaster, despite criminal investigations and environmental law enforcement, the impacts in the Doce River are still indeterminate, although the first studies have been already published (Gomes et al. 2017; Hatje et al. 2017). After rupture, 16 Mm^3 of refuse were left in the tailings dam. To date, 959,000 m^3 have been removed to be treated, and 2020 is the deadline for dam closure (Fundação Renova 2019).

Environmental Governance in Brazil

The Fundão tailings pond rupture raised awareness on the Brazilian environmental protection system. Despite the pressing environmental issues of any given country, citizens rely on the environmental protection systems or agencies to assure and provide both environmental and health security and to act in an emergency crisis. To understand the decision-making process after the disaster, it is necessary to envisage how the Brazilian environmental governance system works. At the national level, the environmental institutions are Ministério do Meio Ambiente (MMA), IBAMA, Instituto Chico Mendes de Conservação e Biodiversidade (ICMBio), ANA, whereas Ministério de Minas e Energia (MME) deals with energy and mineral production. The MME houses the DNPM that supervises and monitors tailings ponds. In the Doce River, there are equally responsible entities at State level: IEMA and Agência Estadual de Recursos Hídricos do Espírito Santo (AGERH) in Espírito Santo and Fundação Estadual do Meio Ambiente – Minas Gerais (FEAM), Instituto Mineiro de Gestão das Águas (IGAM), and Instituto Estadual de Florestas – Minas Gerais (IEF) in Minas Gerais. The Doce River Basin Management Committee was created in 2002 to achieve the goals set through the Integrated Water Resource Management Plan of the Doce River (PIRH-Doce). When a watershed is transboundary, management is supervised at the Federal level but implemented at the regional/state level.

Brazil is a resource-economy country that is highly dependent on commodity exports, with the belief that environment compliance hinders economic growth. This has prioritised mining and weakened environmental industry and regulating agencies. Between the accidents of Fundão (2015) and Brumadinho (2019), instead of changing the political framework to provide effective safety to populations around mining areas, what happened was the flexibilisation of some legal provisions, such as reduction of licensing steps (Armada 2019). Led mainly by the public sector, environmental protection is allocated scarce financial resources or is ill-distributed among the existing bodies. Lack of transparency and communication among state, agencies, institutes, and organisations, may be the culprit for the overall current standstill.

The New Governance Framework

A 'new Governance Framework' was created to deal with the Doce River disaster, a local, transboundary structure that addresses the Doce River disasters and its aftermath. To accelerate the watershed environmental recovery process and prevent delays at the Supreme Federal Court, Samarco made a *Framework Agreement* between Vale S.A., BHP Billiton Brasil Ltda, Federal Government of Brazil (IBAMA, ICMBio, ANA, DNPM, FUNAI), the States of Espírito Santo (IEMA, Instituto de Defesa Agropecuária e Florestal do Espírito Santo—IDAF, AGERH), and Minas Gerais (IEF, IGAM, FEAM), signed on 2 March 2016. This mechanism to remedy the Doce River disaster was created without any consultation or participation of the affected communities (Nabuco and Aleixo 2019). The *Framework Agreement* is formed by three new entities (Fig. 5.4a): the Renova Foundation, Inter-Federative Committee (CIF—Comitê Inter-Federativo) (Figs. 5.3 and 5.4), and technical boards.

The CIF has a multi-level structure, composed of agencies at the Federal, State, and Municipal level. The CIF has authority to conduct the agreement acts; it has been formally designated to approve the projects and activities; and consist of members of Environmental Ministry, Federal Government, State of Espírito Santo, State of Minas Gerais, municipalities impacted, the Doce River Hydrograph Basin Committee, and Public Defenders of the States. By creating the *Framework Agreement*, public agencies assumed that it is a technical issue to guarantee and protect transindividual rights and interests (Santos and Milanez 2017). These bodies have therefore assigned and restricted decision-making powers to the Renova Foundation, environmental agencies, third-party experts, and the state bureaucracy. The *Framework Agreement* also exempted the Federal government from their responsibility in preventing the disaster (Gonçalves e Silva 2019).

The Renova Foundation started operations on 2 August 2016, having responsibility for managing US$6.3 billion and for developing, proposing, enabling, and implementing plans, programmes, and projects that tackle the above-mentioned environmental priorities. Ultimately, the technical groups discuss and deliver socio-environmental and

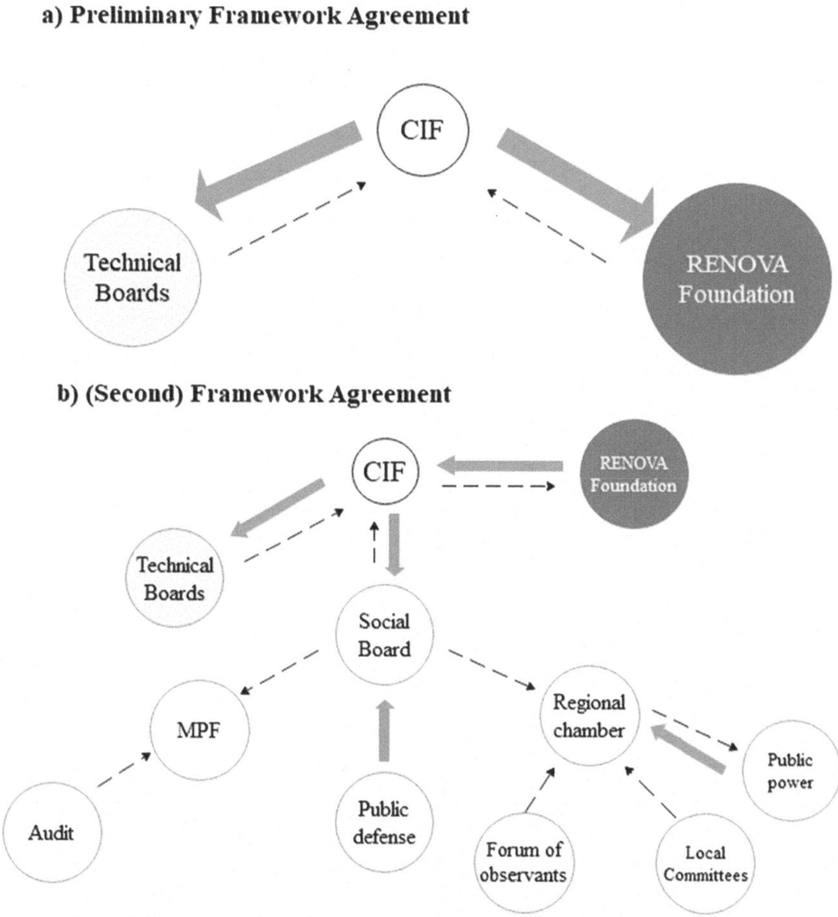

Fig. 5.3 (a) Composition of the preliminary framework agreement and (b) The most recent (second) framework agreement

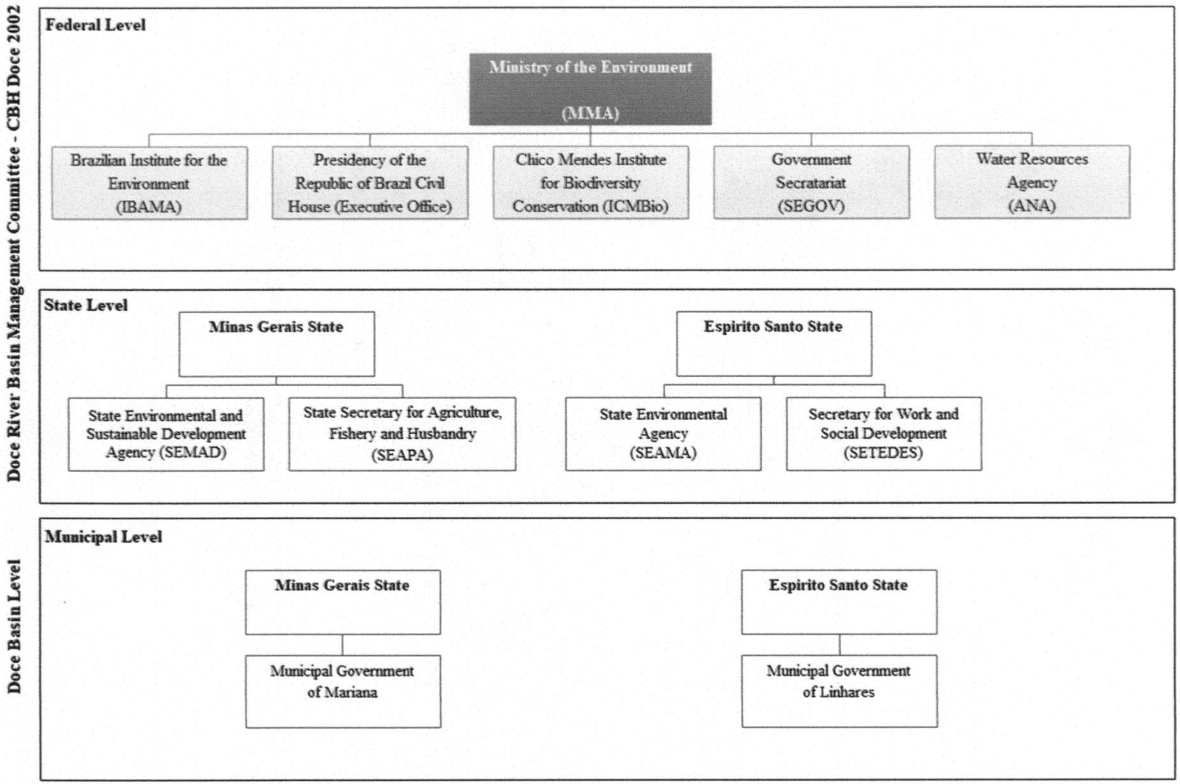

Fig. 5.4 Organogram representing the Inter-Federative Committee (CIF) and its multi-level structure

socio-economic programmes, aiming at impact recovery. Both the technical boards and the Renova Foundation respond to the CIF, in a rigid hierarchical structure, and operate according to its ruling (Fig. 5.4a). Nevertheless, a major player does not take part in the *Framework Agreement*. MPF, the Federal Prosecutors' Office, is a separate administration, focusing on promoting social justice and democratic rights and is the central institution with legitimacy to homologate agreements and other legal protocols. MPF does not participate in this agreement, stating that 'the considerations given by the MPF were not taken into account by the remaining parties of the agreement (…) resulting in partial and incomplete settings, illegitimate/illegal procedures', defining the *Framework Agreement* 'unconstitutional in its merits' (MPF 2016b). Also, civil society organisations, scholars, members of the Public Prosecutor's Office and the Public Defender's Office, representatives of social movements, and the affected communities sharply criticised the *Framework Agreement*, as it did not include the most important stakeholders: the affected communities (Nabuco and Aleixo 2019).

The *Framework Agreement* aims to provide restoration of environmental damage to the communities affected, including the indigenous. A fund of up to US$6.3 billion (BRL 20 billion) is available for clean-up costs and recover damages related to the Samarco dam failure. The *Framework Agreement* establishes a public-private governance system to rehabilitate the Doce River Basin with environment-related incentives, constituting the first hybrid governance system in Brazil. According to Muradian and Rival (2012), hybrid regimes are suited to deal with the governance challenges derived from the characteristics of ecosystem services. The major challenge is the integration of *regulatory processes, mechanisms and organisations through which political actors influence environmental actions and outcomes* (Lemos and Agrawal 2006). To evaluate the stakeholders' role, a stakeholder analysis was carried out to identify weakness and loss of opportunities to improve governance, according to Brown (2006), which defends a narrative approach to the study of organisations. This approach focuses on stories that underpin our cognitive and emotional lives as agents of memory, emotion, and meaning (Brown 2006). The stakeholders were categorised from the authors' perspectives, constructed by a mix of internal participants who had joined CIF meetings, analysing legal and technical documents, and an empirical explorer group that collected data about geographical changes and post-disaster actions. This vast contextualisation allowed comparing how political discussions at CIF's table were being applied in the field. Figure 5.5 was then generated and is central to understanding the reasons why this complex organisation created to manage an environmental disaster has low efficiency although it has the power, hegemony, and control over the Doce River restoration.

The stakeholder's analysis generates 29 stakeholders involved in the new *Framework Agreement*, with different degrees of decision-making power (Fig. 5.5a). Brown (2006) defines the importance of these stakeholder groups as for how their livelihoods were (1) impacted by the outcome of decision-making and (2) their influence over the decision-making process.

Stakeholders were categorised into primary and secondary, depending on decision-making power. The upper left square of Fig. 5.5a describes a stakeholder that is defined as *Demanding* by Mitchell et al. (1997), a group that has no power nor competences but is profoundly impacted by the decision-making process. In the current case, local communities (1) are considered the highest impacted stakeholder, only to be matched by the local professional fisherman (3). Local communities are considered here as the local population living in the river vicinity but also the indigenous people who live and worship the river. For these people, the Doce constituted water, food, shelter, and a belief system. The most impacted and influenced by the disaster, local communities (1) have the lowest influence over the new governance system. The local professional fishermen (3) have higher influence than (1) since they formed official associations that represented their well-being and interests. Both land developers (7) and industries (8) are considered here as small local businesses, like farmers and dairy farms. Both (7) and (8) are currently facing economic and environmental scarcity, in terms of degraded land and river, in the absence of natural resources that sustained their business. AGERH (10), Agência Nacional de Energia Elétrica (ANEEL) (11) and IDAF (12) (*Discretionary* stakeholders) are Federal and State institutions that possess both expertise and legislative power but are neither greatly impacted by

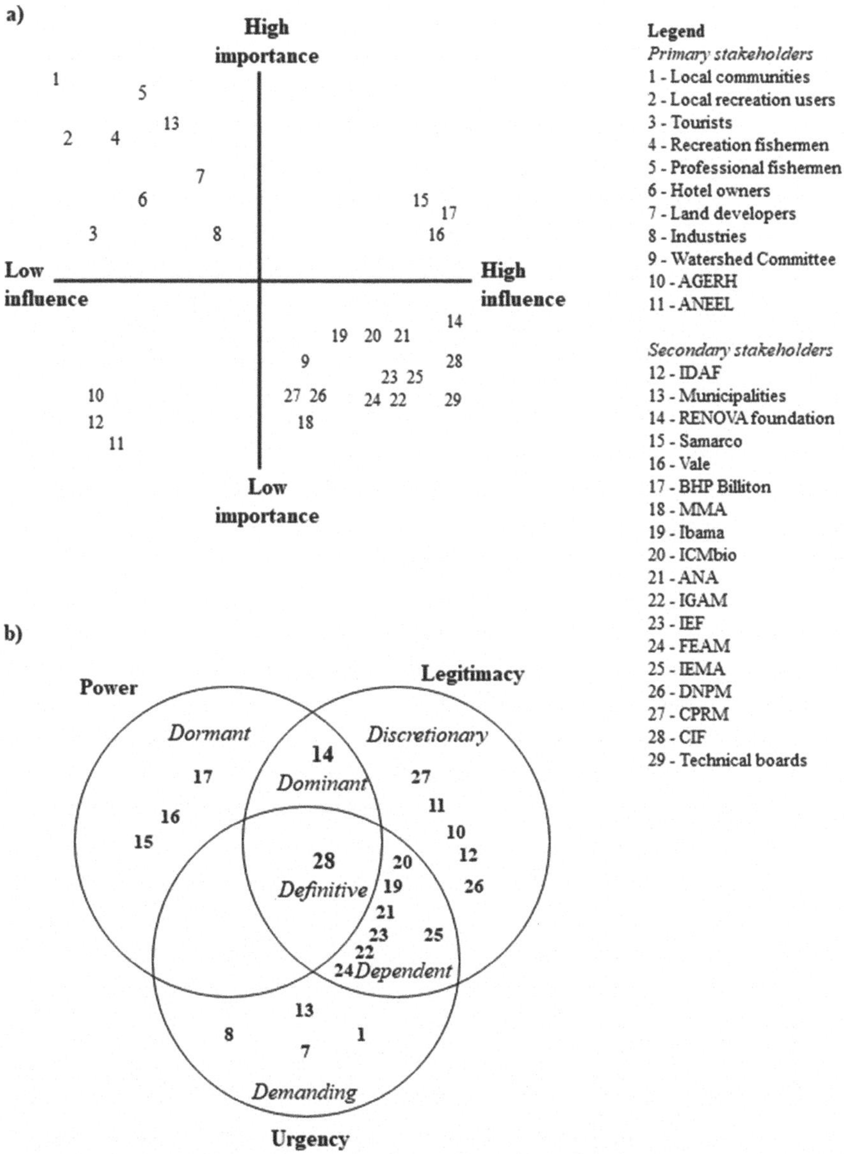

Fig. 5.5 (a) Stakeholder level of importance (Brown 2006) concerning impact by the disaster and level of influence on decision-making on the post-disaster actions following the new framework agreement and (b) stakeholder definition (Mitchell et al. 1997)

the disaster nor have significant influence over the decision process. These institutions have little to no representation at the CIF and are not currently included in the watershed recovery, but they have legislative power at the State level. Therefore, we consider them as *Discretionary* stakeholders (Mitchell et al. 1997).

Another type of stakeholder is the *Dangerous* type, where NGOs take this role regularly. However, Brazil does not host NGOs with sufficient power to influence the decision process as it is. Therefore, we leave this category blank. The Fundação Renova (14) is the sole stakeholder defined as *Dominant*, as it has power with legitimacy to manage the funds that were allocated to the Doce River recovery. The DNPM, the institute supervising tailings dams in Brazil, is indicated as neutral in terms of impact and influence (very close to the origin in Fig. 5.5a and *Discretionary* stakeholder in Fig. 5.5b). The DNPM has all the legitimacy to sanction and stop mining exploitation prior to disaster. Once the tailings dam was ruptured, the DNPM has no competencies relative to ecosystem and environmental restoration. Mitchell (1997) describes the *Dependent* stakeholder as stakeholders who lack power but who have urgent legitimate claims because these stakeholders depend upon others for the power necessary to carry out their will. In this sense, we defined the majority of environmental agencies as dependent (Fig. 5.5b). In the new Framework Agreement, these autonomous agencies that generally have the authority to implement directives, supervise and execute sanctions are now dependent on CIF decisions. The *Definitive* stakeholder is a stakeholder that has all three driving attributes for effective decision-making (Mitchell et al. 1997). Here, we define CIF as the sole *Definitive* stakeholder in the decision process of the Doce recovery (Fig. 5.5b). Empirically, the MPF has all the three main attributes as well, but it removed itself from the Framework agreement early in the process.

Samarco (15), Vale (16), and BHP (17) are considered as powerful stakeholders because they have financial capacity and provide the funds that will be used to recover the ecosystem. We consider that they are *Dormant* stakeholders because they lack the urgency to recover the environment. However, this urgency should increase in a post-disaster scenario, mainly by authority imposition—the MPF. Since the MPF has removed itself from the Framework Agreement, the powerful dormant stakeholders will remain as such. The stakeholder analysis suggests that the new *Framework Agreement* continues to underrepresent the most impacted stakeholders (Fig. 5.5a, b). As the CIF is the *definitive* stakeholder, and the Renova Foundation the *dominant* one, measures should be taken to assure that decision-making is not mainly affected by economic interests but based on the technical and scientific advice provided by the technical boards (Fig. 5.3a).

The jurisdictions of environmental management systems and water resources management systems in Brazil are separate, and the *Framework Agreement* addresses this divide. Efforts to decentralise decision-making from the CIF and to reduce the industrial ruling should be made. Instead, the watershed ecosystem recovery should be prioritised, and concepts of ecological engineering and ecohydrology should be adopted (McClain 2002; Millenium Ecosystem Assessment 2005). For this to take place, a basin approach should be considered, and for that, the watershed committee (9) should assume a central role as *Definitive* stakeholder. This way, environment, energy production, population, sewage treatment, agricultural irrigation, and wetland preservation, to mention a few, will be incorporated in decision-making. An instrumental aspect would also be to engage indigenous people and make them more representative in the process. The recovery of wetlands along the Doce River is instrumental in supporting indigenous people, the Krenak, who support themselves from the river. Also, as the Krenak perceive the Doce River as a deity, working proactively in its recovery would reassure and engage them. For example, the Kagera project, a transboundary watershed between Burundi, Rwanda, Tanzania, and Uganda supported by Food and Agriculture Organization (FAO), can serve as an imprint for the Doce River recovery. Actions portfolio consist of protection of wetlands for water and food supply carried out by local communities in tandem with the technical support. This decentralised approach involves fieldwork and teaching local communities.

The Broken Policy

With diverse and complex governance, Brazil offers a flexible system that may be considered an advantage when risk (such as disasters) arises (Renn et al. 2011). The new *Framework Agreement* represents a new structure in the national governance paradigm, housing members of different governmental bodies in a three-axis structure (Fig. 5.3a). Since its inception, several public meetings have taken place to develop the current structure and coordinate with key stakeholders to the Doce River recovery (IBAMA 2018). The *Framework Agreement* developed a regulatory body—CIF—an independent foundation (Renova Foundation) and technical boards (Fig. 5.4), independent of the established rigid regulatory structure itself. When institutional diversity is taken into account, several benefits can be observed (Renn et al. 2011):

- increased flow of communication across environmental agencies;
- reduced bureaucracy;
- expedite watershed rehabilitation, since communication and decision-making are faster;
- Simplifies decision-making because scientific and technical information is costume made; and
- aggregates information to be provided to the public.

The *Framework Agreement* is viewed as a reasonable governance structure for watershed recovery and management. The structure is horizontal, including members from the main environmental federal and state agencies; funds are available for ecosystem recovery; and there is sound national and international technical expertise available. Like other hybrid governance systems, it is a complex structure, involving a multiplicity of actors and many interrelations between the 'local' and the 'global' (Muradian and Rival 2012). Partners in such governance systems tend to have common environmental issues and therefore coordinate activities and resources towards common research and development (Hardy 2010).

However, the *Framework Agreement* has proven to be lacklustre, maybe due to its complexity or other factors. As pointed out by Moss and Newig

(2010), spatial distance between stakeholders directly influences their level of involvement in collaboration, despite the CIF, and meetings have been held roughly one month apart since its inception (IBAMA 2018). Four years after the disaster, an impact assessment on water quality, aquatic biodiversity, and human health is still missing. There was limited knowledge exchange between researchers and local communities for remedy and collective development (de Abreu and de Andrade 2019). A concise environmental monitoring programme is being planned and was launched in late 2017, involving academia and technical experts from government bodies, among others (IBAMA 2017). According to Hardy (2010), this is significant, as he found that effective agency-based partnerships comprise highly skilled technical experts, government officials, and representatives from regional and State agencies. However, two big *Framework Agreement* caveats consist of:

- The MPF does not participate in the agreement, that is, the national regulatory body does not partake of the CIF. Therefore, the *Framework Agreement* does not hold juridical power to implement and regulate recovery actions. According to Eckersley (2004), management decisions regarding public and common pool goods require that higher-level institutions and organisations be recognised as legitimate.
- The *Framework Agreement* establishes Samarco, the 'polluter', as the creator of the Renova Foundation. This point implies a bias towards the management of financial resources, diminishing effective institutional diversity in the decision-making process.

The *Framework Agreement* places itself between markets and hierarchies constituting a hybrid governance structure, similar to the Chesapeake Bay transboundary watershed management (Just and Netanyahu 1998). In this particular case, an agency-based watershed partnership was created that contained over 300 scientists, 22 different state and federal agencies, as well as representatives from the United States Environmental Protection Agency (USEPA) (Diaz-Kope and Miller-Stevens 2015). In these circumstances, policy decisions regarding restoration and protection of the Chesapeake Bay are made by the members of the Executive

Council (governors of Virginia, Maryland, and Pennsylvania; the Mayor of Washington, D.C.; as well as a representative from the US Environmental Protection Agency (EPA) and the Chair of the Chesapeake Bay Commission) (Chesapeake Bay Program 2009; Diaz-Kope and Miller-Stevens 2015). Policy decisions, however, are taken by the Executive Council, which comprises environmental regulators and high-level government representatives and delegations of federal and state agencies (Chesapeake Bay Program 2009). Indeed, decision-making within the Chesapeake Bay Program is complex, as the different partners respond to several jurisdictional authorities and are affected differently by watershed issues (e.g. pollution affected both water and air). Therefore, the programme incorporates four distinctive decision-making levels that guide governance activities: (1) consensus, (2) unilateral, (3) champion, and (4) voting (Chesapeake Bay Program 2009; Diaz-Kope and Miller-Stevens 2015). Creating levels of decision-making can be the solution to the Doce River *Framework Agreement* delay. By allocating funds and decision-making, a restoration and protection group could speed up the process, bypass the CIF centrality (Fig. 5.5b), and empower the 'technical boards' and consequently environmental restoration.

Diversity is key when complex, uncertain, and ambiguous risk problems need to be addressed (Renn et al. 2011). Although diversity is achieved within the *Framework Agreement* constitution, it is evident that the mining industry has strong influence over the Renova Foundation and the overall decision-making process (Fig. 5.5a). The funds have been allocated to the recovery of the Doce River, but the actions taken so far are not effective for the recovery of the river. As Muradian and Rival (2012) stated, without appropriate incentives or local engagement in rulemaking, there is abundant evidence that state policies might be ineffective. The economic influence of the private sector over governments must then be addressed. In Brazil, political campaigns could be financed by private companies up to a limit of 2% of their gross annual revenue. Specifically, a company that bills US$2 billion a year may donate up to US$3 million to a given political party. This practice of 'lending money', which ended in 2015, may have undermined the integrity of political decision-making. Politicians have been criticised because they focused on the companies' growth to the detriment of the protection of the population and the environment (Westra et al. 2013). Since 2015, by electoral reform, donations by private companies to political parties are prohibited. In the aftermath of the Fundão failure, several close collaborations between the mining industry and politicians were disclosed, some related to bribery, exposing powerful alliances between industry and politicians (Lyra 2019).

However, industries were given flexibility throughout the years, including the compliance with international regulations such as safety and security (e.g. US Occupational Safety and Health Administration OSHA). Brazil mining activities are spread out, with over 3000 listed tailings ponds). Currently, there are a considerable number of dams at risk of rupture in Brazil (DNPM 2016). Basin vulnerability was defined based on tailings dams' risk assessment and its size and on socio-economic data regarding the representation of the indigenous people, GDP, and administration improbity. High-risk mining industries should supervise and regulate locally but should respond to national governmental bodies, where contingency plans are instrumental in preventing and minimising environmental impacts along the entire fluvial-estuarine-marine continuum, and policymaking needs to focus more on prevention at source (Lu et al. 2015).

When we consider dam failure risk alone, Brazil has considerable basin resilience based merely on its basin sizes (Lacerda et al. 2002). Indeed, when we add a social dimension because of local policies to watersheds to basin vulnerability given by risk of dam failure, we see that basins with medium vulnerability rise from two to seven (Fig. 5.6). Policy and its social impacts do add to the risk of watershed management in case of dam failure. In the case of the Doce River, collaboration between the different layers of Federal and State government, academia, industry, and local communities, including indigenous people, is essential to recuperate the basin. Given the scale of mining operations in the Iron Quadrangle, monitoring, contingency plans, and legislative reinforcement need to be of the same scale, if we are not to have a similar event soon. However, we alert to the fact that there are other basins in Brazil in danger of dam failure, and that current local policies would result in the same type of disaster management. Forcing industries to implement contingency plans for possible dam failures now may mitigate uncertainty in the future (like the Canadian Directive 085).

The *Framework Agreement* created in 2018 has yet to prove to be effective, as the affected communities face serious obstacles to have their voice heard in both CIF and Renova Foundation, as well as to have access to equitable remedies (Nabuco and Aleixo 2019). The affected communities should be able to freely choose assistance committees (including legal counselling) to have representatives with the capacity to influence remedy programmes in the decision-making bodies within Renova Foundation and CIF, and have the power to decide on the remedial measures tailored to their needs (Nabuco and Aleixo 2019). For example, the indigenous communities of the Krenak, Tupiniquim, and Guarani; the 'quilombolas', and the artisanal fishermen still lack access to inherent human rights such as water, work, wealth, and an adequate standard of living and health (Gonçalves e Silva 2019). Other issues, such as transparent and trustworthy systems of monitoring, control, and disclosure of results, are also not implemented (Santos and Milanez 2017). Also, the predominance of Executive Branch agencies in the CIF compromises the monitoring activity, as they currently do not have the capacity for effective enforcement of the regulations (Santos and Milanez 2017).

If we need to single out the main reason why the *Framework Agreement* is on a standstill, we would have to address the stakeholder analysis (Fig. 5.5b). The CIF is the *Definitive* stakeholder, the single entity that has effective power over decision-making in this governance system. Concentrating decision-making power in one single entity is the principle of classical governance systems (Hammer et al. 2011; Lemos and Agrawal 2006; Milas and Latif 2000; Muradian and Rival 2012). The new *Framework agreement* (Fig. 5.3b) introduces social stakeholders and both federal and state environmental agencies. However, it is our understanding that the hierarchy does not change. According to Muradian and Rival (2012), solving the problems posed by ecosystem services usually requires that we move from thinking in terms of single, ideal managerial approaches to combining governance structures, scales, and tools. If we want the *Framework Agreement* to be successful, governance requires to be displaced from a single centre of power (McGinnis 2000). Similar to the Chesapeake Bay Program, the *Framework Agreement* should adopt

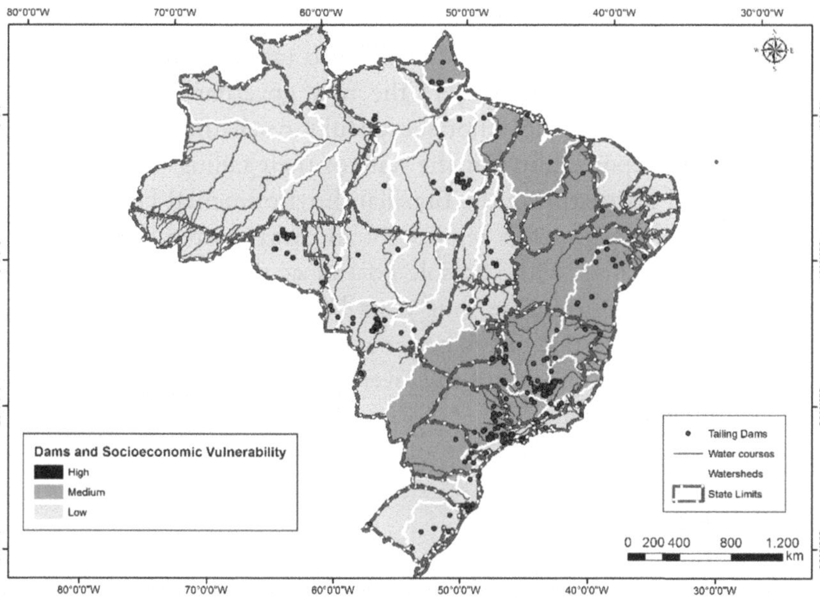

Fig. 5.6 Basin vulnerability based on dam risk failure assessed by DNPM (2016), combined with the ratings attributed to basin size and socio-economic data

distinctive decision-making levels that guide governance activities (Chesapeake Bay Program 2009). Moreover, since most Brazilian watersheds are transboundary, we urge the central government to initiate and implement, a priori, a similar type of Framework Agreement to each national watershed (especially the ones highlighted with medium vulnerability in Fig. 5.6). More policy guidelines for disaster management can be found in, for example PEDRR (2011); Schoenberger (2016).

A second massive disaster in the same region has recently occurred. On 25 January 2019, tailings from a dam in Vale's Córrego do Feijão mine (Belo Horizonte, Minas Gerais) released 12 Mm^3 of iron tailings to the Paraopeba River (Freitas et al. 2019). This rupture had a lower volume of tailings released (and lower environmental impact when compared to Fundão disaster), but higher loss in human lives (almost 300 fatalities), and it is considered a large-scale work accident (the largest in Brazil) (Almeida et al. 2019). The slurry released travelled approximately 7 km downhill before reaching Rio Paraopeba, thereby destroying a bridge of the mine's railway branch, and spreading to Vila Ferteco local community, near the town of Brumadinho. The company responsible for this disaster is one of the owners of Fundão tailings dam. This second disaster highlights two things: (1) the new governance adopted—the *Framework Agreement*—does not successfully reach the mining companies and their responsibility for the state of their tailings ponds and (2) the urgency of restructuring the Brazilian Environmental Policy System. The *Framework Agreement* is temporary and was created outside of the environmental system with the sole goal of accelerating the watershed recovery process. During the four years between the two disasters, no other actions were carried out to better appropriate responsibilities in the mining industry. As a result of this oversight, another 300 human lives were lost. How many more?

Final Remarks

Resource-based countries, such as Canada, Russia, China, Australia, and Brazil, rely heavily on mining activities to support their economy, and they will continue to exploit natural resources in the future (Montesanti 2014). Currently, in Brazil, integrated state participation regarding industrial resource exploitation is non-existent. The new Framework Agreement constitutes the first hybrid governance system in Brazil and presents limitations regarding stakeholder engagement and empowerment. Based on basin vulnerability, we can predict that the Doce River and 'Costeira do Norte Oriental' are potential watersheds to undergo new tailings dam rupture in the future and require special judicial and environmental audit to these basins. In principle, the Framework Agreement in place would be a diverse and well-structured system to oversee ecosystems recovery. However, efforts to decentralise decision-making and minimise the industrial biases should be undertaken. According to the Brazilian law, and the polluters-pay principle, *"the person exploiting mineral resources must recover the degraded environment, in accordance to the technical solution required by the competent public authority"* (Article 225, 2, Brazilian Constitution). To this day, no concrete action plan, goals, and recovery targets have been established.

Acknowledgements This work was partly supported by a Young Talent fellowship funded by the programme Science without Borders of the Brazilian Federal Agency for the Support and Evaluation of Graduate Education (Coordenação de Aperfeiçoamento de Pessoal de Nível Superior—CAPES).

References

Almeida, I. M., Jackson Filho, J. M., & Vilela, R. A. G. (2019). Reasons for Investigating the Organizational Dynamics of the Vale Tailings Dam Disaster in Brumadinho, Minas Gerais State, Brazil. *Cadernos de Saúde Pública, 35*(4), ee00027319. https://doi.org/10.1590/0102-311X00027319.

Alves, R. H., & Rietzler, A. C. (2015). Toxic Effects of Arsenic on *Eisenia andrei* Exposed to Soils Surrounding Gold Mining Operations. *Revista Brasileira de Ciência do Solo, 39*, 682–691.

ANA. (2015). *Noticia da Agência Nacional de Águas*. http://www2.ana.gov.br/Paginas/imprensa/noticia.aspx?id_noticia=12271. Accessed 22 Aug 2017.

ANA. (2016). *Encarte Especial da Bacia do Doce*. http://arquivos.ana.gov.br/RioDoce/EncarteRioDoce_22_03_2016v2.pdf (accessed 10 October 2017)

ANA. (2017). *Sewage Atlas*. http://atlasesgotos.ana.gov.br. Accessed 17 Oct 2017.

Antón, A., Cebrian, J., Heck, K. L., et al. (2011). Decoupled Effects (Positive to Negative) of Nutrient Enrichment on Ecosystem Services. *Ecological Applications, 21*(3), 991–1009.

Armada, C. A. S. (2019). The Environmental Disasters of Mariana and Brumadinho and the Brazilian Social Environmental Law State. *SSRN*. https://doi.org/10.2139/ssrn.3442624.

Biermann, F., Bai, X., Bondre, N., et al. (2016). Down to Earth: Contextualizing the Anthropocene. *Global Environmental Change, 39*, 341–350.

Boger, D. V. (2013). Rheology of Slurries and Environmental Impacts in the Mining Industry. *Annual Review of Chemical and Biomolecular Engineering, 4*, 239–257.

Bourcy, S. C., & Weeks, R. E. (2000). Stream Morphology and Habitat Restoration of Pinto Creek, Gila County, Arizona. In *Proceedings of the Seventh International Conference on Tailings and Mine Waste 2000* (pp. 467–475). Fort Collins, CO, USA. Rotterdam: Balkema.

Bridge, G. (2004). Contested Terrain: Mining and the Environment. *Annual Review of Environment and Resources, 29*, 205–259.

Brown, A. D. (2006). A Narrative Approach to Collective Identities. *Journal of Management Studies, 43*, 731–753.

Cambridge, M., & Shaw, D. (2019). Preliminary Reflections on the Failure of the Brumadinho Tailings Dam in January 2019. *Dams and Reservoirs, 29*, 113–123.

Chesapeake Bay Program. (2009). *Summary Report to the Chesapeake Executive Council.* 2009 State of the Chesapeake Bay Program, Annapolis.

Chiaretti, D. (2017). *Reconstrução Após Desastre da Samarco Esbarra em Entraves.* https://valor.globo.com/brasil/coluna/reconstrucao-apos-desastre-da-samarco-esbarra-em-entraves.ghtml. Accessed 15 Oct 2018.

Costa, A. T. (2001). *Geoquímica das Águas e dos Sedimentos da Bacia do Rio Gualaxo do Norte, Leste-Sudeste do Quadrilátero Ferrífero (MG): Estudo de uma Área Afetada por Atividades de Extração Mineral.* MSc dissertation, Universidade Federal de Ouro Preto.

Costanza, R., d'Arge, R., de Groot, R., et al. (1997). The Value of the World's Ecosystem Services and Natural Capital. *Nature, 387*, 253–260.

CPRM. (2015). *Portal CPRM – Serviço Geológico do Brasil.* http://www.cprm.gov.br. Accessed 22 Aug 2015.

Davies, M. P. (2002). Tailings Impoundment Failures: Are Geotechnical Engineers Listening? *Geotechnical News, 20*(3), 31–36.

de Abreu, M. C., & de Andrade, R. D. J. C. (2019). Dealing with Wicked Problems in Socio-ecological Systems Affected by Industrial Disasters: A Framework for Collaborative and Adaptive Governance. *Science of the Total Environment, 694*. https://doi.org/10.1016/j.scitotenv.2019.133700.

de Oliveira, E. N., Knoppers, B. A., Lorenzzetti, J. A., et al. (2012). A Satellite View of Riverine Turbidity Plumes on the NE-E Brazilian Coastal Zone. *Brazilian Journal of Oceanography, 60*, 283–298.

Diaz-Kope, L., & Miller-Stevens, K. (2015). Rethinking a Typology of Watershed Partnerships: A Governance Perspective. *Public Work Management Policy, 20*, 29–48.

DNPM. (2016). *Classificação de Barragens de Mineração.* www.dnpm.gov.br/assuntos/barragens/cadastro-nacional-de-barragens_2016-_atualizacao_campanha. Accessed 25 Mar 2018.

Eckersley, R. (2004). *Green State. Rethinking Democracy and Sovereignty.* Cambridge, MA: MIT Press.

Escobar, H. (2015). Natural Disasters: Mud Tsunami Wreaks Ecological Havoc in Brazil. *Science, 350*, 1138–1139.

Foley, J., Defries, R., Asner, G. P., et al. (2005). Global Consequences of Land Use. *Science, 309*, 570–574.

Fourie, A. B., Papageorfiou, G., & Blight, G. E. (2000). Static Liquefaction as an Explanation for Two Catastrophic Tailings Dam Failures in South Africa.

In *Proceedings of the Seventh International Conference on Tailings and Mine Waste 2000* (pp. 149–158). Fort Collins, CO, USA. Rotterdam: Balkema.
Freitas, M. B., & Rodrigues, S. C. A. (2014). As Consequências do Processo de Desterritorialização da Pesca Artesanal na Baía de Sepetiba (RJ, Brasil): Um Olhar sobre as Questões de Saúde do Trabalhador e o Ambiente. *Ciência & Saúde Coletiva, 19*, 4001–4009.
Freitas, C. M., Barcellos, C., Asmus, C. I. R. F., et al. (2019). Da Samarco em Mariana à Vale em Brumadinho: Desastres em Barragens de Mineração e Saúde Coletiva. *Cadernos de Saúde Pública, 35*(5). https://doi.org/10.1590/0102-311X00052519.
Fundação Renova. (2019). *Dados da Reparação – Terra e Água*. https://www.fundacaorenova.org/dadosdareparacao/terra-e-agua. Accessed 24 Oct 2019.
Garcia, L. C., Ribeiro, D. B., de Oliveira Roque, F., et al. (2017). Brazil's Worst Mining Disaster: Corporations Must Be Compelled to Pay the Actual Environmental Costs. *Ecological Applications, 27*, 5–9.
Gomes, L. E. O., Correa, L. B., Sá, F., et al. (2017). The Impacts of the Samarco Mine Tailing Spill on the Rio Doce Estuary, Eastern Brazil. *Marine Pollution Bulletin, 120*, 28–36.
Gonçalves e Silva, T. V. (2019). The Mariana Dam Disaster in Brazil: Open Issues About Corporate Accountability under International Human Rights Law. *Educamazônia, 1*, 35–53.
Grangeia, C., Ávila, P., Matias, M., et al. (2011). Mine Tailings Integrated Investigations: The Case of Rio Tailings (Panasqueira Mine, Central Portugal). *Engineering Geology, 123*, 359–372.
Guedes, G. R., Simão, A. B., Dias, C. A., et al. (2015). Risco de Adoecimento por Exposição às Águas do Rio Doce: Um Estudo Sobre a Percepção da População de Tumiritinga, Minas Gerais, Brasil. *Cadernos de Saúde Pública, 31*, 1257–1268.
Hammer, M., Balfors, B., Mörtberg, U., et al. (2011). Governance of Water Resources in the Phase of Change: A Case Study of the Implementation of the EU Water Framework Directive in Sweden. *Ambio, 40*, 210–220.
Hardy, S. D. (2010). Governments, Group Membership, and Watershed Partnerships. *Society and Natural Resources, 23*, 587–603.
Hatje, V., Pedreira, R. M. A., de Rezende, C. E., et al. (2017). The Environmental Impacts of one of the Largest Tailing Dam Failures Worldwide. *Scientific Reports, 7*, 10706. https://doi.org/10.1038/s41598-017-11143-x.
Herard, D. (2010). *The Hungarian Toxic Red Sludge Spill and Determining Public Accountability*. MSc dissertation, Florida International University.
IBAMA. (2016a). Technical Report PAR. 02024.000023/2016-47 NUGEO/RO/IBAMA. Brasília.

IBAMA. (2016b). *Rupture of the Fundão Dam: Documents related to the Samarco disaster in Mariana/MG*. www.ibama.gov.br/informes/rompimento-da-barragem-de-fundao#pareceres. Accessed 25 Oct 2017.

IBAMA. (2017). *Bacia do Rio Doce Recebe Rede para Monitorar Qualidade da Água*. www.ibama.gov.br/noticias/422-2017/1154-bacia-do-rio-doce-recebe-rede-para-monitorar-qualidade-da-agua. Accessed 25 Oct 2017.

IBAMA. (2018). *CIF Meetings*. www.ibama.gov.br/cif/pautas-das-reunioes. Accessed 24 Aug 2018.

IBGE. (2016). *Mudanças na Cobertura e Uso da Terra 2000–2014*. Rio de Janeiro.

IBGE. (2017). *Pesquisa Nacional por Amostra de Domicílios 2015*. www.ibge.gov.br/estatisticas/sociais/educacao/9127-pesquisa-nacional-por-amostra-de-domicilios.html. Accessed 25 Oct 2017.

IEMA. (2015). *Grupo de Técnico de Enfrentamento da Crise Ambiental do Rio Doce* (Processo No 72423994). Cariacica.

IEMA. (2017). *IEMA – Informativo Rio Doce*. https://iema.es.gov.br/informativo-rio-doce. Accessed 17 Oct 2017.

Just, R. E., & Netanyahu, S. (1998). *Conflict and Cooperation on Trans-Boundary Water Resources*. Springer US.

Kossoff, D., Dubbin, W. E., Alfredsson, M., et al. (2014). Mine Tailings Dams: Characteristics, Failure, Environmental Impacts, and Remediation. *Applied Geochemestry, 51*, 229–245.

Lacerda, L. D., Kremer, H. H., Kjerve, B., et al. (2002). *South American Basins: LOICZ Global Change Assessment and Synthesis of River Catchment – Coastal Sea Interaction and Human Dimensions*. Texel.

Lemos, M. C., & Agrawal, A. (2006). Environmental Governance. *Annual Review Environment and Resources, 31*, 297–325.

Lima, A. T., Mitchell, K., O'Connell, D. W., et al. (2016). The Legacy of Surface Mining: Remediation, Restoration, Reclamation and Rehabilitation. *Environmental Science and Policy, 66*, 227–233. https://doi.org/10.1016/j.envsci.2016.07.011.

Lu, Y., Song, S., Wang, R., et al. (2015). Impacts of Soil and Water Pollution on Food Safety and Health Risks in China. *Environment International, 77*, 5–15.

Lyra, M. G. (2019). Challenging Extractivism: Activism over the Aftermath of the Fundão Disaster. *Extractive Industries and Society, 6*, 897–905.

McClain, M. E. (2002). *The Ecohydrology of South American Rivers and Wetlands* (IAHS Special Publication No. 6). International Association of Hydrological Sciences.

McGinnis, M. D. (2000). *Polycentric Games and Institutions: Readings from the Workshop in Political Theory and Policy Analysis*. Ann Arbor: University of Michigan Press.

Medeiros, A. O., Kohler, L. M., Hamdan, J. S., et al. (2008). Diversity and Antifungal Susceptibility of Yeasts from Tropical Freshwater Environments in Southeastern Brazil. *Water Research, 42*, 3921–3929.

Meira, R. M. S. A., Peixoto, A. L., Coelho, M. A. N., et al. (2016). Brazil's Mining Code Under Attack: Giant Mining Companies Impose Unprecedented Risk to Biodiversity. *Biodiversity Conservation, 25*, 407–409.

Milas, S., & Latif, J. (2000). The Political Economy of Complex Emergency and Recovery in Northern Ethiopia. *Disasters, 24*, 363–379.

Millenium Ecosystem Assessment. (2005). *Ecosystems and Human Well-Being: Wetlands and Water Synthesis*. Washington, DC: World Resources Institute.

Miranda, L. S., & Marques, A. C. (2016). Hidden Impacts of the Samarco Mining Waste Dam Collapse to Brazilian Marine Fauna – An Example from the Staurozoans (Cnidaria). *Biota Neotropica, 16*(2). https://doi.org/10.1590/1676-0611-BN-2016-0169.

Mitchell, R. K., Agle, R., & Wood, D. J. (1997). Toward a Theory of Stakeholder Identification and Salience: Defining the Principle of Who and What Really Counts. *Academy of Management Review, 22*, 853–886.

MMSD. (2002). *Breaking New Ground: Mining, Minerals and Sustainable Development, Final Report*. London: Minerals and Sustainable Development Project.

Montesanti, S. (2014). Policy: Count the Social Cost of Oil Sands Too. *Nature, 513*, 172–172.

Morgenstern, N.R., Vick, S.G. and Viotti, C.B. (2016). Fundão Tailings Dam Review Panel: Report on the Immediate Causes of the Failure of the Fundão Dam.

Moss, T., & Newig, J. (2010). Multilevel Water Governance and Problems of Scale: Setting the Stage for a Broader Debate. *Environmental Management, 46*, 1–6.

MPF. (2016a). *Procedimento Investigatório Criminal (PIC)*. MPF n.° 1.22.000.003490/2015-78. Brasilia.

MPF. (2016b). *Acão Civil Pública 155M com Pedido de Liminar Inaudita Altera PARS*.

MPMC. (2015). *Post Event Environmental Impact Assessment Report – Key Finding Report*.

Muradian, R., & Rival, L. (2012). Between Markets and Hierarchies: The Challenge of Governing Ecosystem Services. *Ecosystem Services, 1*, 93–100.

Nabuco, J., & Aleixo, L. (2019). Rights Holders' Participation and Access to Remedies: Lessons Learned from the Doce River Dam Disaster. *Business Human Rights, 4*, 147–153.

Nazareno, A. G., & Vitule, J. R. S. (2016). Pollution: Too Many Mining Disasters in Brazil. *Nature, 531*, 580–580.

Nobre, C. A., Marengo, J. A., Seluchi, M. E., et al. (2016). Some Characteristics and Impacts of the Drought and Water Crisis in Southeastern Brazil During 2014 and 2015. *Journal of Water Resources and Protection, 8*, 252–262.

Oliveira, K. S. S., & da Quaresma, V. S. (2017). Temporal Variability in the Suspended Sediment Load and Streamflow of the Doce River. *Journal of South American Earth Sciences, 78*, 101–115.

Pain, D. J., Meharg, A., Sinclair, G., et al. (2003). Levels of Cadmium and Zinc in Soil and Plants Following the Toxic Spill from a Pyrite Mine, Aznalcollar, Spain. *Ambio, 32*, 52–57.

PEDRR. (2011). *Managing Watersheds for Urban Resilience. Partnership for Environment and Disaster Risk Reduction (PEDRR).* Policy Brief Presented at the Global Platform for Disaster Risk Reduction, Roundtable on "Managing Watersheds for Urban Resilience". Geneva.

Petticrew, E. L., Albers, S. J., Baldwin, S. A., et al. (2015). The Impact of a Catastrophic Mine Tailings Impoundment Spill into one of North America's Largest Fjord Lakes: Quesnel Lake, British Columbia, Canada. *Geophysical Research Letters, 42*, 3347–3355.

Pires, J. M. M., de Lena, J. C., Machado, C. C., et al. (2003). Potencial Poluidor de Resíduo Sólido da Samarco Mineração: Estudo de Caso da Barragem de Germano. *Revista Árvore, 27*, 393–397. https://doi.org/10.1590/S0100-67622003000300017.

Renn, O., Klinke, A., & van Asselt, M. (2011). Coping with Complexity, Uncertainty and Ambiguity in Risk Governance: A Synthesis. *Ambio, 40*, 231–246.

Ribeiro, R. A., Lemos-Filho, J. P., Ramos, A. C. S., et al. (2011). Phylogeography of the Endangered Rosewood *Dalbergia nigra* (Fabaceae): Insights into the Evolutionary History and Conservation of the Brazilian Atlantic Forest. *Heredity, 106*, 46–57. https://doi.org/10.1038/hdy.2010.64.

Rico, M., Benito, G., Salgueiro, A. R., et al. (2008). Reported Tailings Dam Failures. *Journal of Hazardous Materials, 152*(2), 846–852.

Rodrigues, L. (2017). *Governo de Minas Autoriza Pesca de Algumas Espécies na Bacia do Rio Doce.* http://agenciabrasil.ebc.com.br/geral/noticia/2017-05/governo-de-minas-autoriza-pesca-de-algumas-especies-na-bacia-do-rio-doce. Accessed 17 Oct 2017.

Rodrigues, A. S. L., Malafaia, G., Costa, A. T., et al. (2014). Iron Ore Mining Promotes Iron Enrichment in Sediments of the Gualaxo do Norte River Basin, Minas Gerais State, Brazil. *Environmental Earth Sciences, 71*, 4177–4186.

Rooney, R. C., Bayley, S. E., & Schindler, D. W. (2012). Oil Sands Mining and Reclamation Cause Massive Loss of Peatland and Stored Carbon. *PNAS, 109*, 4933–4493.

Samarco. (2008). *Relatório Anual de Sustentabilidade da Samarco SA*. Belo Horizonte.

Samarco. (2014). *Relatório Anual de Sustentabilidade da Samarco SA*. Belo Horizonte.

Santolin, C. V. A., Ciminelli, V. S. T., Nascentes, C. C., et al. (2015). Distribution and Environmental Impact Evaluation of Metals in Sediments from the Doce River Basin, Brazil. *Environmental Earth Sciences, 74*, 1235–1248.

Santos, R. S. P., & Milanez, B. (2017). The Construction of the Disaster and the Privatization of Mining Regulation: Reflections on the Tragedy of the Rio Doce Basin, Brazil. *Vibrant: Virtual Brazilian Anthropology, 14*(2). https://doi.org/10.1590/1809-43412017v14n2p127.

Schoenberger, E. (2016). Environmentally Sustainable Mining: The Case of Tailings Storage Facilities. *Resouces Policy, 49*, 119–128.

Segura, F. R., Nunes, E. A., Paniz, F. P., et al. (2016). Potential Risks of the Residue from Samarco's Mine Dam Burst (Bento Rodrigues, Brazil). *Environmental Pollution, 218*, 813–825. https://doi.org/10.1016/j.envpol.2016.08.005.

Tuck, C. A. (2015). *USGS Minerals Information: Iron Ore*. https://minerals.usgs.gov/minerals/pubs/commodity/iron_ore. Accessed 12 Aug 2017.

UNEP/OCHA. (2000). *Cyanide Spill at Baia Mare, Romania*. UNEP/OCHA Assessment Mission, UNEP/Office for the Co-ordination of Humanitarian Affairs.

van Niekerk, H. J., & Viljoen, M. J. (2005). Causes and Consequences of the Merriespruit and Other Tailings-dam Failures. *Land Degradation and Development, 16*(2), 201–212.

Vick, S. G. (1997). Failure of the Omai Tailings Dam: Closure. *Geotechnical News, 15*(1), 49–55.

Westra, L., Taylor, P., & Michelot, A. (2013). *Confronting Ecological and Economic Collapse : Ecological Integrity for Law, Policy and Human Rights*. New York: Routledge.

World Bank. (2017). *DataBank*. http://databank.worldbank.org/data/home.aspx. Accessed 2 Aug 2017.

Zhang, R., Cuartas, L. A., de Castro Carvalho, L. V., et al. (2018). Season-based Rainfall-runoff Modelling Using the Probability-Distributed Model (PDM) for Large Basins in Southeastern Brazil. *Hydrological Processes, 32*, 2217–2230.

6

What Do We Want to Be When We Grow Up? The Political Dimensions of Climate Change in Brazil, China and Mozambique

Leila da Costa Ferreira, Fabiana Barbi, and Mariana D. Barbieri

Anthropocene and Risk Society in the Climate Change Context

Since the last two decades of the twentieth century, the environmental issue has assumed the condition of global problem, mobilising not only civil society organisations and media sectors, but governments from all

L. da C. Ferreira (✉)
Institute of Philosophy and Human Sciences (IFCH), University of Campinas (UNICAMP), Campinas, Brazil
e-mail: leilacf@unicamp.br

F. Barbi
Centre for Environmental Studies and Research (NEPAM), University of Campinas (UNICAMP), Campinas, Brazil

M. D. Barbieri
University of Campinas (UNICAMP), Campinas, Brazil

A. A. R. Ioris (ed.), *Environment and Development*,
https://doi.org/10.1007/978-3-030-55416-3_6

regions of the planet. In this scenario, it has been interesting to observe, especially in recent decades, how the environmental issue has become institutionalised all over the world, a process which has increasingly attracted the attention of the scientific community. This chapter presents a synthesis of the main strategies, actions and current trends of environmental policies in Brazil, China and Mozambique. These examples have been chosen for their international importance and, above all, because of the significance of the environmental issue at the heart of these countries' political process like not pay attention in environmental problems. Their strategies for responding to the risks associated with climate change can provide insights with the potential to reshape international negotiations and debates about the problem. The text is focused on global environmental issues and, in particular, on the issue of climate change, which is recognised as one of the main challenges of twenty-first-century global society and a central element in discussion of the so-called Anthropocene (Biermann et al. 2012). Climate change is ongoing, and it is now accepted that increases in the average surface temperature of the Earth are a result of human activities that produce excessive emissions of greenhouse gases (GHGs), mainly from the burning of fossil fuels, combined with natural variations in the global climate system (IPCC 2007).

Risks from climate change can be minimised by reducing GHG emissions, that is, through mitigation strategies. But states also need to develop strategies to adapt to climate change impacts. Therefore, mitigation and adaptation are not mutually exclusive actions; on the contrary, they must be complementary (IPCC 2007; Renn and Klinke 2015). Approaches to such a multifaceted challenge as climate change must also be comprehensive, incorporating various areas of human activity and various stakeholders and sectors of society including multilateral agencies, governments, the private sector, research institutes and organised civil society groups in order to elucidate the factors contributing to climate change risk and find ways to confront them (Bulkeley and Newell 2010; Di Giulio et al. 2017; Ferreira 2018; Jacobi and Sulaiman 2016). Current global environmental changes, unprecedented in human history, are challenging the adaptability and responsiveness not only of the planet but also of societies, as their impacts are yet not completely clear (Young et al. 2006). This situation persists despite various scientific initiatives aimed at providing a better understanding of these processes on an integrated planetary scale (Rockström et al. 2009). The complexity of our condition at this point in history is so unique that the environmental changes brought about by human intervention are understood to be the foundation of a new historical-geological age: the Anthropocene (Crutzen and Stoermer 2000; Spring 2016; Will et al. 2011).

The Anthropocene is characterised by the cumulative impact of human civilisation, which has changed the Earth as much as ice ages but in a much shorter period of time. Although there is no consensus, there are two proposed beginnings to the Anthropocene: the period after the Industrial Revolution, when economic growth and industrial expansion increased GHG emissions, and the post-World War II period, when GHG emissions increased across the globe and profound changes in lifestyles led to higher fossil fuel consumption, urbanisation and processes of industrialisation all over the world (Crutzen and Stoermer 2000; Spring 2016; Steffen et al. 2011; Syvitski and Kettner 2011; Zalasiewicz et al. 2011). Climate change is just one result of these changes. For Giddens (2005), the Anthropocene describes a 'world out of control', where humans have unleashed processes more powerful than their abilities to control them. Discussion about the transition to this new era is relevant as it reflects on the influence of human activities on processes that alter and destabilise the earth system as a whole, thus calling into question the cultural, social and political aspects of societal development processes.

Living in a global society in the Anthropocene means facing multiple risks shaped by various 'stressors', meaning that there are great uncertainties in terms of what the world will be like 20, 50 or 100 years from now. In relation to climate change, it is only possible to present probabilities and project possible scenarios that could materialise in the form of social imaginaries. In other words, the consequences of new risks like climate change are uncertain, mainly because the social, economic and political spheres are so tightly interwoven. Moreover, it is not possible to predict exactly what will happen in the future because the implementation of appropriate mitigation and adaptation policies or lack thereof will impact the results of changes produced during the Anthropocene. This means that contemporary society is being directly forced into the challenge of managing the growing risks that define the Anthropocene as a new era, a geological age in which human action directly influences the various terrestrial systems, altering the stability of the planet (Giddens 2009).

In the risk society, as theorised by Beck (1992), Dunlap and Brulle (2015) suggest the need to reconsider the nature of the current ecological crisis, given that the side effects of industrial production are considered to result from a profound institutional crisis in industrial society. Climate change risks result from the development process of industrial society and are directly influenced by globalisation. Although they are invisible and cause no immediate, observable harm, when 'development products' such as methane (CH_4) and carbon dioxide (CO_2) (from industrial agriculture, increased energy production and consumption, etc.) are released into the atmosphere and change their composition, the consequences are incalculable (Bulkeley 2001). The Anthropocene, in this sense, would have risk as its main horizon, as suggested by Beck (2009); that is, living in this age means living with probabilistic calculations, perceptions and materiality of catastrophic future scenarios in a way that highlights the deleterious effects of the contemporary social order. Another key point relating to risk theory (Beck 1992, 2009) is that the risks posed by climate change go beyond sensory perception, that is, they are situated in the abstract realm of physical equations and chemical formulas, remaining invisible to everyday senses. They can thus be altered, amplified, exaggerated or minimised within the limits of existing knowledge, meaning that they are open to definition and social construction. Also, these risks are both local and global, so their causes and consequences go beyond geographical boundaries (Beck 1992, 1995, 2000, 2009, 2010; Dunlap and Brulle 2015; Ferreira 2018; Giddens 1991, 2005, 2009).

In risk society and in the Anthropocene era, technical-industrial development causes unpredictable, invisible and incalculable risks that require societies to confront and reflect on the foundations of their social cohesion and their rationality. The Anthropocene and risk society, consequently, have an interesting elective affinity that creates multiple connections of meaning: risk seems to be a key element in the ways of acting, thinking and feeling that will shape the social forms created in the Anthropocene. Scientists in all areas share the great concern that our planet is approaching irreversible biophysical tipping points. Scientific research in the natural sciences is fundamental for a better understanding of these processes. However, the results of a worldwide survey by the International Human Dimensions Program on Global Environmental Change (IHDP) have shown that the human dimensions of global environmental change, including social, political and cultural aspects, are just as important as the biophysical dimensions, but severely underestimated (ISSC and UNESCO 2013). In addition, the organisational and institutional arrangements through which human beings govern their relationships with the natural environment and global biochemical systems have been shown to be inadequate and insufficiently understood in the face of the challenges brought about by global environmental changes (Biermann et al. 2012). This is one of the justifications for the Earth System Governance Project launched by IHDP, one of the leading scientific efforts on this issue.

A better understanding of the social and political dimensions of climate change can contribute to minimising the impacts of global environmental change. The social and political dimensions of climate change have proven to be important aspects of research on this topic, given the complexity of the relationship between knowledge, policy recommendations and changes in societal development patterns (Zehr 2015). The discussion covered in this chapter contributes to these debates, focusing on how social groups and society as a whole respond to the risks posed by climate change through policy strategies and instruments. In terms of responses to climate change, it has been argued that governments play a key role in shaping norms, institutions and appropriate modes of governance to address these risks at different levels and scales (Bulkeley and Kern 2006; Bulkeley and Newell 2010; Giddens 2009). It is in this sense that the challenge of the Anthropocene imposes the need for dialogue, reactions and changes at a global scale, capable of balancing different local interests in favour of a common future.

While climate change is a risk for all countries, developing countries are considered the most vulnerable. Brazil, China and Mozambique are cases where climate risks may deepen vulnerabilities and undermine development prospects. It is estimated that developing countries will be impacted with between 75% and 80% of the cost of climate change damage. Even a conservative warming scenario of 2 °C above pre-industrial temperatures could result in permanent GDP (current US$) from The

World Bank: Data reductions of 4–5% for Africa and Southeast Asia. Most developing countries lack sufficient financial and technical capacity to cope with increasing climate risks. In addition, they rely more directly on climate-sensitive natural resources for income generation and well-being, and most are in tropical and subtropical regions already subject to a highly variable climate (World Bank 2010). On the other hand, countries such as China and Brazil, to varying degrees and across different sectors, are major contributors to global GHG emissions. Chinese emissions, mainly from burning fossil fuels, are the highest in the world, while Brazilian emissions come primarily from the land-use change sector, more specifically resulting from Amazon deforestation (see Graph 6.1).

Next, we present an analysis of key climate change strategies and action in Brazil, China and Mozambique since the 1970s, when the environmental issue emerged onto the world political agenda following the 1972 United Nations Conference on the Human Environment in Stockholm, and which is therefore considered a starting point for political internalisation of the environmental issue in most countries. Our analysis focuses on the institutional arrangements and political strategies that have

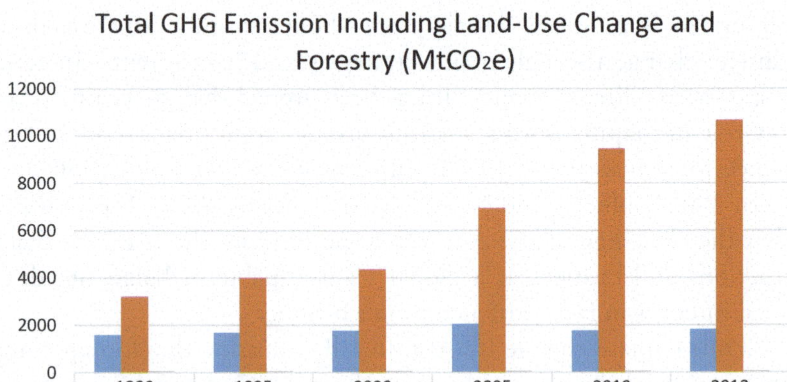

	1990	1995	2000	2005	2010	2012
Brazil	1606.59	1688.95	1778.98	2056.99	1773.77	1823.15
China	3218.45	4019.32	4371.05	6974.05	9473.05	10684.29
Mozambique	54.86	55.92	54.06	60.66	59.2	57.66

Graph 6.1 Total GHG emissions in China, Brazil and Mozambique (1990–2012). (Source: Boden et al. 2015)

emerged since then in Brazil, China and Mozambique and how these countries have addressed the climate issue in their political processes until 2019.

Policy Responses to Climate Change Risks in Brazil, China and Mozambique

In this section, we explore key environmental actions and strategies in Brazil, China and Mozambique, paying particular attention to policies related to climate change. These locations face several challenges in relation to the set of problems that make up the environmental issue in a world characterised by high modernity, the risk society and global environmental changes. Given the scope of possibilities for studies on the scenario of global environmental changes in the south-south axis, the analytical approach is specifically positioned to study the problem of climate change. The analysis is not intended to be comparative, but seeks to identify different political, economic and social contexts of the Global South in order to capture the dynamics related to climate change.

Brazilian Strategies in Response to Climate Change

Brazil is known for having faced difficulties in making environmental concerns a priority in national politics. Analysis of the political-administrative dimension of environmentalism in Brazil reveals a legal and institutional framework defined at various levels of public administration, which has become increasingly complex since the 1970s. In fact, several government environmental agencies have been created, together with a considerable number of environmental regulations (Ferreira and Tavolaro 2008). Therefore, there have been various policies aimed at improving quality of life in urban centres and natural areas over the past 40 years or more at federal, state and local levels.

The internalisation process of the climate issue in Brazil can be divided into three main moments (Barbi et al. 2015). The first phase of political and scientific mobilisation, from 1992 to 2002, was marked by the establishment of political-institutional and scientific structures focused on climate change. At this early stage, the problem entered the national political agenda. It was a moment of articulation between international and national negotiation levels and among institutions and actors, and of identifying political and scientific needs. During this phase, in 2000, the Brazilian Forum on Climate Change (FBMC) was created, representing one of the first national movements related to climate change. Its purpose was to present, educate and mobilise society to debate issues related to the problem of global warming (Viola et al. 2013).

The second phase lasted from 2003 until 2008, and was marked by a greater understanding of the problem of climate change in Brazil. It was notable for the development of a political-scientific agenda around the theme and for the planning of actions at national level. Building this agenda was essential for the national climate change policy. The Inter-Ministerial Committee on Climate Change was created in 2007, with the main task of planning and implementing the National Policy on Climate Change through a National Climate Change Plan. The third phase, starting in 2009, was marked by the creation, approval and implementation of the national policy on climate change, as well as the strengthening of the scientific agenda on climate change. The National Policy on Climate Change was approved in 2009, two weeks before the 15th Conference of the Parties of the United Nations Framework Convention on Climate Change in Copenhagen (COP 15). Brazil shifted its historic position in negotiations and adopted a voluntary target to reduce GHG emissions by between 38% and 36.1%, or 9% of total projected emissions, by 2020 (Brasil 2009). The national policy stipulated sector plans, which were developed in 2012 and 2013 to guide implementation of the policy.

During the process of designing and promulgating the national policy, some research institutions and civil society organisations were established at federal and state levels. Notable examples include *Rede Clima* (i.e. Climate Network), the National Institute of Science and Technology for Climate Change, a network of civil society organisations called the Climate Observatory, the launch of the Research Program on Global Climate Change in the State of São Paulo by São Paulo Research Foundation (FAPESP) and the creation of the Brazilian Panel on Climate Change. The Brazilian Panel, launched in 2009 by the Ministry of

Science, Technology and Innovation and the Ministry of the Environment, aimed to strengthen the links between climate science and politics. It was structured on the basis of the Intergovernmental Panel on Climate Change (IPCC) to provide scientific assessments of climate change impacts and vulnerabilities. The First National Assessment Report was presented in 2012 during Rio+20 (PBMC 2013). The importance of this report lies in the guidance it provided for the development of climate policies in the country, related to both mitigation and adaptation. More recently, in 2015, as part of the international negotiations on the Paris Agreement under the Climate Convention, the Brazilian government approved its Intended Nationally Determined Contributions (INDC). The Brazilian INDC proposes a 37% reduction in GHG by 2025 and a 43% reduction by 2030, compared to 2005. To this end, it establishes some targets to be reached by the year 2030, which include increasing the use of biofuels; reducing illegal deforestation to zero; restoring and reforesting 12 million hectares of lost forest; increasing renewable energy to make up 45% of the energy matrix; increasing wind, solar and biomass energy production; increasing energy efficiency by 10%; and expansion of low-carbon agriculture (Itamaraty 2015).

Throughout these three phases, and particularly since 2000, it is possible to observe a significant increase in the production and systematisation of studies and reports, which could potentially contribute to the development of climate policies and reduce uncertainty related to climate change risks; for example, scientific knowledge is critical to the development of GHG emissions inventories that guide the implementation of emission reduction targets and plans. Regarding political responses, it is notable that mitigation has always been present in the plans informing national policy. Specific GHG emission reduction targets were a highlight of the Brazilian policy. Although emissions in Brazil had previously been decreasing for some time, the fact that they have now been increasing since 2012 represents a new set of challenges for the present political and institutional framework (see Graph 6.2).

Analysing emissions from 2015 to 2016, it is possible to recognise that the increases are mainly due to the 23% growth in emissions linked to land use change, resulting from increases in deforestation and agribusiness expansion. These data show that Brazil's emission pattern differs

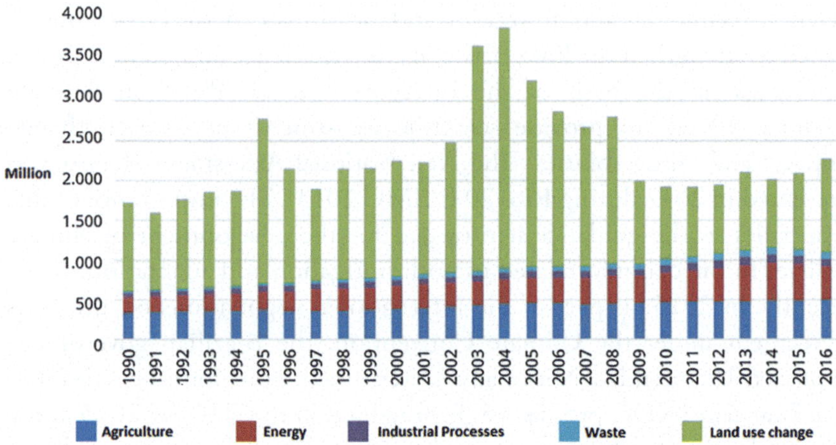

Graph 6.2 Total GHG emissions in China, Brazil and Mozambique (1990–2012). (Source: OC 2018)

from that of other major climate players such as China and the United States in that increases are not linked to energy production or industrial processes. Policies aimed at reducing deforestation and preserving forest areas can have a rapid result, dramatically reducing Brazil's GHG emissions, while in China, as we shall see below, reducing emissions requires a complete restructuring of the country's energy and industrial production. In addition to strengthening GHG mitigation strategies, Brazil must develop extensive adaptation policies. The country's population is concentrated along its 8000-kilometre coastline, and coastal cities are considered the most vulnerable to climate change impacts, mainly due to sea level rise. Other regions, such as the Northeast and North, are expected to experience reduced rainfall, which will affect agricultural production and power generation through hydroelectric dams, as well as affecting the well-being of the population (PBMC 2013). The National Adaptation Policy was published in 2016, showing that this aspect of climate policy has taken longer to develop in the country.

A fourth phase, starting in 2019, is characterised by the current Brazilian federal government's climate change denial. In this regard, given the government's backtracking on environmental issues and its threat to leave the Paris Agreement in 2019, the World Meteorological Organisation

(WMO) called on the Brazilian government to continue to engage in issues related to climate change. The theme has become less of a priority in the national political scenario, meaning that new arrangements are needed to keep attention focused on this issue, as has happened in the United States following the country's withdrawal from the agreement, with a number of civil society and private sector organisations, state and local governments mobilising through the 'We Are Still In' movement to ensure that the American INDC is met. If at the beginning of the 2010s there was an incentive for civil society, non-governmental organisations (NGOs) and scientists to participate in discussions about climate and environmental policies, Brazil is now experiencing setbacks in this regard due to public policy changes at federal level. Nevertheless, within universities and research institutes, there is a growing amount of research related to climate change, as can be seen in the FAPESP Global Climate Change Research Program, which has funded projects in various fields of knowledge since 2011.

Climate Action in China

China is a key global actor in international climate governance because of its status and influence in the G77, a group that brings together developing countries to increase their negotiating capacity in the UN system (Basso and Viola 2014; Moreira and Ribeiro 2016). In addition, it has also been the world's largest GHG emitter since 2007, when it overtook the United States (Edgar 2017). China can be considered the engine of the world's economic growth because of its rapid development and its economic and industrial expansion. This scenario has been accompanied by dramatic growth in the population density of its urban areas, a factor directly linked to the emergence of megacities since the 1990s. As a result, energy consumption has grown significantly, along with emissions of air pollutants and GHG. Thus, the size of the country, its demographics and the current scenario of rapid development are causing environmental problems to spread not only at the local level, but throughout the world, and not only through macroeconomic impacts, disturbing the financial equilibrium of different countries, but also with regard to transboundary pollution in the form of acid rain and dust storms. The country is therefore a major contributor to the global climate change scenario (Chan and Yao 2008; Huang et al. 2010; Macbean 2007; Xie 2011).

Politics in China faces a constant dilemma between prioritising either economic growth or environmental protection, especially since the introduction of reforms by Deng Xiaoping in 1978 (the Four Modernisations and the opening of China to the international market). For most of China's history, nature has been considered a constraint to be mastered or overcome rather than something to be harmonised and preserved. Hung and Tsai (2012) highlight the challenges China faces as a result of climate change, and their costs. These include melting glaciers, especially in Tibet and Tiansham; loss of agricultural production, which may be reduced by up to 10% by 2030; increased numbers of droughts, storms, floods and natural disasters caused by extreme weather; sea level rises, which will affect 67 million people; declining resources; and severe pollution problems. These all represent typical considerations in development dilemmas, and the potential cost of climate change for China is extremely high. In this context, China began to develop environmental legislation in the 1970s, as well as to build institutions aimed at protecting the environment. China's environmental protection laws are considered more progressive than those in any other developing country, at least on paper; for example, there is an institutional arrangement exclusively dedicated to the promotion and enforcement of environmental laws, The State Council of Environmental Protection. There is also an active presence of non-governmental organisations (NGOs) concerned with environmental protection. Lawsuits against polluters have been filed in the courts, some successfully. However, at the provincial level, where important layers of city political administration are located, conflicting objectives, as well as substantial corruption in state management, severely weaken the effectiveness of policies and legal measures directed at environmental protection (Ferreira 2018; Macbean 2007; Oliveira 2011).

In order to address a growing range of global environmental problems, the Chinese government has become more internationally engaged over the last two decades. In 1990, the State Council announced a statement on 'Problems and positions related to global environmental issues', whose principles guided China's position in global climate negotiations. This declaration emphasises the responsibility of developed countries for the environmental crisis, the harmony between environmental protection and economic development, the right of countries to develop, the sovereign equality of all countries and the need to create funds for developing countries (Oliveira 2011). China has engaged in the international climate change debate by developing appropriate domestic institutions and collaborating with the Climate Convention in the international negotiations that led to the Kyoto Protocol. In addition, as a board member of the United Nations Environment Program (UNEP), China adopted and signed 50 international treaties, 15 conventions and 27 bilateral agreements related to environmental protection during the 1990s. However, once the established international negotiations were being consolidated and had to be absorbed internally, the intentions of the Chinese representatives came into conflict with international aspirations. In this context, the internal position was that global policies could not slow down the pace of growth in China.

Although openness to international debate is a reality, with the Chinese government increasingly participating in international forums, GHG emissions in China continue to grow. In this context, due to pressure from the international community, the country has begun to play a more proactive role in climate change negotiations in terms of commitments. In 2007, for example, China announced the National Climate Change Program and created the National Climate Change Leadership Group. In 2009, China made a pledge to the international community to cut GHG emissions significantly by 2020 by increasing the use of 'clean' fuel sources by 15% and increasing the pace of reforestation and the development of a 'green economy'. More recently, in 2015, China committed to the Paris Agreement, setting through its INDC the target of reducing its GHG emissions by up to 65% per unit of GDP by 2030 compared to 2005. To this end, it is expected that an investment of US$6 billion will increase renewable energy capacity and increase forest cover in the country (PRC 2015). The central actor influencing China's climate change strategies is the National Development and Reform Commission (NDRC), which is responsible for economic planning and climate policy formulation and implementation through its Department of Climate Change. This commission has absorbed the National Leading Group to Address Climate Change.

At the same time, the China Meteorological Administration (CMA) is the leading scientific body on climate change in China and is also a key player in formulating climate-related strategies and international negotiations. CMA leads the National Executive Group on Climate Change, leads China's IPCC-related activities, co-chairs the National Climate Committee, hosts the National Panel on Climate Change and is the focal point of the China Global Climate Observation System Committee. Through its Climate Change Centre, it is responsible for preparing annual reports on climate change and GHG. China has adopted a series of policies and programs aimed at mitigating the problem of climate change. However, the position of Chinese leaders is that emission reduction targets should not compromise the country's economic growth, which would have an impact on the pace of social development currently observed in China (Huang et al. 2010; Oliveira 2011).

More recently, following the reform of China's environmental legislation in 2015, the country has adopted mechanisms that ensure greater data transparency, fines and government action to increase control over GHG emissions and improve the country's environmental quality. The Chinese civil society is starting to become incorporated as a social actor, able to participate in environmental governance, especially through the activities of environmental NGOs, which act as partners of the central government and seek to work with society on environmental improvement and legislation compliance. Despite this, the number of environmental protests has grown exponentially as a result of greater awareness of environmental degradation among the population. Although NGOs are increasingly involved, the state acts in a centralised manner on climate and environmental policy. With the announcement of US withdrawal from the Paris Agreement, China has the opportunity to consolidate itself as one of the main players in the international arena, striving to fulfil agreements and achieve stipulated goals.

The Mozambican Perspective on Climate Change

The third country analysed in this chapter is Mozambique, which has a completely different history and development trajectory from Brazil and China in terms of environmental political strategies. Mozambique's GHG emissions are considered low, but the country faces the direct consequences of global emissions and climate change, which are destabilising the country's climate system, intensifying the occurrence of floods, droughts and cyclones. Studies of environmental history (e.g. Sedrez 2002) have emphasised the existence of a literature gap and therefore a lack of knowledge in terms of academic production on environmental issues in the African continent. This is very much the case for Mozambique, whose academic development around environmental issues is little explored (Muhale 2018). Mozambique faces difficulties in estimating its emissions, and the last official emissions inventory that took place was published in 1994. Nevertheless, Mozambique is estimated to contribute about 0.025% of global emissions, a figure which is unlikely to be significantly reduced over the next 10 years. Emissions are largely due to deforestation, fires, environmental degradation and energy production from wood and coal (Zolho 2010).

Mozambique is one of the African countries most vulnerable to the effects of climate change due to its geographical location in the intertropical zone; its long coastline downstream from large river basins; and a vulnerability to floods, droughts, cyclones, landslides, epidemics and extreme cold and heat waves. The country experiences frequent extreme weather events that have caused numerous deaths, injuries and homelessness. Rising sea levels in cities such as Maputo and Beira are a threat to life in these populous and important regions of the country. Climate change in African countries will intensify existing pressures related to limited financial resources, poor infrastructure and degraded ecosystems, exacerbating food insecurity, disease, poverty and social inequality (Zolho 2010). The floods in 2000 were a historic natural disaster in Mozambique, responsible for the deaths of 800 people and losses estimated at about US$400 million. In 2012, the country was hit by cyclones Dando and Funso, when floods, rains and tropical depressions affected 81,200 people in central and southern parts of country, leaving 37 people dead, 41 injured and just over 81,000 homeless families (Artaxo 2014). All of this contributed to increasing the country's vulnerability to the risks arising from climate change. Moreover, the country lacks adequate governmental, institutional and financial structures to deal with these climatic events in a region of significant economic importance (Muhale 2018).

After gaining independence in 1977, Mozambique adopted a socialist model of development, whereby FRELIMO, the only ruling party in the country's one-party system, planned to end underdevelopment within a period of ten years. To this end, it chose a strategy of modernisation, in which agriculture was positioned as the 'engine' of the economy (Abrahamsson and Nilsson 1995). This model was replaced by a multiparty democracy through the approval of the 1990 Constitution, which established the separation of powers, political pluralism and decentralisation of power (Osório 2010), and is considered the major driver for legal institutionalisation of the environmental issue (Muhale 2018; Serra 2012). In addition, some documents and studies point to the 1980s as the starting point for political institutionalisation of the environmental issue in Mozambique (e.g. MICOA 1996).

Political mobilisation around environmental issues in Mozambique is very much based on the need to deal with environmental risks and disasters, as exemplified by the Coordinating Council for the Prevention and Combat of Natural Disasters and the Department for the Prevention and Combat of Natural Disasters, established in 1980. The main environmental management instruments in Mozambique are: Resolution 5/95, which created the National Environmental Policy in 1995; the Environment Law, passed in 1997; the Regulation on the Environmental Impact Assessment Process of 1998; and the Ministry for Coordination of Environmental Action, created in 1994 (Serra Júnior 2003). Legislation around environmental management is varied and extensive in Mozambique, a result of political and economic transformations and extreme events in the country's history (Amaral 2018). There are a number of legal instruments relating to risk and disaster management, including the 1999 National Disaster Management Policy, the National Disaster Management Institute, established in the same year, the Emergency

Operations Centre, founded in 2006, and the Organic Statute of the National Institute of Disaster Management of 2007.

In 2013, the government published its National Strategy for Adaptation and Mitigation of Climate Change, 2013–2025, which includes strategic and priority lines of action for the country. In this document, it is recognised that climate change is a determining factor for development processes, and the greatest risk to the achievement of set goals, especially for the least developed countries (Amaral 2018). The strategy sets climate risk adaptation and reduction as national priorities, while recognising the need to seize the opportunities that the country has to reduce the impacts of climate change through a set of low-carbon mitigation and development actions. Despite a track record of low accumulated emissions, it is crucial for Mozambique to review its energy choices, as in recent years, there has been an increase in the volume of energy generated via coal and oil, which are highly polluting methods. The country's large coal reserves, if exploited, will lead to increased emissions from both coal use and deforestation in mining areas, contributing directly to climate change.

Thus, since the 1980s, concrete actions have been developed in the country to reduce climate change risks, and the impacts and variability resulting from these. They include extending an early warning system for floods, increasing water resource management capacity and building the capacity of agrarian producers to better manage the process of reducing the impact of environmental change on coastal zones. Reliable monitoring and recording of information relating to natural disasters has become increasingly important in national development plans. In this context, citizen participation in the socio-economic life of the country is now considered vital to the process of seeking optimal solutions to the environmental problems that local communities face (Amaral 2014; Muhale 2018). There has been substantial criticism from NGOs and Mozambican civil society groups that the government has made little effort to mitigate and adapt to climate change. Although Mozambique is a signatory to the Paris Agreement, 'the country was unable to promote a debate and inform Mozambicans about the failed Kyoto protocol, now replaced by the Paris Agreement. After two decades, Mozambicans are unaware of their government's positions on the negotiations that take place every year in the

Conference of the Parties (COP) within the United Nations Framework Convention on Climate Change (UNFCCC)' (Monjane 2015: 1).

During COP 21, the World Bank reported a US$50 million investment in forest reform in Mozambique in order to increase carbon sequestration and offset GHG emissions. Nonetheless, there is a strong belief among much of the population that Africa will continue to pay dearly and that African countries will remain dependent, colonized and powerless, unable to secure mechanisms to avert environmental damage in their increasingly resource-intensive countries, where highly polluting companies use cheap labour for production and leave trails of environmental destruction across Africa (Monjane 2015).

Three National Experiences: Lessons in Terms of Environment and Development

With human activities being the major contributors to the worsening of global environmental changes, especially climate change, understanding of the social and political dimensions of these changes is fundamental to finding strategies to cope with their impacts. Thus, the main challenges in the Anthropocene are how to respond to unpredictable, invisible and often incalculable contemporary environmental risks. This chapter has discussed some of these responses, which are ongoing in Brazil, China and Mozambique. This was not intended to be a comparative analysis, but a referenced analysis of how different societies have undertaken actions and strategies to address climate change, focusing particularly on the governmental sphere. Both Brazil and China have focused much of their efforts on mitigating actions, that is, the stabilisation or reduction of greenhouse gas emissions. Their climate agenda is motivated by treating climate risk in this way, while the adaptation agenda is incipient in these countries. On the other hand, in Mozambique, political action on climate change is largely based on the need to respond to the risks and disasters that are intensified by these changes.

Climate change constitutes an issue that permeates not only the three countries analysed, but all other countries on the globe. Responses to this challenge are distinct, directly related to the political, economic and social structure of each country. In analysing three countries in the Global South, we seek a shift from Western-centric perspectives (Stuenkel 2018), that is, we are looking at different realities and experiences from those experienced by European and North American countries, which are commonly found in climate change analyses. The Global North discourse is already well known and recognised, but we must go further to grasp the complexity of today's risk society. The Brazilian, Chinese and Mozambican cases present three distinct situations in terms of emission patterns, impacts, political and scientific responses to climate change. In the Brazilian case, GHG emissions are mainly linked to the land use change sector, with an increase in emissions directly linked to increases in deforestation and agribusiness expansion. In 2016, Brazil ranked sixth among the largest GHG emitters in the world. Our analysis has shown that Brazilian strategies related to climate change have not been enough to reduce the country's emissions. Furthermore, all development in terms of institutional arrangements and climate strategies has been compromised since the country entered the climate change denial phase.

In the case of China, emissions are directly linked to energy production, due to the use of coal as the main source of energy. Burning coal has increased air pollution, posing a visible threat to the inhabitants of large cities, who throughout the 2000s experienced many days of poor air quality. The government's response to climate change arises from international pressure, domestic pressure for improved air quality and the recognition of the high costs of climate change and high pollution rates. China's challenge is to maintain economic growth while reducing GHG emissions and the vulnerability of its population to the impacts of climate change. Mozambique's situation is completely different from those of Brazil and China. Like many African countries, Mozambique has very low GHG emissions, but it is crucial to recognise that responses to climate change in Mozambique have resulted from the intensification of extreme climate phenomena that have hit the country, causing thousands of deaths and economic losses, and making the survival of the population, which already struggles with poverty and poor quality of life, even more difficult.

The global character of climate change, and its risks as a phenomenon with universal effects, not restricted to a particular social class or geographical location, increases the vulnerability of those who are already suffering from it. Thus, women, the elderly, children and the poor may find the difficulties they already experience in everyday life intensified by climate change, particularly as responses to the problem are uneven. Residents of vulnerable areas in Mozambique, Brazil and China are much more likely to suffer consequences related to the intensification of extreme phenomena than residents of Europe, where levels of education, income and social and economic development are generally higher. The cases of Brazil, China and Mozambique exemplify the need to respond to climate change considering both aspects of climate policy: mitigation of GHG emissions, in the cases of Brazil and China, and adaptation, in all three cases. The three countries are vulnerable to the consequences of climate change, especially due to their extensive coastlines, which are likely to suffer from sea level rises, and areas predisposed to drought and flooding, as well as intense waves of heat and cold, which decimate lives, as well as increasing poverty, and hunger, reducing water availability and impacting quality of life.

Conclusions

Risks are the result of multilevel processes, problems that arise from the organisation and structure of society, reflecting organisational and developmental choices made in societies. Thus, climate change represents a multilevel challenge, simultaneously relating to the local and global scales. National governments are, in fact, key actors in developing climate change mitigation and adaptation strategies. Moreover, when dealing with such a multidimensional and complex Anthropocene challenge, it would be naive to believe that only one group of actors would be able to solve the climate crisis. Involvement of non-governmental actors, civil society organisations, the private sector, universities and research institutions is essential for producing efficient and successful responses to the problem.

Finally, climate change risks are products of the development processes of contemporary societies themselves, and confronting them involves questioning these processes. This is where climate change policies should be targeted. As Beck (2010) warns, climate policy is not precisely about climate, but about the transformation of the basic concepts and institutions established throughout the development of contemporary society. Addressing such a challenge requires comprehensive action, in not only areas of human activity such as policy, science and the economy, but also various sectors of society including governments, businesses, companies, research institutes and civil society groups in order to find the necessary conditions for confronting climate change risks.

References

Abrahamsson, H., & Nilsson, A. (1995). *Mozambique: The Troubled Transition: from Socialist Construction to Free Market Capitalism*. London/Atlantic Highlands: Zed Books.

Amaral, G. A. (2014). *O Processo de Institucionalização da Questão Ambiental em Moçambique 1980–2000*. MSc Dissertation, Universidade Federal Fluminense.

Amaral, G. A. (2018). *Mudanças Ambientais, Percepções de Risco e Estratégia de Adaptação aos Eventos Extremos em Moçambique: Estudo de Caso em Machanga*. PhD Thesis, UNICAMP.

Artaxo, P. (2014). Uma Nova era Geológica em Nosso Planeta: O Antropoceno? *Revista USP, 103*, 13–24.

Barbi, F., Mozzer, G. B., Martins, R. D., & Ferreira, L. C. (2015). Mudanças Climáticas na Rio+20: Uma Reflexão sobre os 20 Anos de Negociações Climáticas no Brasil. In F. Castro & C. Futemma (Eds.), *Governança Ambiental no Brasil: Entre o Socioambientalismo e a Economia Verde* (pp. 53–83). Jundiaí: Pacco Editorial.

Basso, L., & Viola, E. (2014). O Progresso da Política Energética Chinesa e os Desafios na Transição para o Desenvolvimento de Baixo Carbono, 2006–2013. *Revista Brasileira de Política Internacional, 57*, 174–192.

Beck, U. (1992). *Risk Society*. Beverly Hills: Sage.

Beck, U. (1995). *Ecological Politics in an Age of Risk*. Cambridge: Cambridge University Press.

Beck, U. (2000). Risk Society Revisited: Theory, Politics and Research Programmes. In B. Adam, U. Beck, & J. V. Loon (Eds.), *The Risk Society and Beyond: Critical Issues for Social Theory* (pp. 211–229). London: Sage.
Beck, U. (2009). *World at Risk*. Cambridge: Polity Press.
Beck, U. (2010). Climate for Change, or How to Create a Green Modernity? *Theory, Culture and Society, 27*(2–3), 254–266.
Biermann, F., et al. (2012). Navigating the Anthropocene: Improving Earth System Governance. *Science, 335*(6074), 1306–1307.
Boden, T. A., Marland, G., & Andres, R. J. (2015). *Global, Regional, and National Fossil-Fuel CO2 Emissions*. http://cdiac.ornl.gov/trends/emis/overview_2011.html. Accessed 15 Feb 2020.
Brasil. (2009, December 29). *Lei n° 12.187, de 29 de dezembro de 2009*. Institui a Política Nacional sobre Mudança do Clima – PNMC. Diário Oficial da União. Brasília, DF, n° 248, S. 1, p. 109.
Bulkeley, H. (2001). Governing Climate Change: The Politics of Risk Society? *Transactions of the Institute of British Geographers, 26*(4), 430–447.
Bulkeley, H., & Kern, K. (2006). Local Government and the Governing of Climate Change in Germany and the UK. *Urban Studies, 43*(12), 2237–2259.
Bulkeley, H., & Newell, P. (2010). *Governing Climate Change*. New York: Routledge.
Chan, C. K., & Yao, X. (2008). Air Pollution in Mega Cities in China. *Atmospheric Environment, 42*(1), 1–42.
Crutzen, P. J., & Stoermer, E. F. (2000). The "Anthropocene". *Global Change Newsletter, 41*, 17.
Di Giulio, G. M., Bedran-Martins, A. M., da Vasconcellos, M., & Ribeiro, W. C. (2017). Mudanças Climáticas, Riscos e Adaptação na Megacidade de São Paulo, Brasil. *Sustentabilidade em Debate, 8*(2), 75.
Dunlap, R., & Brulle, R. (2015). *Climate Change and Society: Sociological Perspectives*. New York: Oxford University Press.
EDGAR (Emissions Database for Global Atmospheric Research). (2017). *Fossil CO2 & GHG Emissions of All World Countries, 2017*. http://edgar.jrc.ec.europa.eu/overview.php?v=CO2andGHG1970-2016. Accessed 27 Feb 2019.
Ferreira, L. C. (2018). *The Sociology of Environmental Issues: Theoretical and Empirical Investigations*. Curitiba: CRV.
Ferreira, L. C., & Tavolaro, S. B. F. (2008). Environmental Concerns in Contemporary Brazil: An Insight into Some Theoretical and Societal Backgrounds (1970s–1990s). *International Journal of Politics, Culture, and Society, 19*(3–4), 161–177.

Giddens, A. (1991). *As Consequências da Modernidade*. São Paulo: UNESP.
Giddens, A. (2005). *Mundo em Descontrole: O Que a Globalização Está Fazendo de Nós*. Rio de Janeiro: Record.
Giddens, A. (2009). *The Politics of Climate Change*. Cambridge: Polity Press.
Huang, X., Zhao, D., Brown, C. G., & Wu, Y. (2010). Environmental Issues and Policy Priorities in China: A Content Analysis of Government Documents. *China: An International Journal, 8*(2), 220–246.
Hung, M., & Tsai, T. C. (2012). Dilemma of Choice: China's Response to Climate Change. *Revista Brasileira de Política Internacional, 55*(Special Edition), 104–124.
IPCC (Intergovernmental Panel on Climate Change). (2007). Climate Change 2007: Synthesis Report. In R. K. Pachauri & A. Reisinger (Eds.), *Contribution of Working Groups I, II and III to the Fourth Assessment Report of the Intergovernmental Panel on Climate Change* (pp. 1–52). Cambridge: Cambridge University Press.
ISSC and UNESCO. (2013). *World Social Science Report 2013, Changing Global Environments*. Paris: OECD and UNESCO.
Itamaraty. (2015). *Pretendida Contribuição Nacionalmente Determinada (iNDC) para a Consecução do Objetivo da Convenção-Quadro das nações Unidas sobre Mudança do Clima*. http://www.itamaraty.gov.br/images/ed_desenvsust/BRASIL-iNDC-portugues.pdf. Accessed 6 Mar 2019.
Jacobi, P., & Sulaiman, S. (2016). Governança Ambiental Urbana em Face das Mudanças Climáticas. *Revista USP, 109*, 133–142.
Macbean, A. (2007). China's Environment: Problems and Policies. *The World Economy, 30*(2), 292–307.
MICOA (Ministério para a Coordenação Ambiental). (1996). *Programa Nacional de Gestão Ambiental*. Maputo.
Monjane, B. (2015, December 10). *Moçambique e o Desastre do Novo Acordo Climático de Paris*. Pressenza International Press Agency. www.pressenza.com/pt-pt/2015/12/mocambique-e-o-desastre-do-novo-acordo-climatico-de-paris. Accessed 27 Feb 2019.
Moreira, H. M., & Ribeiro, W. C. (2016). A China na Ordem Ambiental Internacional das Mudanças Climáticas. *Estudos Avançados, 30*(87), 213–234.
Muhale, I. J. (2018). *O Papel da Produção Científica na Institucionalização da Política Ambiental em Moçambique: 1980–2014*. PhD Thesis, USP.
OC. (2018). Análise das Emissões Brasileiras de Gases de Efeito Estufa e suas Implicações para as Metas do Brasil. Observatório do Clima.

Oliveira, A. P. (2011). A China Abraça a Causa Verde. In A. Moreira & P. Ramalho (Eds.), *Estratégia* (pp. 55–68). Lisboa: Instituto Português da Conjuntura Estratégica.

Osório, C. (2010). *Gênero e Democracia: As Eleições de 2009 em Moçambique*. Maputo: WLSA Moçambique.

PBMC (Painel Brasileiro de Mudanças Climáticas). (2013). *Base Científica das Mudanças Climáticas* (Primeiro Relatório de Avaliação Nacional, vol. 1). Rio de Janeiro: PBMC.

PRC (Popular Republic of China). (2015). *Enhanced Actions on Climate Change: China's Intended Nationally Determined Contributions*. www4.unfccc.int/sites/submissions/INDC/Published%20Documents/China/1/China's%20INDC%20-%20on%2030%20June%202015.pdf. Accessed 27 Feb 2019.

Renn, O., & Klinke, A. (2015). Risk Governance and Resilience: New Approaches to Cope with Uncertainty and Ambiguity. In U. Fra Paleo (Ed.), *Risk Governance: The Articulation of Hazard, Politics and Ecology* (pp. 19–41). Dordrecht: Springer.

Rockström, J., et al. (2009). Planetary Boundaries: Exploring the Safe Operating Space for Humanity. *Nature, 461*, 472–475.

Sedrez, L. (2002). Historia Ambiental de América Latina: Origen, principales interrogantes y lagunas. In G. Palacio & A. Ulloa (Eds.), *Repensando la Naturaleza: Encuentros y Desencuentros Disciplinarios en Torno a lo Ambiental* (pp. 99–109). Bogota: Universidad Nacional de Colombia – Sede Leticia/Instituto Amazónico de Investigaciones Imani/Instituto Colombiano de Antropología e Historia/Conciencias.

Serra, C. M. (2012). Quadro Jurídico-ambiental em Moçambique: Pontos Fortes, Constrangimentos e Desafios. In C. M. Serra, D. Dondeyne, & T. Durang (Eds.), *O Meio Ambiente em Moçambique: Notas para a Reflexão sobre a Situação Atual e os Desafios para o Futuro* (pp. 11–34). Maputo: Governent Report.

Serra Júnior, C. (Ed.). (2003). *Coletânea de Legislação do Ambiente*. Maputo: CIEDIMA.

Spring, U. O. (2016). Perspectives of Global Environmental Change in the Anthropocene. In G. Sosa-Nunez & E. Atkins (Eds.), *Environment, Climate Change and International Relations* (pp. 29–41). Bristol: E-International Relations.

Steffen, W., Grinevald, J., Crutzen, P., & Mcneill, J. (2011). The Anthropocene: Conceptual and Historical Perspectives. *Philosophical Transactions of the Royal Society A, 369*, 842–867.

Stuenkel, O. (2018). *O Mundo Pós- Ocidental: Potências Emergentes e a Nova Ordem Global.* Rio de Janeito: Ed. Zahar.

Syvitski, J. P. M., & Kettner, A. (2011). Sediment Flux and the Anthropocene. *Philosophical Transactions of the Royal Society A, 369*, 957–975.

Viola, E., Franchini, M., & Ribeiro, T. L. (2013). *Sistema Internacional de Hegemonia Conservadora: Governança Global e Democracia na Era da Crise Climática*. São Paulo: Annablume.

Will, S., Grinevald, J., Crutzen, P., & Mcneill, J. (2011). The Anthropocene: Conceptual and Historical Perspectives. *Philosophical Transactions of the Royal Society A, 369*(1), 842–867.

World Bank. (2010). *Relatório sobre o Desenvolvimento Mundial de 2010: Desenvolvimento e Mudança Climática*. Washington, DC.

Xie, L. (2011). China's Environmental Activism in the Age of Globalization. *Asian Politics & Policy, 3*(2), 207–224.

Young, O., et al. (2006). The Globalization of Socio-ecological Systems: An Agenda for Scientific Research. *Global Environmental Change, 16*, 304–316.

Zalasiewicz, J., Williams, M., Haywood, A., & Ellis, M. (2011). The Anthropocene: A New Epoch of Geological Time? *Philosophical Transactions of the Royal Society A, 369*(1), 835–841.

Zehr, S. (2015). The Sociology of Global Climate Change. *WIREs Climate Change, 6*, 129–150.

Zolho, R. (2010). *Mudanças Climáticas e as Florestas em Moçambique*. Maputo: Centro de Integridade Pública Moçambique.

7

Colombia's Developmental and Socioecological Trajectory and the Mounting Risks Associated with the 2016 Havana Accord

Rafael R. Ioris and Antonio Augusto Rossotto Ioris

A Country Ravaged by War[1]

In light of the peace accord between the Colombian government and the *Fuerzas Armadas Revolucionarias de Colombia* (FARC) in 2016, this chapter scrutinises present-day dilemmas and obstacles to a sustainable settlement capable of addressing the long-term developmental inequities of Colombia. It is expected that as Colombia becomes more integrated in the globalised economy and influenced by global public opinion, the

[1] An earlier version of this chapter was published as Ioris, R.R. and Ioris, A.A.R. (2018). Consolidating the Past and Risking the Future: Colombia's Developmental Trajectory and the Prospects for a Lasting Peace in the Wake of the Havana Accord. *Journal of Global South Studies* 35(1): 155–173.

R. R. Ioris (✉)
History Department, University of Denver, Denver, CO, USA
e-mail: Rafael.Ioris@du.edu

A. A. R. Ioris
School of Geography and Planning, Cardiff University, Cardiff, UK
e-mail: IorisA@cardiff.ac.uk

A. A. R. Ioris (ed.), *Environment and Development*,
https://doi.org/10.1007/978-3-030-55416-3_7

many parties involved in the conflict will increasingly realise that they had more to gain with peace than otherwise and that this will pave the way to a renewed effort towards peace. Yet despite its many promises, as of now, the peace negotiations have primarily served as a means of potentially boosting the economy and facilitating access to leading business groups and associated foreign investors to Colombian resources rather than as a transformative move towards much-needed political and economic inclusion. In fact, disturbing continuities between the inequalities that deepened during the protracted civil war and elements of peace agreement are noticeable, as will be indicated below. In a late-night announcement, on 12 November 2016, the then Colombian president Juan Manuel Santos declared, 'We have reached a new final accord to end the armed conflict that integrates changes and propositions suggested by the most diverse sectors of society'. With these negotiations, President Santos was trying to address the crisis that had been created by the 'No' vote that had stunned everyone in the country and around the world the previous month.[2] Later in November, the National Colombian Congress endorsed the revised agreement and, this time around, despite some criticism from 'No' supporters (particularly from former president Alvaro Uribe), Santos did not risk putting the document through a second referendum that would risk another defeat.[3]

It remained to be seen how this revised agreement will be implemented over the years and the long-term repercussions, but at least it signalled the possibility that Latin America's longest military conflict in modern history would come to an end. The rejection of the so-called Havana Accord signed between the Colombian government and the FARC (*Fuerzas Armadas Revolucionarias de Colombia*, that is, the Revolutionary Armed Forces of Colombia, the country's main remaining armed insurgent group) in the referendum of 2 October sent a clear signal that though formal peace is essential, it is not the only measure the country needs. In

[2] 'Colombian Government and FARC Reach New Peace Deal,' *The Guardian*, 13 November 2016, https://www.theguardian.com/world/2016/nov/13/colombian-government-and-farc-reach-new-peace-deal. Accessed 21 June 2017.

[3] 'Colombia's Congress Approves Peace Accord with FARC,' *New York Times*, 30 November 2016, https://www.nytimes.com/2016/11/30/world/americas/colombia-farc-accord-juan-manuel--santos.html?_r=0. Accessed 23 June 2017.

addition, the success of the negotiations that led to the peace accord and its eventual approval by the Colombian Congress are attributable to both the willingness of President Santos (in office between 2010 and 2018, succeeded by Iván Duque Márquez who was himself backed by the hard-liner ex-president Alvaro Uribe, the main voice against the agreement with FARC) and the widespread war fatigue in Colombian society.

To paraphrase the title of a novel by Gabriel García Márquez, the agreement was a 'chronicle of a necessary accord foretold'; neither the government nor the guerrillas had any more appetite for a pointless civil war. Colombia had paid a high price for a protracted conflict that had ravaged Colombian society for decades and had made it impossible for more sustainable economic growth to occur. Although since the mid-2000s the activities of insurgent groups had decreased (Salas-Salazar 2016), this did not necessarily translate into actual state rule or at least a state presence in many areas of the country (Ríos Sierra 2016). In several rural and more isolated regions, the threat of a return to overt violence is still very present, and many still find their main means of subsistence in illegal activities that often involve profitable connections with drug traffic dealers. In effect, cocaine, which is often the only economic alternative in remote corners of the country, may account for around 3–5% of Colombia's GDP. This commerce fosters violence, corruption, illegality, and uncertainty (Franz 2016). Having said that, it is also the case that the connections between the civil war, drug production, and economic development are not clear and the findings are contradictory. Some regions and sectors of the economy seem to have thrived even in the most violent periods of the long civil war. Moreover, hopes that the October referendum could lead to a lasting peace or at least to some sort of social normalcy were dashed by the victory of the 'No' vote. Domestic public opinion proved to be deeply polarised. And whereas regions that had been most ravaged by war-related activities exhibited some of the highest levels of support, broad urban coalitions opposed the Ceasefire agreement.[4] In the end, the main winners of the 'No' were conservative

[4] "Resultado del Plebiscito: Un País Dividido entre Periferia e Interior", *El Tiempo*, 2 October 2016, http://www.eltiempo.com/politica/proceso-de-paz/resultado-por-ciudades-de-votacion-en-plebiscito-2016-46988. Accessed 2 February 2018.

politicians (under the leadership of Álvaro Uribe and public prosecutor Alejandro Ordóñez), the belligerent section of the army, and the military industry.[5] More importantly, the difficulty the Santos administration had in leading the nation to a clear victory with the peace referendum reveals a great deal about the limitations of Colombia's fragile democracy.

Colombia continues to face serious socio-economic challenges, many of which were recently aggravated by rising levels of organised crime activity in both the countryside and the urban areas. These developments have not occurred in a void; they have emerged in the context of an already deeply unbalanced land tenure system that has fuelled the continuous operation of various guerrilla groups over the past 60 years (Ross 2007). Throughout the country, land distribution is extremely inequitable: in the year 2002, 0.4% of the landowners owned 61.2% of the registered properties in Colombia (IGAC 2012). In addition, land ownership is notoriously insecure, especially for indigenous peoples, peasants, and subsistence farmers. The attempts of successive governments to advance agrarian reforms have largely been ineffective due to corruption and limited resources, and in recent years, the focus has shifted from land reform to agribusiness-based rural development (Richani 2010). Thus, although President Santos has tried to move beyond an overly militaristic approach to conflict in the country, his political overtures will need to be complemented by economic measures aimed at reintegrating local communities into the national economy. This seems particularly relevant today, as the country's mainstream economic policies have become ever more aligned with those of free trade agreements, such as the *Alianza del Pacífico* (Pacific Alliance) and the Trans-Pacific Partnership. These international agreements are likely to aggravate social inequality because their priorities are market-based.

This chapter examines present-day obstacles to a more socially inclusive and politically sustainable peace settlement that would be capable of addressing Colombia's many developmental shortcomings. The chapter particularly considers the oligarchic developmental trajectory of Colombia

[5] 'Estos Son los Principales Actores que Ganan con el "No" del Plebiscito,' *El Tiempo*, 2 October 2016, http://www.eltiempo.com/politica/proceso-de-paz/quien-gano-el-plebiscito-en-colombia-50450. Accessed 2 February 2018.

and the ability of its national elites to continuously reinvent themselves and reaffirm their power during both times of peace and conflict. This is directly related to the political economy of protracted war that Richani (2001) describes, which is characterised by the accumulation of economic assets by 'armed actors', including members of the army and political and commercial elites. However, as discussed below, the benefits the war brought to the military and the guerrillas in the 1970s and 1980s and particularly in the early 1990s have been increasingly constrained by shrinking political and economic returns since the emergence of paramilitary groups in the 1990s.

Development in Colombia: A Long and Tortuous Road to (No) Peace

Although Colombia is a country of abundant natural resources and economic potential, it is mired in problems that have arisen from major socio-economic disparities between regions, peoples, and economic sectors. The nation has followed an intensely painful path in recent decades, which included a resilient leftist guerrilla movement, drug-trafficking activities, violence in urban and rural areas, and aggressive pro-business policies invigorated by growing exports of natural resources, particularly oil.[6] Colombia's history has been defined by sharp socio-economic contrasts and a succession of fast-paced development initiatives, such as the country's remarkable insertion into the global economy as an important exporter of natural resources at the end of the nineteenth century and significant levels of industrialisation in the middle of the twentieth century. Although these economic activities have been successful, they have not been enough to ameliorate, let alone address, the severe inequality that is to this day a defining feature of the country. Colombia managed to maintain relative economic stability, largely avoiding the

[6] Before the 2003 energy reforms, the Colombian oil and gas industry was controlled by Ecopetrol S.A., a state-owned oil and gas company and industry regulator. The reforms removed regulatory functions from Ecopetrol and opened Colombia to international competition. These reforms made investments in energy exploration and production more attractive to international companies (more than US$ 4.8 billion in 2014, about 30% of total foreign direct investment).

boom-and-bust and debt crises of other Latin American countries. Colombia's ability to achieve this stability has depended largely on maintaining and deepening sociopolitical and sociospatial inequalities.

This is a particularly disheartening reality, considering that for much of the twentieth century, there was a substantial degree of economic and political experimentation and change in Colombia. The nation's economy grew quickly and significant levels of industrial development took place in several parts of the country. And given the context of dictatorial regimes throughout much of the region in the 1960s and 1970s, Colombia's successful experiences with peaceful transitions of power among the two main political parties (Liberal and Conservative) in the wake of the creation of the National Front in 1957 seemed to present an example of a functioning formally liberal political system. However, what looked like a peaceful transfer of power was in fact a highly undemocratic and oligarchic electoral system, structured after *La Violencia*, the ten-year civil war that scarred the country, particularly the countryside, from 1948 to 1958. Peace had become possible only because of the threat of renewed violence, which forced ruling parties to agree on a minimal agenda of rotating power among themselves, while excluding other social sectors from partaking in the process. Thus, rather than viewing Colombia as an example of a successful democracy, the country's experiences since at least the middle of the twentieth century should be seen as an example of the shortcomings of formal democracy and of the failure of top-down attempts to modernise.

The oligarchic nature of the Colombian state was further reinforced by two central features of the political evolution of the country: a weak central state structure and strong regional elites. The Colombian elite, who have promoted economic development through a very conservative social platform, are not united on many issues and they maintain their leadership through a complex system of public and private initiatives (Dent 1974). Colombian regional oligarchies have managed to advance modernisation and economic growth while having to accommodate multiple interests and groups. The central Colombian state has more than once proven unable to promote broad social reform because of powerful and entrenched provincial ruling groups (Lopez-Alves 2000). These features of Colombia's development fostered rising levels of dissatisfaction, particularly during periods of rapid demographic growth, such as in the post-war era. The shortcomings of Colombia's political institutions were also aggravated by the frustration of even middle-income Colombians about the possibility of significant socio-economic reform. Even when land reform was attempted in a more concerted fashion, such as during the sponsorship of the Alliance for Progress in the early 1960s, when the country was chosen to be a showcase for the initiative, the programme achieved far less than both the Colombian and US governments expected and was certainly much less than what was needed (Fajardo 2003; Lopez 2014).

Sadly, this experience was not unique in the history of Colombian land tenure. Large landholders have not allowed traditional small farmers to grow coffee on their own plots because that would have freed them from the economic and political rule of local bosses (Hirschman 1963). In much the same way, the industrialisation of the middle of the twentieth century was largely attributable to an alliance between traditional landowning families and regional industrialists. Instead of creating a new social sector that would be capable of breaking away from the traditional holders of political power, these elites worked to consolidate traditional oligarchs' control of the decision-making process. Thus, industrialisation preserved restricted access to power, though industrial workers did then and still do their best to push the boundaries of local and national spheres of economic and political power (Farnsworth-Avelar 2000; Palacios 1983). During this period, the National Front operated through clientelistic politics that ensured that the labour movement would be divided and weak (Hartlyn 2008). It may be true that the National Front proved successful in preventing overt military intervention in the political realm for much of the second half of the twentieth century. Unfortunately, this outcome seems to have been reached at the cost of repressing those who demanded more social and economic progress, such as students and intellectuals who promoted a socialist agenda. In the long run, this dynamic exacerbated the already present political apathy of growing numbers of politically and economically disenfranchised groups. The guaranteed sharing of state funds, which assured the democratic election of four presidents from the Liberal and Conservative parties in the period 1958–1970, was conducive to corruption and the patronage of favoured supporters (Palacios 2006).

The second half of the twentieth century in Colombia was marked by sustained economic growth and the rise of a domestic manufacturing industry, as well as by continued and growing problems with internal violence and the displacement of many thousands of people. Especially during the putative economic success of the late 1960s and 1970s, the country benefited from rising international prices for coffee, its main export item. Even though Colombia was affected by the regional debt crisis of the early 1980s, it withstood that experience significantly better than most of its neighbours, at least in part due to rising prices for oil exports. Although the formal economy has been strengthened in the last few decades, its benefits were distributed in deeply uneven ways. The level of poverty remained significantly high in the period 1979–1991, while wealth remained concentrated in the hands of the elites (Delgado et al. 1999). In addition, an important element of this economic revival was the production and exportation of illegal drugs.

In the late 1980s, there was an upsurge of widespread violence on a new scale, marked by rise of powerful and brash drug lords (e.g. Pablo Escobar) who controlled enormous economic resources and exercised extensive political influence in the country. These developments precipitated a movement for a new constitution, which for many people symbolised a new start for the country. The 1991 constitution further strengthened judicial powers, recognised the country's cultural pluralism, broadened the scope of individual and political rights, and expanded the autonomy of municipalities and regions. At the same time, however, the document was largely silent about the status of the police and military and their relationships to civil authority. Military spending, which was unchecked by the new legislators, increased five times from 1985 to 1990, the size of the army doubled, and the national police remained under the control of the Ministry of Defence (Palacios 2006). Alongside these legal reforms, Colombia's government has engaged in negotiations with revolutionary groups throughout the period of brutal conflicts. For instance, President Betancur negotiated a cease-fire in 1982–1985, which would have granted amnesty to some guerrillas and released some political prisoners. These negotiations paved the way for the creation of the

Unión Patriótica (Patriotic Union), a political arm of the FARC. However, these negotiations fell apart when state security and paramilitary forces systematically gunned down many members of the Patriotic Union and another revolutionary group, M-19, seized the Supreme Court building in Bogota. When the military tried to retake the building, 11 justices and more than 100 people were killed.

New rounds of peace negotiations took place under the presidency of Andrés Pastrana Arango (1998–2002), but these ended abruptly when the guerrillas kidnapped a member of Congress and other political figures. Pastrana introduced the unpopular and controversial Plan Colombia, which the United States supported and financed. The main goal of this plan was to stop the flow of drugs (especially cocaine and heroin) to the United States over the Mexican border. The plan failed to decrease production, generate economic development, or, ultimately, reduce violence. Since the late 1990s, the US government has invested more than US$25 billion on international drug-control programmes, but these initiatives have largely failed to reduce the supply of cocaine and heroin entering the US market and have generated widespread and profoundly damaging consequences in Latin America (Youngers and Rosin 2015). Although Plan Colombia may have reduced the number of coca plants, it has not reduced cocaine processing in Colombia (now estimated at around 70% of global production). What it has done is exacerbate the humanitarian crisis in Colombia and forced internal displacements (Franz 2016). In 2004, with the direct help of the United States, Álvaro Uribe Vélez launched the *Plan Patriota* (Patriotic Plan), which sought to displace guerrilla groups, occupy remote areas traditionally controlled by the guerrillas, and promote social programmes. New legislation passed in 2005 included reduced punishments for paramilitary and guerrilla fighters if they surrendered their arms, renounced violence, and returned assets they had appropriated illegally. The FARC's response was the Resistance Plan, an offensive that it organised to counteract military action and undermine the perception that security had improved under Uribe. In 2006, some preliminary peace talks between the government and National Liberation Army took place in Cuba but did not produce concrete results.

A succession of political scandals involving associations between politicians and the paramilitary, clashes with Ecuador and Venezuela, and the continuous war with the FARC meant that the situation in Colombia remained unstable. At the dawn of the twenty-first century, the FARC launched the Rebirth Plan, which intensified warfare with the controversial use of land mines, snipers, and bombings in urban areas. This plan spurred the government to take the offensive again against the guerrillas. Despite investments to improve the army's ability to face the guerrillas, violence and dissatisfaction continued in the countryside because of the increasing concentration of land ownership and serious levels of inequality. Guerrilla groups were easily able to recruit more dissatisfied individuals (Montañez-Gómez 2016).

In 2010, Juan Manuel Santos, who had been Uribe's secretary of defence, was elected president after the courts decided that Uribe could not run for a third term. In 2012, Santos announced that the government would engage in talks with the FARC to explore ways to end the long conflict. Santo pledged to learn from past mistakes and from unsuccessful attempts to resolve the issues. Former president Uribe criticised this level of commitment as a sign of weakness and as an effort to seek peace at any cost. In a climate of great national and international expectation, Santos announced the plan of settlement of the dispute between the government and the FARC in September 2016, which included a comprehensive truth and reconciliation process. The agreement included admissions of guilt and the requirement that the perpetrators of violence during the conflict do community service. Although it was not a legal requirement, President Santos promised in 2015 to submit the Peace Accord to a public referendum in order to legitimise its goals and its mechanism of retributive justice. Considering the tumultuous history of the county, it is no surprise that peace negotiations have been so protracted and late in coming. In fact, it seems to be largely because of the idiosyncratic place President Santos is trying to occupy vis-à-vis his predecessor and the weakened state of the FARC that a path towards a lasting peace opened up (Javier Orjuela 2015). Yet when most Colombians rejected the first version of the accord, it was clear that the terms of the agreement were far from universally accepted.

One of the champions of the 'No' vote was Uribe, who articulated a fierce campaign against what his group considered a soft plan of accommodation. The former president has close ties to powerful groups of landowners and other conservative forces that are likely to lose economic and political ground from any new plans to distribute land to former guerrilla members.[7] Just a few days after the frustrating outcome of the referendum, another unexpected development surprised the country and the international community: the Nobel Committee awarded the Novel Peace Prize for 2016 to President Santos, even though no concrete solution to the conflict had been reached. The prize seems to have been a bet on the potential for future peace, even though this decision was criticised as direct interference in national negotiations because it suddenly increased the popularity and the visibility of Santos.

In fact, the Nobel Committee was the first to declare that their decision was intended to encourage 'all those who are striving to achieve peace, reconciliation and justice in Colombia'. The president had made it clear that he would continue to work for peace right up until his last day in office, and the committee hoped that the Peace Prize would give him strength to succeed in this demanding task.[8] The negative result at the ballot boxes forced the parties involved to go back to the negotiation table. It was not clear what would happen after the 'No' vote prevailed, but both the government and the FARC had already travelled a long way and there was a strong will to try to secure peace. And though it is still unclear whether this peace deal will indeed be implemented, it will likely serve as a benchmark for more lasting and much-needed reforms in the realm of development.

[7] 'Álvaro Uribe Vélez: Ex presidente de la República y Senador,' *La Silla Vacía*, 10 May 2017, http://lasillavacia.com/quienesquien/perfilquien/alvaro-uribe-velez. Accessed 24 June 2017; Aldo Cívico, "Uribe y una Justicia de Doble Rasero," *El Espectador*, 13 September 2016, http://www.elespectador.com/opinion/opinion/uribe-y-una-justicia-de-doble-rasero-columna-654573. Accessed 24 June 2017.

[8] *The Nobel Peace Prize for 2016*. https://www.nobelprize.org/nobel_prizes/peace/laureates/2016/press.html. Accessed 15 November 2016.

The Havana Accord

Peace negotiations between the government of Colombia and the FARC started taking place in Havana, in 2015, a location seen by the guerrillas as a safe and reliable neutral ground. This time around, despite many setbacks along the way, it seemed that things could come to a fruitful end after more than 50 years of conflict. In contrast to similar attempts in the early 1980s and late 1990s, most observers saw this round of negotiations as the culmination of a process of national political restructuring that could finally produce a lasting peace. The process deviated from the conservative path, the *Política de Defensa y Seguridad Democrática* (Democratic Security and Defence Policy), that the Uribe administration (2002–2010) had put together in the wake of the national crisis of the 1990s. Responding largely to the latest wave of drug-related violence, the Democratic Security Policy was aimed chiefly at restructuring the national economy through pro-market reforms that included new, more strict security measures to curb the FARC's presence in the territory, an effort to buttress the traditional oligarchic rule of Colombia's elites.

The country continues to face many socio-economic challenges, including urban-rural inequalities, concentration of land ownership, modest rural development, and a restrictive electoral democracy. The uncertainties of global trade and the depressed price of commodity exports (such as oil) also present challenges. To many, especially international and domestic investors, the repressive policies undertaken in the mid-2000s seemed to be the best way to restructure the country's economy, especially in a time when much of Latin America was ruled by leftist regimes pursuing nationalist economic policies (Ioris and Ioris 2013). As Colombia pushed for market-led reforms to insert itself more fully in regionally and globally integrated commodity markets, the Colombian government intensified its efforts to contain guerrilla insurgents, particularly under the auspices of Plan Colombia, which provided for an increase in the number of paramilitary forces. Even though some of this violence was eventually curbed, largely in response to international pressure, militarism and increased military expenditures have become continuous and permanent trends (Palacios 2006). And although the activities of insurgent groups have decreased in the late 2000s, this has not necessarily translated into actual state rule or even a state presence in many areas of the country.[9]

The violence perpetrated by armed groups sympathetic to the interests of the oil industry (e.g., right-wing paramilitaries) has facilitated investments in that sector because forced displacements and violence against civilian groups have operated to protect oil industry infrastructures and guaranteed land for oil exploration (Maher 2015). The Colombian state could not develop the country without establishing a monopoly on violence because of a symbiotic relationship between government officials and non-state actors (e.g. guerrillas) who exercised power in several rural areas (Acemoglu et al. 2013). In the latter decades of the twentieth century, widespread violence reached disturbing levels. From 1985 to 1998, an average of 1420 Colombians were killed each year. During this period, there were 14,000 violent incidents, which included attacks on civilians, sabotage, and kidnappings. Twenty-seven per cent of the victims were members of the civil society, 28% were members of the army, and 46% were members of guerrilla groups (Martínez Ortiz 2001). Now that some form of peace is looming, there are emerging pressures for permission to extract natural resources (primarily oil, but also minerals, timber, biodiversity goods, among others) from Colombia. Such initiatives are likely to benefit enormously from Colombia's pursuit of pro-market institutional reforms (e.g. royalty reductions). President Santos's political goals seek to integrate many communities into the national economy, particularly now that Colombia's mainstream economic policies have become more closely aligned with those of free trade agreements, such as the Trans-Pacific Partnership.[10]

Land reform and restrictions on foreign involvement in key industries, such as mining and oil exploration, were two of the main demands of FARC negotiators. The Santos administration's growing acceptance of land reform has been seen as one of the most promising aspects of the peace negotiations (Ince 2013). But formal peace is prone to directly

[9] Fundación Paz y Reconciliación, http://www.pares.com.co. Accessed 20 November 2016.

[10] *100 Preguntas de la Alianza del Pacífico*. http://www.mincit.gov.co/tlc/publicaciones/7180/100_preguntas_de_la_alianza_del_pacifico. Accessed 21 June 2017.

benefit foreign companies interested in Colombia's abundant natural resources. One of the new frontiers is the Pacific coast in the Department of Choco, where the British embassy in Bogota sees opportunities for UK corporate miners and timber millers (based on meetings held at the embassy in 2016 by the second author). It is likely that, despite the elegant rhetoric of the peace treaty and its emphasis on 'integral rural development' and respect for local cultures, much-needed elements of effective development will be pursued only tentatively. The importance of the agrarian question is not surprising in a country ravaged by historical conflicts related to land tenure. Even when the government pursued an aggressive land reform programme, as in the 1960s, it did little to curb violence in the countryside. Over the last six decades, such violence has been responsible for the deaths of 220,000 Colombians, most of whom were the rural poor (Sanchez Torres 2007). Even though issuing land titles has been conducted in many parts of the country at different times over the last 70 years, it has been implemented in piecemeal fashion and politically powerful large landholders blocked such efforts. This has created widespread and high levels of insurgent activity (Albertus and Kaplan 2012). Colombia remains one of the most unequal countries in the world, with a Gini index of 0.535 (higher in the countryside), low levels of investment in social programmes, and a growing concentration of income in the hands of the richest 1% of the population between 1993 and 2014 (CEPAL 2016).

If land reform indeed is to lead to a lasting peace, it must be conducted in a consistent, democratic manner and on a significant scale. These two conditions continue to pose serious political challenges in a country where promoting mobilisation related to legal, political, and labour issues means a real risk of death. At the same time, it is important to understand the complexity and uniqueness of contemporary social and economic circumstances. Columbia's economy relies on exports of a small number of primary goods and commodities (oil, coffee, bananas, etc.), while its population is becoming increasingly urbanised. Family farmers [*campesinos*] now account for only 12% of the population (Montañez-Gómez 2016). It is also crucially important to consider the broad range of disputes and the deeply politicised interactions involved in the armed conflict now that peace seems a viable prospect.

According to Jaime Torres González, the focus of liberal Colombian administrations on globalisation and market-based solutions over the last two decades has done little to improve infrastructure, literacy rates, or openness to trade (Torres González 2014). Capitalism in Colombia has developed in an intrinsically violent way that produced poverty and led to selective capital circulation and accumulation (Thomson 2011). Although sustained violence has facilitated the transformation of the subsistence economy, it also promoted forcible land grabbing and the expansion of commercial agriculture (Gómez et al. 2015). If the former FARC fighters enter parliament and become integrated into the national economy, thus legitimising Santos as an effective leader, it is likely that it will be easier for both sides to approve a package of liberalising measures and demonstrate to the rest of the planet that Colombia is open for business.

The Prospects for a Lasting Peace and the Need for Inclusive Development

The referendum of October 2016 was an unnecessary and poorly conducted course of action aimed at buttressing popular legitimacy of the decisions set in motion by the Santos administration. It is possible to conclude that even though the nation voted against the referendum, the majority of the Colombian people are still on the side of peace. Key areas of the country had a historically low turnout because the referendum coincided with a hurricane along the Caribbean coast, and it is clear that Uribe's 'vote no' campaign largely relied on the tactics of fear. However, the point remains that both sides need to quickly find a new way forward if an agreement for lasting peace is to be implemented. A key question that will become gradually more urgent is which sectors and groups in society will win or lose as the conflicts wane and peace begins to reign. During the period of belligerence—the 'war system' from the1970s to the 1990s, as Richani (2001) superbly describes it—there was a tacit, even comfortable, arrangement between the guerrillas and the national state, which allowed both sides to extract rents from various groups behind the scenes of a low-intensity war. When paramilitaries intervened in the second part of the 1990s, the situation became less stable and it became increasingly more difficult to maintain the status quo. Almost all parties involved demanded some form of change. This led to a serious national debate and the eventual approval of the Havana Accord. There were many reasons for the peace agenda, but probably the attempt to maintain the overall balance of political and economic power was the main one. Crucially, being on the side of peace meant, in practice, being on the side of a strong socio-economic agenda that depends on and is likely to flourish from the peace agreement that was signed.

However, about two-thirds of eligible voters abstained from participating in the 2016 referendum. Most Colombians either did not agree with the terms of the agreement or did not feel motivated to participate in the process. One thus needs to go beyond the specifics of the agreement to understand the structural reasons behind this widespread refusal to participate. Johan Galtung's concept of socio-economic violence is a useful tool for understanding this dynamic. It captures the reality of lasting socio-economic and political factors that prevent specific segments of society from accessing goods needed to meet their basic needs but benefit others (Galtung 1969).

Colombia is a nation defined by forced displacements, economic insecurity, and a structural lack of opportunities for most Colombians. It is clear thus that peace will require subtantial and enduring changes in the country's economic structures, political institutions, and legal framework so that systemic violence historically ravaging Colombian society can be overcome. A viable agenda for lasting, transformational peace needs to include a significantly different basis of development that includes key elements of social justice, understood as the reforming of the institutional conditions that sustain and reproduce poverty and inequality. This would include including moving away from the historical, and more recently deepened, strategy of inserting the country into the global economy by means of commodity export, especially in the field of carbon-based energy production.

It is important to remember that the focus of Uribe's plan to end the violence under his Democratic Security Policy was the creation of a more welcoming environment for foreign investors instead of actually addressing the deep-rooted causes of the continued conflicts in the country (Mora Cortes 2013). Concrete measures should include total demilitarisation of all armed groups in a process that assigns responsibility to all actors involved in acts of violence. Similarly, the Santos administration needs to make a much better case for the value of any type of transitional justice. This would include efforts to incorporate former guerrillas into the political system. Effective and peaceful integration of former FARC members into the political arena has to be conditioned on concrete ways to prevent what happened with the Patriotic Union of the mid-1980s, when more than 3000 former members of the guerrillas who had joined the political process were murdered with total impunity (Walch 2016).

Though essential, these are not sufficient elements of a lasting peace, understood in larger terms than the mere absence of overt violence. To be sure, effective representation of most communities affected by the violence of insurgency, and especially communities that need land, must be a central element of any new move towards a sustainable peace. Columbia cannot afford to repeat the errors of the 1970s, when the National Peasant Association was formally allowed to exist, but conservative administrations consistently undermined it (Zamosc 1986). Any peace plan should also reassess the market-led land policies of the 1990s, which mostly benefited larger landholders (Giugale et al. 2003). Perhaps even more challenging, the formalisation of a peace accord needs to be seen as only the first step towards comprehensive reforms that are capable of addressing the root causes that gave rise to such a devastating and destructive era in Colombian history. As it stands, the peace accords may create new opportunities for trade with China and open access to areas that were once inaccessible due to guerrilla activity. The Colombian government has adhered closely to Washington, and so far, the free market economy has attracted little Chinese attention. China's investment in Colombia is comparable to its investments in the much smaller Ecuadorian market.[11] However, much more is required to address the low quality of the 'actually existing democracy', the ineffectual national state, and the reliance

[11] "China and Latin America in 2016," *Council on Hemispheric Affairs*, 16 August 2016, http://www.coha.org/china-and-latin-america-in-2016. Accessed 16 November 2016.

on a narrow spectrum of natural resources such as oil, coal, nickel, and gold (Robinson 2016).

It should be clear, therefore, that in the short term, the Santos administration needs to prove capable of preventing the derailment of the commitments both sides made in the Havana Accord, largely by steering things away from the rising conservative rhetoric of his right-wing predecessor. Although, in recent years, a slowdown in the global demand for many of Colombia's exports has posed additional challenges to finding an effective path to peace, an agenda that includes a more inclusive national development needs to be pursued through incorporating the most traditionally excluded social segments, particularly rural populations, in the political system and in the nation's economy. These actions need to be coupled with a clear rejection of the neoliberal agenda, designed to integrate the country into economic trends shaped by the United States. Tragically, considering the long tradition of Colombian politics and the ongoing conservative turn in the region and around the world, it is likely that the promise of peace will soon be undermined by growing tensions and micro-scale wars that are difficult to resolve. Colombia must effectively surmount its painful history in order to realise a brighter future.

More Recent Developments

Just a few years after the historic signing of Havana Accord, which sought to bring to a halt violence between the Colombian repressive forces and the country's most important and consequential guerrilla group, FARC, thus ending a small-scale civil war that ravaged that Andean nation for over 50 year, key leaders of the FARC, now converted into an official political party, under the same acronym (which now stands for *Fuerza Alternativa Revolucionaria del Común*, Alternative Revolutionary Force for the Common), announced in September 2019 that they would go back to armed resistance against a new administration that was now living up to the commitments of the Accord. It seems that three years down the road, the aura of hope that initially surrounded it has depleted and the country now has to face a bitter situation in which armed groups are again fighting primarily for the maximisation of wealth and territory

(Rochlin 2020). Many of the key aspirations of the peace agreement were too ambitious and generated expectations that exceed the capacities of existing political and economic institutions (Rettberg 2020) and failed to satisfy the also important requirements of sustainability and land justice (Genoud 2020).

To understand how these foreboding developments come to be, we need to retrace the process of approval of the very Accord. First, the original version of what was called in Colombia el *Acuerdo Final de Paz* (Final Peace Accord) was signed in Cartagena on 26 September 2016. Immediately after, on 2 October, Juan Manuel Santos, then president of the country, asked Colombians to vote, on a general plebiscite, whether they agreed or not with the accord, on what proved to be a historic and legally unnecessary blunder, since opponents to the deal narrowly rejected it on the basis of alleged lack of punishment to members of the guerrilla, a conservative stance promoted by right-wing former president (later senator) Alvaro Uribe Velez. This unexpected but significant outcome did not immediately undermine the accord as internal actors, namely from the United Nations, exerted enough levels of influence to prevent a rapid escalation of violence. Moreover, between October and November the same year, all parties involved in the negotiation, including international observers, worked hard to introduce amendments to the accord in order to make it more palatable to the punitive logic underpinning the 'No' vote. Among the most important changes, military members were excluded from legal prosecution under the auspices of the, to be created, Special Jurisdiction for Peace (JEP), now restricted to decommissioned FARC members, and the decision to place the very Accord could be revised by the Colombia Congress, a decision that effectively weakened the treaty by making it conditional on ad hoc approval of a non-signatory body. Despite these changes, the revised version of the Accord was signed in Bogota in November, 2016, a major step that was followed by the total demilitarisation of the FARC in late June 2017, when the former guerrilla turned in all weapons it possessed to the Colombian government, under the supervision of a special commission of the United Nations. In November the same year, FARC held its first official congress as a political party, though internal divisions about the course of implementation of the Accord, now stalled in the Colombia Congress, were becoming visible.

Especially relevant among those who decided to return to the armed struggle in mid-year, 2018, Ivan Marquez, who was the main FARC negotiator for over four years in Havana, denounced the election of Uribe-anointed new president Ivan Duque in June the same year and his attempts to stall the implementation of the Accord. Even more concerning, from the government side, a major bombardment against regrouped FARC dissidents by official forces in the south of the country early in November, 2019, which amounted to the killing of minors present in the encampment were so violent that Guillermo Botero, Duque's right-hand Minister of Defence, the country's most important cabinet position, was forced to resign. Furthermore, based on the based on the report titled State of Implementation of the Colombian Final Accord, produced by the Kroc Institute for International Peace Studies at the University of Notre Dame, organisation in charge of monitoring the implementation of the Accord, published in April 2019, only one-third of the commitments, mostly related to the demilitarisation of the conflict, had been completed.

Namely relevant insofar as their state of completion (or lack thereof), initiatives related to land reform and to the replacement of illicit drugs in conflict areas have either not been pursued, or initial actions taken have been reversed. It is indeed clear that under Duque's executive control, reintegration of former combatants and restitution to victims, especially in the countryside, have been consistently underfunded. Further escalating things, echoing traffic events of the failed attempt to become a political party in the 1990s, the FARC members joined a political project under the label of Patriotic Union (UP, *Unión Patriótica*); in the last two years, 167 former guerrilla members, who joined civilian life have been assassinated, a process that is augmented by the continued killings of human rights activists, a consistent feature of Colombia life. What is even more worrying is that though the majority of FARC members have, for now, remained decommissioned, other guerrilla forces, namely ELN (the National Liberation Army) and the new, FARC dissident, AUC (Self-Defence Units of Colombia) are still operating, including in areas formerly controlled by the FARC. In short, notwithstanding the historic significance of the initial commitment to bring an end to Latin America's longest civil war, Ivan Duqe's right-wing government is currently undermining the path towards the implementation of the Havana Accord, a tragic trend for Colombia and Latin America as a whole.

References

Acemoglu, D., Robinson, J. A., & Santos, R. (2013). The Monopoly of Violence: Evidence from Colombia. *Journal of the European Economic Association, 11*(1), 5–44.

Albertus, M., & Kaplan, O. (2012). Land Reform as a Counterinsurgency Policy: Evidence from Colombia. *Journal of Conflict Resolution, 57*(2), 198–231.

CEPAL. (2016). *Panorama Social de América Latina, 2015*, doc. LC/G.2691-P. Santiago de Chile: Comisión Económica para América Latina y el Caribe.

Delgado, L. C., Reyes Gonzalo, C. E., & Sarmiento Gómez, A. (1999). Colombia. In *Gasto Público en Servicios Sociales Básicos en América Latina y el Caribe: Análisis desde la Perspectiva de la Iniciativa 20/20*, doc. LC/R.1933 (pp. 289–333). Santiago de Chile: Comisión Económica para América Latina y el Caribe Santiago de Chile.

Dent, D. W. (1974). Oligarchy and Power Structure in Urban Colombia: The Case of Cali. *Journal of Latin American Studies, 6*(1), 113–133.

Fajardo, L. E. (2003). *From the Alliance for Progress to the Plan Colombia: A Retrospective Look at U.S. Aid to Colombia* (Working Paper 28). London: Development Centre, London School of Economics.

Farnsworth-Avelar, A. (2000). *Dulcineia in the Factory: Myths, Morals, Men, and Women in Colombia's Industrial Experimentation, 1905–1960*. Durham: Duke University Press.

Franz, T. (2016). Plan Colombia: Illegal Drugs, Economic Development and Counterinsurgency-A Political Economy Analysis of Colombia's Failed War. *Development Policy Review, 34*(4), 563–591.

Galtung, J. (1969). Violence, Peace, and Peace Research. *Journal of Peace Research, 6*(3), 167–191.

Genoud, C. (2020). Access to Land and the Round Table on Sustainable Palm Oil in Colombia. *Globalizations* (forthcoming).

Giugale, M. M., Lafourcade, O., & Luff, C. (2003). *Colombia: The Economic Foundation of Peace*. Washington, DC: World Bank.

Gómez, C. J. L., Sánchez-Ayala, L., & Vargas, G. A. (2015). Armed Conflict, Land Grabs and Primitive Accumulation in Colombia: Micro Processes, Macro Trends and the Puzzles in Between. *Journal of Peasant Studies, 42*(2), 255–274.

Hartlyn, J. (2008). *The Politics of Coalition Rule in Colombia*. Cambridge: Cambridge University Press.

Hirschman, A. O. (1963). *Journeys Toward Progress: Studies in Economic Policy Making in Latin America*. New York: Twentieth Century Fund.

IGAC. (2012). *Atlas de la Distribución de la Propiedad Rural en Colombia*. Bogotá: Instituto Geográfico Agustín Codazzi.

Ince, M. (2013). Defeating Colombia's Oldest Insurgency: Prospects for Peace and Reconciliation with the FARC. *The RUSI Journal, 158*(1), 20–27.

Ioris, R. R., & Ioris, A. A. R. (2013). The Brazilian Developmentalist State in Historical Perspective: Revisiting the 1950s in Light of Today's Challenges. *Journal of Iberian and Latin American Research, 19*(1), 157–172.

Javier Orjuela, L. (2015). Uribe versus Santos: Del Enfrentamiento Personal a la Diferenciación Ideológica. *Revista de Estudios Sociales,* April–June 52, 201–207.

Lopez, R. (2014). From Middle Class to Petit Bourgeoisie: Cold War Politics and Classed Radicalization in Bogotá, 1958–1972. *Estudios Interdisciplinarios de America Latina y el Caribe, 25*, 99–130.

Lopez-Alves, F. (2000). *State Formation and Democracy in Latin America, 1810–1900*. Durham: Duke University Press.

Maher, D. (2015). The Fatal Attraction of Civil War Economies: Foreign Direct Investment and Political Violence, a Case Study of Colombia. *International Studies Review, 17*(2), 217–248.

Martínez Ortiz, A. (2001). Análisis Económico de la Violencia en Colombia: Una Nota sobre la Literatura. *Cuadernos de Economía, 20*(34), 157–187.

Montañez-Gómez, G. (2016). Territorios para la Paz en Colombia: Procesos entre la Vida y el Capital. *Bitácora Urbano Territorial, 26*(2), 11–28.

Mora Cortes, A. F. (2013). Conflicto, Violencia Socioeconómica y Desplazamiento Forzado en Colombia. *Cuadernos de Economía, 61*, 721–754.

Palacios, M. (1983). *El Café en Colombia (1830–1970): Una Historia Económica, Social y Política*. Bogotá: Editorial Presencia & FEDEDESAROLLO.

Palacios, M. (2006). *Between Legitimacy and Violence: A History of Colombia, 1875–2002*. Durham: Duke University Press.

Rettberg, A. (2020). Peace-Making Amidst an Unfinished Social Contract: The Case of Colombia. *Journal of Intervention and Statebuilding, 14*(1), 84–100.
Richani, N. (2001). The Political Economy of Colombia's Protracted Civil War and the Crisis of the War System. *Journal of Conflict Studies, 21*(2), 2.
Richani, N. (2010). Colombia: Predatory State and Rentier Political Economy. *Labour, Capital and Society, 43*(2), 119–141.
Ríos Sierra, J. (2016). Dinámicas de la Violencia Guerrillera en Colombia. *Revista de Ciencias Sociales, 22*(3), 84–103.
Robinson, J. A. (2016). La Miseria en Colombia. *Desarrollo y Sociedad, 76*(1), 9–90.
Rochlin, J. (2020). Colombia and the Transformation of Warfare: The End of Ideology? *Global Change, Peace and Security, 32*(1), 79–101.
Ross, E. B. (2007). Clearance as Development Strategy in Rural Colombia. *Peace Review, 19*(1), 59–65.
Salas-Salazar, L. G. (2016). Conflicto Armado y Configuración Territorial: Elementos para la Consolidación de la Paz en Colombia. *Bitácora Urbano Territorial, 26*(2), 45–57.
Sanchez Torres, F. (2007). *Las Cuentas de la Violencia*. Bogotá: Grupo Editorial Norma.
Thomson, F. (2011). The Agrarian Question and Violence in Colombia: Conflict and Development. *Journal of Agrarian Change, 11*(3), 321–356.
Torres González, J. (2014). Colombia: Logros en dos Décadas de Modelo de Desarrollo Aperturista: Análisis Según Resultados de Balanza Comercial. *Análisis Político, 27*(82), 212–235.
Walch, C. (2016). Rethinking Ripeness Theory: Explaining Progress and Failure in Civil War Negotiations in the Philippines and Colombia. *International Negotiation, 21*(1), 75–103.
Youngers, C., & Rosin, E. (Eds.). (2015). *Drugs and Democracy in Latin America: The Impact of U.S. policy*. Boulder/London: Lynne Rienner Publishers.
Zamosc, L. (1986). *The Agrarian Question and the Peasant Movement in Colombia*. Cambridge: Cambridge University Press.

8

Cerca del Rio y Lejos del Agua: Water, Autonomy, and Hope in the Ecuadorian Andes

Geoff Goodwin

Introduction[1]

Water is abundant in Latin America. It cascades from mountains, crashes through valleys, and surges through forests. Yet, millions of Latin Americans face water scarcity. The paradox of scarcity amid abundance is not unique to water. Indeed, it is one of the defining features of capitalism. However, water scarcity is particularly perverse because of the

[1] This chapter is based on a presentation I gave at the Agrocultures conference in Porto Velho, Brazil, August 2019. I am grateful to participants at the conference for their constructive comments and to Antonio Ioris for inviting me to participate in this event. I am also indebted to the numerous people who shared their thoughts and experiences with me while I conducted fieldwork in Ecuador, and to the Department of International Development at the London School of Economics for supporting this research through RIIF funding. Thanks also to Ana Dinerstein, Jonathan Alderman, and Antonio Ioris for their comments on earlier drafts of this chapter.

G. Goodwin (✉)
Oxford Department of International Development, Queen Elizabeth House, University of Oxford, Oxford, UK
e-mail: geoff.goodwin@qeh.ox.ac.uk

A. A. R. Ioris (ed.), *Environment and Development*,
https://doi.org/10.1007/978-3-030-55416-3_8

central role water performs in supporting human and nonhuman life.[2] The paradox is especially acute in rural contexts as rural families often live close to major water sources but nonetheless lack access to potable and irrigation water.

In this chapter, I use the phrase *cerca del rio y lejos del agua* to capture this predicament and highlight struggles to overcome socially constructed water scarcity across rural Latin America.[3] Since the 1960s, community water associations have been the main protagonists in this struggle in the Ecuadorian Andes.[4] These organisations, which supply potable and irrigation water, are highly heterogeneous, with their own dynamics, histories, and futures. However, they share certain traits, most notably, structures and practices that enable collective decision making over various aspects of water services and management, including distribution, tariffs, and infrastructure. Hence, water associations have some degree of autonomous control over local water services and supplies. The extent of their autonomy varies; however, even in cases where they have developed considerable autonomous capacity, autonomy is never complete. Full autonomous control of water is impossible and water associations have to interact and negotiate with various actors, including state bureaucracies, local governments, international development agencies, and non-governmental organisations. Thus, water association autonomy is a deeply relational process.

I will explore this process through the lens of Ana C. Dinerstein's open Marxist approach to collective autonomy (Dinerstein 2015, 2019). Her conceptual framework rests on four modes of autonomous praxis—negating, creating, contradicting, exceeding—and the principle of hope, which she takes and adapts from the Marxist philosopher, Ernst Bloch.I will illustrate this by tracing the evolution of water association autonomy from the 1960s, arguing that autonomy responds to particular social and

[2] On water scarcity, see, for example, Bakker (2010), Ioris (2012), and Swyngedouw (2007).

[3] I borrow the phrase "*cerca del rio y lejos del agua*" from a documentary produced by the *Foro de los Recursos Hidricos* (FRH): *Despojo del agua en la cuenca del Rio Chimbo* (2014). I would like to take this opportunity to thank the FRH, and especially Carlos Zambrano, for their tremendous support during my fieldwork in Ecuador.

[4] In this chapter, I use the term "community water association" to refer to a highly diverse group of organisations that take various names in Ecuador, including *juntas de agua* and *sistemas comunitarios de agua*.

historical conditions, creates political opportunities and tensions, and drives political and social change.[5] The chapter will seek to show that this process has been extremely diverse, with water association autonomy responding to multiple factors, taking various forms, and travelling in alternative directions. Regardless of the particular form autonomy has taken it is ultimately a source of hope, for it provides rural families and communities with a tool to challenge water scarcity, forge new social and political practices, and reconfigure human-nature relations.

The chapter starts by outlining Dinerstein's critical approach to autonomy, where she connects Bloch's principle of hope to collective autonomous struggles, emphasising the need to capture the utopian content of social resistance. This section also discusses the influential work of John Holloway (2002/2010), showing how Dinerstein overcomes some of the limitations of his approach to autonomy. The chapter then explores water association autonomy in the Ecuadorian Andes through the lens of Dinerstein's conceptual framework, seeking to show the potential and limits of the process, and offer insights into collective autonomous struggles elsewhere in Latin America.

Theorising Autonomous Struggles in Latin America

Debates over the theory and practice of collective autonomy have taken multiple forms but have tended to boil down to the nature and capacity of the state. Put simply, while some critical scholars and activists see the capture of state power as a precondition for progressive or revolutionary change, others stress the transformative potential of organising outside the state apparatus and formal political institutions.

[5] The chapter draws on research conducted during multiple trips to Ecuador since 2015 and connects to exploratory fieldwork undertaken in 2010 and 2011. Data includes (i) interviews with community water associations, indigenous and peasant movement leaders, non-governmental organisations, politicians, and bureaucrats; (ii) field notes and observations; (iii) local and national newspaper archives; (iv) laws and legislative proposals (v) official documents related to water services, management and infrastructure. For a fuller methodological discussion, see Goodwin (2019).

Political upheavals in Latin America at the end of the twentieth century energised these debates. The Zapatista uprising in Chiapas, Mexico was pivotal. The rebellion, which commenced in January 1994, demonstrated the capacity of communities and movements to transform politics without toppling or seizing the state. The subsequent development of autonomous local governments showed the possibility of alternative forms of political organisation and practice existing within capitalist societies, challenging conventional thinking about the need to capture state power to effect radical change. Events elsewhere in Latin America at the turn of the century pointed in a similar direction, with a diverse range of organisations and movements striving for autonomy, creating novel practices, and reconfiguring social and political relations (Alderman 2018, 2019; Dinerstein 2013; Esteva 2015; González 2015; Merino 2020; Zibechi 2007).

Taking inspiration from these struggles, a growing number of critical scholars started to explore the radical potential of collective autonomy. John Holloway's interventions have been particularly influential and provocative. In his landmark book, *Change the World Without Taking Power*, he rejects state-centric views of social change and political praxis, and calls for the re-centring of radical politics towards everyday acts of resistance and autonomous forms of organisation (2002/2010).[6] For Holloway, the pursuit of radical change through the state is futile because states are embedded in a global web of capitalist relations and lack the autonomy to break with capitalism. Thus, autonomy performs a dual role in his analysis: states lack the autonomy to effect radical change hence social organisations must develop autonomous practices to challenge and transcend capitalism. Not only does this require challenging the state but confronting alienated, commodified forms of labour, developing new ways of working, producing and exchanging, and decommodifying activities and social relations (2002/2010: 238–242; 2010: 27–37).

Holloway argues that full collective autonomy is impossible within capitalist societies; however, the movement towards autonomy creates the possibility of alternative futures, "a project that continually takes us

[6] Putting it bluntly, Holloway claims: "To struggle through the state is to become involved in the active process of defeating yourself" (2002/2010: 214).

against-and-beyond the barriers of capitalism" (2002/2010: 219, see Bohm et al. 2010).[7] Through this process, autonomous struggles can create "cracks" in the capitalist system which plant the seeds for movements toward a post-capitalist world: "a single total transformation of world capitalism to world socialism or communism is unthinkable, so the only possible way to think about radical social transformation is interstitially" (2002/2010: 220; 2010: 3–9, see Dinerstein 2017, 2019). Holloway notes that these cracks are frequently territorial: territories provide a material base for autonomous movements and create space for the formation of new subjectivities, relations, and practices (2010: 27–37; see Zibechi 2007; Halvorsen et al. 2019). Yet, as Holloway notes, autonomous struggles cannot be reduced to territorial struggles and cracks are not simply "spatial ruptures".

To drive radical change, Holloway argues that autonomous organisations must reject hierarchical relations and structures and establish direct rather than representative democratic practices and institutions (2002/2010: 228–238). Eschewing one-size-fits-all organisational models, he proposes organising principles instead, claiming, for example, that "the drive towards self-determination implies the assumption of mutual respect and shared responsibility as organisational orientations" (2002/2010: 230). Such principles, he argues, can be found in a diverse range of organizational forms, including neighbourhood councils and community assemblies.

Holloway opens space for critical and creative thinking about autonomy; however, his approach has some important limitations. Here I will briefly highlight three issues. First, he tends to see all autonomous struggles as anti-capitalist or as cracks in the capitalist system (e.g. 2010: 28). In doing so, he appears to fall into one of the traps he cautions against: assigning particular actors "hero" status in the revolutionary struggle (2002/2010: 104–5). Whereas some autonomous movements clearly challenge capitalist relations and practices, others enthusiastically embrace them, or sit somewhere in the middle. Characterising them all as

[7] Further: "The drive towards self-determination is not instrumental: we do not start from an aim and deduce from that aim the path we must follow to reach that aim. It is rather a movement outwards, a path that is made in the process of walking – walking in the dark, guided only by the light provided by the utopian star of our own projections" (2002/2010: 221).

inherently or potentially anti-capitalist ignores the complexity and diversity of autonomous organising, and risks misinterpreting some autonomous struggles. Second, he tends to homogenise the state, effectively collapsing all state forms into a single capitalist state (2002/2010: 11–18). The potential for individual and collective agency within the state is seemingly rejected, and the state is portrayed as a monolithic entity that uniformly acts in favour of capital and against autonomy. Conceptually, this is problematic as states are best understood as fluid and fragmented rather than static and singular. Tactically, advocating the rejection of state power and party politics runs the risk of abandoning the state to the right, which, as recent experiences in Latin America demonstrate, can result in disastrous outcomes. Third, his rejection of the state leads him to separate out state actors and autonomous organisations and reinforce the state-autonomy binary. Consequently, he provides little insight into how autonomous movements interact with the state or the consequences of these interactions for autonomous struggles and radical politics.

In her critique of Holloway's notion of the "crack", Dinerstein (2015, 2019) overcomes some of these limitations, and creates a conceptual framework that transcends the state-autonomy dichotomy. She achieves this by viewing the struggle for autonomy as more than a crack or an opening: the quest for autonomy is an antagonistic and contradictory process which takes place within the social relations of capital and is full of setbacks and disappointments. She identifies four modes of praxis that are present in these autonomous struggles.[8] The first—*negating*—is understood as the rejection of existing realities—including, patriarchy, racism and oppression—and the embrace of negativity as political practice (2015: 34–37). This entails working against capitalist structures, processes and relations, and rejecting conventional, institutionalised forms of politics. It involves viewing politics as disagreement rather than consensus and using antagonism to redefine and transform political practices and relations. The other side of this coin—*creating*—is orientated towards the construction of "concrete utopias", radical practices, relations, and

[8] Dinerstein explores these modes of autonomous praxis through the analysis of autonomous struggles in Argentina, Bolivia, Brazil, and Mexico. In this chapter, I focus solely on her conceptual framework.

dreams that exist in the present (2015: 37–41).[9] Creating requires seizing opportunities, confronting fears, and stepping into the unknown. Thus, imagination and daring are central to this mode of autonomous praxis. The third—*contradicting*—stems from the intrinsic but creative tension between autonomism and capitalism (2015: 41–46). Autonomous movements must interact with the state, law and capital and these interactions mediate social resistance insofar as the state aims to "translate" autonomous struggles into the grammar and institutions of power (Dinerstein 2015, 2017, 2020). In so doing, autonomous movements constantly run the risk of capture, appropriation, and subordination. Here, autonomy most clearly emerges as a contradictory and relational process which is internal to capitalism and intrinsically connected to commodification and accumulation. The contradicting mode highlights the fact that autonomous movements must navigate complex terrain, and face the possibility of disappointment and failure. Yet, autonomous struggles also create something that cannot be contained or subordinated (Bohm et al. 2010). This is captured in the fourth mode of autonomous praxis—*exceeding* (Dinerstein 2015: 47–48). Excess—or "surplus"—is the residual or overflow that is produced through cooperation and solidarity, creating possibilities of alternative futures beyond the state.

Hope is the central thread that connects these modes of autonomous praxis. In connecting autonomy and hope, Dinerstein draws on the writing of Ernst Bloch, the influential Marxist philosopher who views utopia as a future that is actively constructed—that is, concrete praxis. His understanding of hope is rooted in the idea of the "not yet", the always present possibility of a (more) humane future. Dinerstein argues that it is only when autonomy is placed in the "key of hope" that its meaning and potential can be fully grasped (2015: 58–75, see Dinerstein 2019). Indeed, for her, autonomous struggles are fundamentally about hope "i.e. about realising something that is not yet – by trying, exploring, rehearsing, anticipating different – better – worlds" (2015: 23). Hence, autonomy has a prefigurative dimension, a quality rooted in the boundless possibility of the not yet. Crucially, this is not dependent on the likelihood of it becoming a reality but merely on its existence in the present as

[9] See Penna (2020) for an illuminating analysis of utopian autonomous struggles in Brazil.

possibility (Dinerstein 2015: 71–74; Holloway 2002/2010: 7). Dinerstein argues that this sense of possibility is a source of hope that propels autonomous struggles forward against the odds. Moreover, her own message is one of hope: she encourages critical scholars to embrace the possibility of the not yet rather than wallow in the fatalism of the not ever (Goodwin 2016).

Dinerstein claims that the four modes of autonomous praxis and the unifying principle of hope are widely found in autonomous struggles in Latin America but take different forms in indigenous and non-indigenous contexts (2015: 48–56, also 212–221). She identifies several factors that distinguish indigenous from non-indigenous autonomy. Indigenous collective memories of oppression and exclusion are fonts of resistance, providing a basis to connect past and present struggles to construct alternative futures. Meanwhile, indigenous cosmologies create possibilities for challenging and transcending capital. The contested and partial integration of indigenous peoples into nation states through colonialism gives indigenous autonomous struggles another distinct twist. For centuries, indigenous communities and nations have fought to defend and expand autonomous spaces. However, they have also attempted to integrate into and transform the state. Novel political demands have emerged out of this contradictory process, including efforts to establish plurinational states, most notably in Bolivia and Ecuador. This indicates the capacity of autonomous indigenous struggles to create excess which spills over into wider political processes and reconfigures local and national politics (Dinerstein 2015: 212–221).

These insights are crucial for understanding autonomous struggles in Latin America. However, creating a separate, singular category for indigenous autonomy runs the risk of homogenising indigenous struggles (Goodwin 2016).[10] Indigenous cultures and identities, though rooted in the precolonial past, are dynamic and diverse and defy simple binary categorisations. Class differentiation has added further layers of complexity as deeper engagement in labour and land markets and capitalist

[10] My point here is that there is the risk of Dinerstein's conceptual framework being employed crudely and failing to capture the diversity and complexity of indigenous struggles by, for example, labelling all autonomous indigenous struggles anti-capitalist or ignoring social and class differentiation within indigenous movements.

production and relations has increased socioeconomic difference within indigenous populations, diversified indigenous cultures, and complicated indigenous struggles. Creating a separate category for indigenous autonomy also suggests indigenous and non-indigenous struggles are separate, which is rarely the case. For example, across Latin America indigenous peoples have worked with white and mestizo leftist and religious activists to build indigenous autonomy and fight racism and exploitation. Moreover, the internal composition of autonomous movements often stretches across ethnic lines, as the Zapatistas and other autonomous movements illustrate. Dinerstein provides conceptual tools and intellectual insights to deal with some of this complexity, and her analysis of indigenous struggles in Latin America reveals some of the diversity within indigenous movements (e.g. 2015: 144–170). However, with her gaze fixed firmly on the autonomy-state-capital nexus, her conceptual framework tends to overlook diversity within autonomous struggles and how difference is negotiated.

Like Holloway, Dinerstein also tends to characterise the state as a singular, monolithic entity (2015, 2020). While this is conceptually and empirically problematic, her critical perspective is important.[11] Capitalist states, in their multiple guises and historical forms, are compelled to perform a set of core functions, including establishing and protecting private property rights, creating and sustaining markets, and promoting and enabling capital accumulation. Viewing autonomous struggles from this perspective highlights some of the fundamental constraints and challenges they face when engaging with the state, and foregrounds the constant threat of state repression and violence. The history of community water associations in Highland Ecuador illustrates these obstacles and dangers, but also shows that state policies and actors move in diverse directions and have the potential to support as well as undermine autonomous organising.

[11] Dinerstein argues that the "state is not a state in a capitalist society, i.e. a neutral arena on which the common good is decided, but a capitalist state" (Dinerstein 2015: 167–8, 2020).

Constructing Autonomy in the Ecuadorian Andes: Water Association Autonomy in Historical and Theoretical Perspective

Community water associations perform a crucial role in the delivery of water services in the rural Ecuadorian Andes.[12] The water they deliver to houses and farmers is the fruit of a historical process of collective organisation that has united rural families and communities in a struggle to overcome water scarcity and take greater control of local water supplies. Through this process, new political relations and subjectivities have been constructed and diverse forms of organisation have been fashioned. Community water associations are extremely diverse; however, they share one important characteristic: collective autonomous capacity. This enables them to influence various aspects of water service delivery, including the conservation of water supplies, the distribution of water, and the construction and maintenance of hydraulic infrastructure. It also allows them to develop and maintain their own organisational practices and structures and renegotiate their relationship with the state. I will illustrate this by tracing the evolution of water association autonomy from the 1960s, arguing that I will argue that autonomy responds to particular social and historical conditions, creates political opportunities and tensions, and drives political and social change. This section will also highlight some of the risks and limits of autonomous organising. However, following Dinerstein, I will ultimately argue that autonomy is a source of hope as it gives rural families and communities a tool to challenge water scarcity, forge new social and political relations and practices, and reconfigure human-nature relations.

The historical trajectory of water association autonomy is depicted below (Fig. 8.1).[13] Autonomy-building started to gather pace in the 1960s and 1970s, before accelerating in the 1980s and 1990s as

[12] On a national level, water associations are estimated to deliver drinking water to around 30% of the population and irrigation water to around 80% of irrigation users (FRH 2011, 2013).

[13] The actual historical process of autonomy building was not nearly as linear or uniform as suggested by Fig. 8.1. The main point I seek to convey in this chart is that the aggregate autonomous capacity of rural community water associations increased from the 1960s and this historical process is crucial for understanding contemporary political struggles in Ecuador. Individual cases often

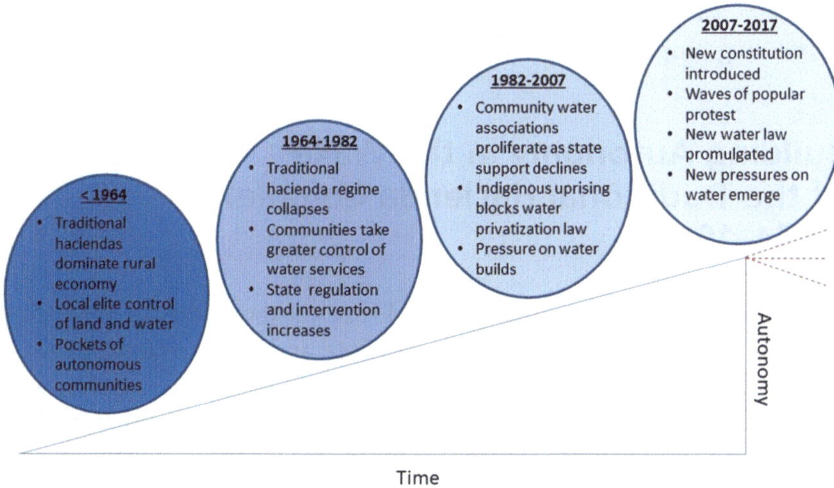

Fig. 8.1 Historical evolution of community water associations in the rural Ecuadorian Andes

governments implemented structural adjustment and neoliberal reform. The reconfiguration of the state and the scaling back of investment and regulation created more space for autonomous organizing and enabled water associations to develop new practices and relations. When the state attempted to increase control of water in the 2000s and 2010s, water associations resisted to protect the autonomous spaces they had constructed in earlier decades. Waves of mass protests derailed the state's plans. However, following a protracted period of protest and negotiation, a new legal framework was established, which threatened water association autonomy by concentrating decision making within state agencies and subjecting water associations to a raft of new regulations. Once the law was introduced, implementing it became a new site of struggle as water associations set about limiting state control and defending their autonomy. I will discuss this historical process in greater depth below, briefly explaining each of the phases depicted in Fig. 8.1, while critically engaging with Dinerstein's

diverged from this general trend, with many water associations, for example, losing autonomous capacity or collapsing altogether.

approach to autonomy, especially her four modes of autonomous praxis: negating, creating, contradicting, and exceeding.

Building Autonomy in the Wake of the Traditional Hacienda Complex (1964–1982)

Communal water organising has long historical roots in rural Highland Ecuador. However, it was not until the 1960s and 1970s that water associations began to proliferate and perform a lead role in water services and management.[14] During this period, the traditional hacienda complex, which had dominated the political landscape for centuries, collapsed, creating greater space for more autonomous forms of communal organisation among indigenous and non-indigenous peasants. Under the traditional hacienda complex, landowning elites monopolised land and water, leaving most peasants reliant on them to access water. Irrigation networks were mainly constructed to cater for the needs of haciendas and potable water was typically reserved for landowning elites and wealthy and middle-class urban households. Some largely autonomous peasant communities existed on the fringes of traditional haciendas and were able to establish collective rights to irrigation water. *Ley de Comunas*, which was first introduced in 1937, provided them with a legal framework to register and protect communal property, including irrigation systems. However, most highland peasants lacked the capacity to manage water independently of landowning elites, especially indigenous peasants who were often tied to haciendas through traditional peasant-patron relations, like *huasipungo*.

The collapse of the traditional hacienda complex, which started to gather pace in the 1950s, was triggered by a confluence of social, political, and economic factors, but land reform, which commenced in 1964

[14] In this section, I primarily draw on the following studies: Andolina (2012), Armijos (2012, 2013), Boelens and Doornbos (2001), Boelens and Vos (2014), FRH (2011, 2013), Hoogesteger (2013, 2014, 2015, 2016) and Pacari (1998). I also use Goodwin (2017, 2019, 2020) and the works cited in these essays.

and ended in 1994, was the most decisive. The process, while weak in terms of the regulation and redistribution of land, gradually eliminated traditional peasant-patron practices which undermined the traditional hacienda regime and supported the capitalist modernisation of agriculture. This transformation was largely overseen by military dictatorships who governed throughout much of the 1960s and 1970s, following the general trend in Latin America.

The unravelling of the traditional hacienda complex created space for rural communities to take greater control of local water supplies. In some cases, new organisations were established to deliver potable or irrigation water. In others, services were delivered through existing communal structures and practices, especially in indigenous communities. In most cases, community assemblies, which took diverse forms, became the site for collective decision making over water services and struggles.

One important feature of this process of collective organisation was the active involvement of water association members in the construction of hydraulic infrastructure, especially digging tunnels and canals for irrigation and drinking water. The bulk of this work, which often lasted weeks or months, was conducted through *mingas*, a collective labour practice, which is deeply rooted in Andean culture, and varies within and between indigenous and non-indigenous communities. The practice is based on principles of solidarity and reciprocity, and involves community members working without wages. However, food, drink, and entertainment are frequently provided. Importantly, *mingas* were not one-off events to construct water systems in the 1960s and 1970s, but the start of a process of collective labour that sustained water systems over the long run. Once water systems were constructed, *mingas* were organised to maintain them, with work ranging from cleaning and repairing irrigation canals to clearing and protecting water sources. The practice provided a basis for the strengthening of collective identities and the exchange of knowledge. Water rights were also frequently established through *mingas*, with participation required to enjoy access to water. The 1960s and 1970s saw an explosion of collective work through *mingas*, which created the foundation for the autonomous control of local water supplies and the formation of new social and political relations.

Although rural communities were the main protagonists in this struggle, they were generally reliant on outside support to develop water systems. Local governments and state agencies typically provided raw materials and technical support, while international development agencies and non-governmental organisations also frequently contributed. The legal framework encouraged autonomous organisation, with water legislation introduced in the 1970s promoting the creation of water associations but without imposing tight restrictions or regulations. *Ley de Comunas*, which was originally introduced to govern and control highland indigenous peoples, provided a tool for rural communities to develop water systems and build autonomy in the late twentieth century. However, these legal instruments also placed constraints on water association autonomy, and bureaucrats and planners often imposed standardised organisational models on water associations, which sometimes undermined existing communal practices and authorities.

Thus, the proliferation of community water associations in the 1960s and 1970s responded to multiple factors and involved diverse practices and relations. Negating, creating, and contradicting were the most prominent modes of autonomous praxis during this period (Dinerstein 2015). Negating involved peasant families and communities rejecting the traditional hacienda complex which amounted to a "total social fact" and encompassed virtually every aspect of social life (Mauss 1925/2002). Protests, strikes, and invasions contributed to the downfall of this regime. Traditional haciendas were embedded in a web of capitalist relations and structures. However, negating in this context was primarily aimed at challenging the peasant-patron practices that regulated labour, land, and water in the traditional agrarian economy. Escaping these practices ultimately resulted in indigenous and non-indigenous peasants becoming more enmeshed in labour and land markets and capitalist production and relations. Hence, at this stage, negating supported rather than undermined capitalist development.

Yet, the negating practices that rural communities and water associations developed and employed echoed some of the features Dinerstein identifies in more contemporary Latin American settings. For example, indigenous communities and movements rejected the exploitation and racism of the traditional hacienda complex and harnessed ethnic identity to challenge colonialism and confront the state. However, these negating practices rarely involved the total rejection of the state. Indeed, state power, expressed most clearly through land reform, opened space for autonomous organising, and indigenous and peasant movements consistently pushed for integration into the state bureaucracies that managed land reform from the 1960s. As noted above, rural communities and water associations also relied on the state to provide raw materials, technical knowledge, and machinery to construct water systems. Hence, the negating, creating, contradicting modes of autonomous praxis overlapped during this period as water associations simultaneously operated with and against the state, developing autonomous relations and practices while facing the risk of being incorporated into state structures and logics.

These early experiences illustrate the problem of conveying the state as a single, monolithic entity that uniformly challenges autonomy. During the 1960s and 1970s, bolstering the autonomous capacity of water associations supported the state's efforts to bring water services to rural communities without having to invest huge resources in water systems. In turn, improving water service delivery reduced the possibility of rural unrest, which was a significant political concern during the Cold War. Hence, a certain degree of water association autonomy was desirable for the state. However, this created space for water associations to establish relations and practices that challenged the state in following decades, indicating the tendency for autonomous organising to overflow and drive politics in unexpected directions.

Deepening Autonomy Amid Neoliberal Reform and Economic Collapse (1982–2007)

Ecuador experienced a brutal economic decline in the 1980s as the debt crisis and structural adjustment saw rising unemployment, plummeting real wages, and increasing poverty and inequality. Structural adjustment programmes laid the foundation for neoliberal reforms which were implemented through the 1980s and 1990s. Further neoliberal restructuring took place in the wake of another massive recession in 1999 and the rushed dollarisation of the economy in 2000. Elected governments from the right, centre-right, and centre-left oversaw structural adjustment and neoliberal reform. From the late 1990s to the early 2000s, a succession of presidents came and went as mass mobilisations and political infighting removed them from office. The period saw significant declines in public investment in water infrastructure and the weakening of public water systems. The privatisation of water services was also promoted and implemented in some cases. However, the full-blown privatisation of the water sector was blocked by mass indigenous mobilisations in the 1990s. Though the legal framework established in the 1970s to regulate water remained intact, the policy and bureaucratic framework was refashioned to loosen state regulation and decentralise water services and management. State bureaucracies were also run into the ground, which further limited state regulation and control of water at the national and local levels.

These changes brought opportunities and threats to community water associations. On the one hand, fiscal constraints and neoliberal reforms reduced the capacity of local governments and state agencies to provide the raw materials and technical support most rural communities needed to build or expand potable and irrigation water systems. Neoliberal restructuring also created greater space for capital intensive agroexporters to operate in the highlands which increased pressures on water supplies and intensified water scarcity. The rapid expansion of flower plantations in the northern highlands was the starkest illustration of this general trend. On the other hand, the lack of state support and regulation created greater space for rural communities to take control of water and build autonomy. The autonomous control of local water services spread as the prospect of the state providing adequate services to rural households diminished. Meanwhile, the regulation of water associations weakened as water bureaucracies declined. New opportunities also emerged for rural communities to take greater control of irrigation water as state-run irrigation networks were transferred to water associations to manage.

Important changes to the internal composition of water associations occurred during this period. Whereas associations that were established in the 1960s and 1970s to deliver irrigation water largely comprised indigenous and non-indigenous peasants with small plots of land, the transfer of public irrigation networks and the expansion of capitalist agriculture incorporated a wider range of agricultural producers in the 1980s and 1990s. In extreme cases, this ultimately resulted in small-scale farmers and agroindustrial enterprises belonging to the same water association. Hence, the class composition of some irrigation organisations changed, which altered the dynamics of autonomous organising. The internal composition of potable water associations also diversified as a broader spectrum of the rural population became involved in water service delivery and social and class differentiation increased within rural communities.

Despite the decline of state support and regulation in the 1980s and 1990s, water associations still engaged with the state. Local governments and state agencies continued to contribute to the construction and maintenance of community water systems, albeit less frequently and more frugally. State agencies also granted formal water rights, which often required renegotiation over time. However, given the paucity of public resources, water associations increasingly turned to overseas development agencies, multilateral institutions, and non-governmental organisations for support. This was not novel: this diverse group of actors were heavily involved in the development of water systems in Ecuador in the 1960s and 1970s, and before in several cases. However, they took on a more decisive role in the 1980s and 1990s, following the general trend in Latin America. Water associations forged alliances with these actors to secure the financial resources and technical expertise they required to construct or expand water systems, with some developing long-term relationships and others engaging fleetingly to access resources and knowledge.

Though the state played a less prominent role in this process in the 1980s and 1990s, the general pattern established in earlier decades remained unchanged. Outside actors supplied raw materials and technical support while community members contributed the bulk of the labour power required to construct water systems and manage them over the long run. *Mingas* remained vital to water association autonomy. However, the practice evolved during this period as a more heterogeneous group of actors became involved in water associations and involvement in labour markets and capitalist production deepened. While participation in *mingas* was obligatory in many cases, members could pay a fine or hire a replacement to avoid carrying out the work in others. Hence, the practice was effectively monetised in some water associations, which weakened ties of solidarity and reciprocity and increased internal differentiation.

Negating, creating and contradicting remained the most prominent modes of autonomous praxis during this period (Dinerstein 2015). Water associations used the space created by structural adjustment and neoliberal reform to build autonomous capacity. However, most still negated neoliberal policies and practices. For the most part, this stopped short of outright rejection of the state. Water associations demanded the state redistribute water and invest in infrastructure to overcome water scarcity. These demands went well beyond simply calling for the rolling back of structural adjustment and neoliberal reform. Struggles to create entirely new political structures and relations emerged. Indigenous movements moved most decisively in this direction, including articulating demands to convert Ecuador into a plurinational state. This demand reflected the *doble intencionalidad* of indigenous struggles in Ecuador in the late twentieth century, with indigenous movements simultaneously striving for greater autonomy from and greater integration into the state (Gutiérrez Aguilar 2008). The plurinational state offered a possible route out of this contradictory process by strengthening indigenous control of resources and politics at the local level while increasing indigenous influence over laws, policies, and practices at the national level. Their demand was therefore not to integrate into the liberal state but to transcend it by creating a new state form. Indigenous struggles over water were part of this wider political process which, as will be discussed below, came to a head in the early twenty-first century.

The contradictory mode of autonomous praxis was also evident in water association interactions with non-state actors. Overseas development agencies, multilateral institutions, and non-governmental organisations performed diverse roles in water services and management in the 1980s and 1990s. While some promoted communal over private water service delivery, others moved in the opposite direction or in both directions simultaneously. The World Bank, for example, enthusiastically supported water privatisation, but also promoted the transfer of state-operated irrigation networks to water associations. Both sought to reduce the state's role in water services and management but the latter created space for greater communal control of water. This highlights a degree of convergence between the objectives of development agencies and the desires of rural communities for greater autonomy. Discourses of empowerment and participation, which underpinned much development programming in the 1980s and 1990s, fitted neatly into the idea that rural communities should take greater control of water service delivery at the local level (Bohm et al. 2010; Dinerstein 2015). However, when viewed from a neoliberal perspective autonomous organising was seen as nothing more than a collection of rational economic individuals coming together to tackle market or state failures. The organisational models some development agencies, multilateral institutions, and non-government organisations imposed on water associations reflected this vision, which undermined existing autonomous practices and structures.

The creating and contradicting modes of autonomous praxis involved navigating this complex terrain. Although interacting with state and non-state actors run the risk of incorporation and disappointment, water associations were able to exert considerable influence over the autonomy-building process and challenge water scarcity in extremely difficult conditions. It was during this period that autonomy as a process of "organising hope" was perhaps most evident, with water associations "using knowledge creatively and politically to weave dreams out of misery, against the odds, amidst brutal state violence, endemic poverty, desperate hunger and social devastation" (Dinerstein 2015: 26).

Negotiating Autonomy During the Commodity Boom and State Expansion (2007–2017)

Autonomous struggles in the 1980s and 1990s created excess which overflowed in the opening decades of the twenty-first century as Ecuador entered a new phase in its political history. Rafael Correa, one of several leftist leaders to emerge in Latin America at the turn of the century, oversaw a period of profound social and political change between 2007 and 2017. Soon after taking office, he set about transforming the constitutional, legal, and bureaucratic framework and bolstering the state. His efforts were aided by the commodity boom, which provided his governments with huge oil revenues, and supported increased public spending, especially in the first half of his presidency. Encouraged by elevated commodity prices, his government also actively promoted medium and large-scale mining.

Under Correa, the state increased in size and capacity, and socioeconomic conditions improved after decades of decline during structural adjustment and neoliberal reform. The 2008 constitution, which was enthusiastically supported in a national referendum, incorporated several important dimensions, including defining Ecuador as a plurinational state, proscribing the privatisation of water, and entrusting water service delivery to public and community organisations. These constitutional declarations, while not fully reflecting the demands of water associations and social movements, were the outcome of the historical process outlined above, with rural highland communities taking greater autonomous control of local water supplies, resisting the privatisation of water sources and services, and demanding greater public investment in hydraulic infrastructure.

The constitution declared that a new water law should be introduced within twelve months. Eleven months later the executive branch of the government and the national indigenous movement sent legislative proposals to the National Assembly setting out their competing visions of water services and management. The two proposals varied in multiple ways but one of the key differences was the degree of decision-making power vested in the state. Whereas the government proposal aimed to concentrate decision-making in the hands of politicians and bureaucrats, the indigenous proposal set out to create political institutions that provided genuine space for water associations and social movements to influence the regulation and control of water at the local and national level. The centrepiece of the indigenous movement proposal was a plurinational water council, which was to comprise representatives of the state, social movements and water associations.

The government proposal threatened the autonomy of water associations by seeking to incorporate them into a centralised legal and bureaucratic framework and subjecting them to a raft of new regulations and controls. The threat prompted water associations and indigenous movements to launch a wave of protests, which forced the government to hold a national consultation and reform its proposals. The final version of the law, which was introduced five years after the initial proposals were sent to the National Assembly, was modified. Nevertheless, it remained within the spirit of the original government proposal and provided the legal foundation for a new state-centric water regime. The legislation gave the water associations legal status and promoted their role in water service delivery. It also recognised their administrative and financial autonomy. Yet, it posed serious threats to autonomy, particularly in relation to the standardisation of organisational structures, the bureaucratisation of communal water services, and centralisation of decision making within state agencies. Indigenous movements feared the government would use the law to reduce opposition to its policies and practices and resisted its implementation. Many water associations also challenged the law and pushed for changes that would protect their autonomy.

The end of Rafael Correa's decade in power started a new phase in this struggle as his successor, Lenin Moreno, set about dismantling Correa's political project and steering policies back towards the neoliberal orthodoxy. Following the classic recipe, this included slashing public spending and reducing the size of the state. In efforts to build bridges with water associations and indigenous movements, Moreno recruited prominent indigenous leaders into water bureaucracies, which created opportunities for water associations to protect their autonomy and create alliances with state actors. However, the brutal state repression of protests against the Moreno government in late 2019 saw indigenous bureaucrats resign as indigenous-state relations once again deteriorated. With the prospect of meaningful state support dwindling, water associations started the new decade even more reliant on autonomous practices and relations to deliver water services and overcome water scarcity.

Contradicting was the most prominent mode of autonomous praxis during these final phases, with water associations confronted with a water regime that simultaneously threatened and supported their autonomy (Dinerstein 2015). On the one hand, the new regime threatened incorporation into state structures and logics and created the prospect of reduced space for genuine autonomous organising. On the other hand, it granted water associations a legal status and recognised their administrative and financial autonomy. The regime also introduced new legal instruments that water associations could use to continue their struggle to overcome water scarcity, including recognition of individual and collective rights to water and the principle of food sovereignty. Following decades of declines, public investment in water also increased during Correa's presidency, which created opportunities for some associations to develop hydraulic infrastructure and improve water services.

During this period, the issue of "translation" became particularly pronounced (Dinerstein 2015, 2017, 2020). That is, the challenges and tensions autonomous movements face when their ideas and demands are translated into the grammar of law and policy. Decades of mobilising resulted in a water regime that recognised important elements of autonomous water struggles, but also created a legal and bureaucratic framework that sought to standardise and deradicalise water association practices and activities. As Dinerstein argues, this was a common feature of autonomous struggles in Latin America during this period, as governments "rendered invisible everything that does not fit into the state's existing parameters of legibility" (Dinerstein 2020).

Consolidating rather than creating was perhaps a more accurate term for a second prominent mode of autonomous praxis during this period. Still, plenty of creating and innovating was on show. For example, as collective acts of resistance and defiance, some water associations rewrote their own internal statutes against the new water law, declaring the community assembly rather than the state bureaucracy the maximum water authority. Communal practices and relations were also rejuvenated to foster solidarity and reciprocity and prefigure alternative futures beyond the capitalist vision promoted by Correa. *Mingas* were central to this process, and provided opportunities to build communal ties, transfer knowledge across generations, and create new collective memories. Water associations that largely or totally comprised indigenous peoples were the most active in this domain, blending traditional practices and beliefs with new ideas and visions to challenge the state. Increasing communal control of *páramos*, high-altitude grassland and vital sources of water in Ecuador, was one important dimension of this autonomous organising, with some associations collectively purchasing tracts of the land to increase control of local water sources and protect water supplies for future generations.

Negating was also on show as water associations and indigenous movements challenged Correa's centralising plans and repressive discourses and practices. Yet, as during earlier phases of autonomous organising, this was not a uniform rejection of state power, and numerous water associations welcomed the prospect of greater state support. Negation was strongest among indigenous water associations. Indigenous leaders, who were singled out for persecution by Correa, challenged his government's plans most vociferously, as they saw it as part of a colonising project which trampled over the plurinational state. His government's refusal to place a plurinational water council at the centre of the new water regime illustrated its lack of interest in transforming Ecuador into a plurinational state and providing indigenous peoples with genuine space to transform the state.

The prominent role indigenous leaders and movements played in challenging the state during this and earlier phases provides some support for Dinerstein's demarcation of indigenous and non-indigenous autonomy (2015: 48–56). Drawing on Raquel Gutiérrez Aguilar (2008), she identifies *doble intencionalidad* as a core element of indigenous autonomy and, as previously noted, this was a fundamental feature of highland indigenous struggles from the 1960s, taking its fullest expression in demands for a plurinational state from the 1990s. Indigenous movements therefore generally pushed to transform rather than reject the capitalist state. The fact autonomy often took more radical forms in water associations that comprised entirely or largely indigenous peoples also lends additional support to Dinerstein's claim that indigenous autonomous struggles take distinct forms. The relations, practices and ideas indigenous associations and movements generated most clearly reflected the utopian and prefigurative dimensions of autonomy and offered the greatest potential of challenging "the universalising force of capital" (Dinerstein 2015: 212). Yet, water association autonomy also exposes some limitations of her framework. Numerous highland water associations comprise both indigenous and non-indigenous members, indicating the challenge of separating out some indigenous and non-indigenous autonomous struggles. Moreover, the highland indigenous population is extremely diverse, with distinct local customs and practices and social and class differences complicating broad generalisations. Identifying a singular, distinct form of indigenous autonomy runs the risk of overlooking these differences and homogenising indigenous struggles.

Conclusions and Lessons

Six decades of autonomous organising have enabled rural communities in Highland Ecuador to challenge water scarcity, take greater control of local water supplies, and create new social and political relations. During this period, community water associations have modified their practices and tactics according to changing conditions, revealing autonomy as a historical process that is perpetually in motion. Drawing on Dinerstein (2015, 2017, 2019, 2020), I have shown that this process has involved distinct modes of autonomous praxis—negating, creating, contradicting, exceeding—and has been a source of hope. What insight does this analysis offer into autonomous struggles elsewhere in Latin America? Here, I will highlight three points.

First, water associations have developed autonomy while working with and against the state and law, highlighting the importance of conceptualising autonomy as a relational, contradictory process rather than a binary outcome (e.g. autonomy v incorporation). Interacting with the state involved the risk of incorporation and subordination, but also created opportunities to develop autonomous capacity and challenge water scarcity. Thus, the case illustrates the importance of not viewing capitalist states as static, singular entities that uniformly challenge autonomy.

Second, the internal composition of water associations changed over time as social and class differentiation increased within rural communities and a broader spectrum of rural society became involved in community water services and management. These changes brought opportunities and challenges. In some cases, autonomy increased as water associations crossed class and ethnic boundaries and united a diverse range of actors in a common struggle to challenge water scarcity. In others, internal tensions escalated as class and ethnic differences undermined unity and solidarity and reduced the radical potential of autonomy. The case therefore highlights the importance of exploring the internal dynamics of autonomous organisations and the ways in which internal differences are negotiated.

Third, collective labour practices have been fundamental to the construction and maintenance of water association autonomy. *Mingas* have bound members of water associations together, strengthened ties of solidarity and reciprocity, and provided the basis for collective ownership of "hydraulic property" (Boelens and Vos 2014). This, in turn, has reconfigured political relations as water associations have challenged the state's authority to regulate and control the water systems that they have built and developed. Not only has this process rewired relationships with the state, but also with nature. Without romanticising a practice that involves considerable toil and sacrifice, *mingas* create opportunities for water association members to interact with water and exchange knowledge and stories about it. This shows that autonomous forms of local water service delivery offer the prospect of a more intimate and meaningful connection to water than private or public systems. Hence, hope also lies in the potential of autonomous struggles to reconfigure human-nature relations, and contribute to efforts to overcome the ecological crises inherent to capitalism.

References

Alderman, J. (2018). Indigenous Autonomy and the Legacy of Neoliberal Decentralization in Plurinational Bolivia. *Latin American and Caribbean Ethnic Studies, 13*(1), 1–21.

Alderman, J. (2019). Whose Autonomy Is It anyway? Tensions Between Class and Ethnicity in the Formation of Collective Citizenship and Self-determination in Plurinational Bolivia. In I. Goudsmit, K. McClean, & W. Moore (Eds.), *Papers from Conference Arranged by the Anglo-Bolivian Society and the Institute of Latin American Studies* (pp. 49–59). London: University of London.

Andolina, R. (2012). The Values of Water: Development Cultures and Indigenous Cultures in Highland Ecuador. *Latin American Research Review, 47*(2), 3–26.

Armijos, M. T. (2012). *Negotiating Citizenship Through Communal Water Management in Highland Ecuador*. PhD Thesis, University of Sussex.

Armijos, M. T. (2013). They Cannot Come and Impose on Us: Indigenous Autonomy and Resource Control Through Collective Water Management in Highland Ecuador. *Radical History Review, 116*, 86–103.

Bakker, K. (2010). *Privatizing Water: Governance Failure and the World's Urban Water Crisis*. Ithaca/London: Cornell University Pres.

Boelens, R., & Doornbos, B. (2001). The Battlefield of Water Rights. Rule-making Amidst Conflicting Normative Frameworks in the Ecuadorian Highlands. *Human Organization, 60*(4), 343–355.

Boelens, R., & Vos, J. (2014). Legal Pluralism, Hydraulic Property Creation and Sustainability: The Materialized Nature of Water Rights in User-Managed Systems. *Current Opinion in Environmental Sustainability, 11*, 55–62.

Bohm, S., Dinerstein, A. C., & Spicer, A. (2010). (Im)possibilities of Autonomy: Social Movements in and Beyond Capital, the State and Development. *Social Movement Studies, 9*(1), 17–32.

Dinerstein, A. C. (Ed.). (2013). *Movimientos Sociales y Autonomía Colectiva: La Política de la Esperanza en América Latina*. Buenos Aires: Capital Intelectual.

Dinerstein, A. C. (2015). *The Politics of Autonomy in Latin America: The Art of Organizing Hope*. Basingstoke: Palgrave Macmillan.

Dinerstein, A. C. (2017). Co-construction or Prefiguration? The Problem of the 'Translation' of Social and Solidarity Economy Practices into Policy. In P. North & M. Scott Cato (Eds.), *Towards Just and Sustainable Economies*. Bristol: Policy Press-Bristol University Press. Accepted version pre-publication.

Dinerstein, A. C. (2019). A Critical Theory of Hope: Critical Affirmations, Beyond Fear. In A. C. Dinerstein, V. A. Garcia, E. González, & J. Holloway (Eds.), *Open Marxism 4. Against a Closing World* (pp. 35–46). London: Pluto Press.

Dinerstein, A. C. (2020, May 20). Why Does the Political Left Fail Grassroots Movements? *Open Democracy*.

Esteva, G. (2015). The Hour of Autonomy. *Latin American and Caribbean Ethnic Studies, 10*(1), 134–145.

FRH. (2011). *Transferencia de Competencias de Riego para el Desarrollo*. Quito: Foro de los Recursos Hídricos.

FRH. (2013). *La Gestión Comunitaria del Agua para el Consumo Humano y el Saneamiento en el Ecuador: Diagnóstico y Propuestas*. Quito: Foro de los Recursos Hídricos.

González, M. (2015). Indigenous Territorial Autonomy in Latin America: An Overview. *Latin American and Caribbean Ethnic Studies, 10*(1), 10–36.
Goodwin, G. (2016). Review of *The Politics of Autonomy in Latin America: The Art of Organizing Hope* by Dinerstein, A.C. (2015). *Radical Americas, 1*(1), 85–88.
Goodwin, G. (2017). The Quest to Bring Land under Social and Political Control: Land Reform Struggles of the Past and Present in Ecuador. *Journal of Agrarian Change, 17*(3), 571–593.
Goodwin, G. (2019). The Problem and Promise of Coproduction: Politics, History, and Autonomy. *World Development, 122*, 501–513.
Goodwin, G. (2020). Fictitious Commodification and Agrarian Change: Indigenous Peoples and Land Markets in Highland Ecuador. *Journal of Agrarian Change* (forthcoming).
Gutiérrez Aguilar, R. (2008). *Los Ritmos del Pachakuti Movilización y Levantamiento Popular-Indígena en Bolivia (2000–2005).* Buenos Aires: Tinta Limón.
Halvorsen, S., Mançano Fernandes, B., & Valeria Torres, F. (2019). Mobilizing Territory: Socioterritorial Movements in Comparative Perspective. *Annals of the American Association of Geographers, 109*(5), 1454–1470.
Holloway, J. (2002/2010). *Change the World Without Taking Power: The Meaning of Revolution Today*. London/New York: Pluto Press.
Holloway, J. (2010). *Crack Capitalism*. London/New York: Pluto Press.
Hoogesteger, J. (2013). Trans-forming Social Capital around Water: Water User Organizations, Water Rights, and Nongovernmental Organizations in Cangahua, the Ecuadorian Andes. *Society and Natural Resources, 26*(1), 60–74.
Hoogesteger, J. (2014). *Los Nuevos Sujetos del Agua: Organización Social y la Democratización del Agua en los Andes*. Quito: Abya-Yala.
Hoogesteger, J. (2015). Normative Structures, Collaboration and Conflict in Irrigation; a Case Study of the Pillaro North Canal Irrigation System, Ecuadorian Highlands. *International Journal of the Commons, 9*(1), 398–415.
Hoogesteger, J. (2016). NGOs and the Democratization of Ecuadorian Water Governance: Insights from the Multi-Stakeholder Platform el Foro de los Recursos Hídricos. *Voluntas: International Journal of Voluntary and Nonprofit Organizations, 27*(1), 166–186.

Ioris, A. A. R. (2012). The Geography of Multiple Scarcities: Urban Development and Water Problems in Lima, Peru. *Geoforum, 43*, 612–622.
Mauss, M. (2002 [1925]). *The Gift: The Form and Reason for Exchange in Archaic Societies*. London: Routledge.
Merino, R. (2020). Rethinking Indigenous Politics: The Unnoticed Struggle for Self-Determination in Peru. *Bulletin of Latin American Research, 39*(4), 513–528.
Pacari, N. (1998). Ecuadorian Water Legislation Analyzed from the Indigenous-Peasant Point of View. In R. Boelens & G. Davila (Eds.), *Searching for Equity: Conceptions of Justice and Equity in Peasant Irrigation* (pp. 281–287). Amsterdam: Van Corcum.
Penna, M. A. (2020). Movimento das Comunidades Populares: A Brazilian Uchronic Utopia. *Ephemera, 20*(1), 51–89.
Swyngedouw, E. (2007). Water, Money and Power. *Socialist Register, 43*, 195–212.
Zibechi, R. (2007). *Autonomías y Emancipaciones América Latina en Movimiento*. Lima: Universidad Nacional Mayor de San Marcos.

9

'The Best-Laid Schemes o' Mice an' Men': Transformative Agency Towards Ecosocialism

Elisabeth Cremona
and Antonio Augusto Rossotto Ioris

The General Crisis of the Twenty-First Century?

The history of great civilisations is often marred by a destabilising and sometimes destructive crisis; recall, for instance, the crisis of the third century, which contributed to bringing the Roman Empire to its knees or the general crises that took hold in Europe in the seventeenth and eighteenth centuries, some of which were highly associated with the transition to urban and industrial capitalist development. Each of these periods was characterised by a wide host of interconnected problems, ranging from war, disease and environmental pressures to economic and political destabilisation. As these crises developed, they had significant transformative impacts on the affected societies, markedly changing the very

E. Cremona
University of Malta, Msida, Malta

A. A. R. Ioris (✉)
School of Geography and Planning, Cardiff University, Cardiff, UK
e-mail: IorisA@cardiff.ac.uk

A. A. R. Ioris (ed.), *Environment and Development*,
https://doi.org/10.1007/978-3-030-55416-3_9

foundation of these civilisations. For instance, Eric Hobsbawm observed that the crisis of the seventeenth century constituted the final break from the feudal system, which had been in place for centuries, to a new capitalist economy (Hobsbawm 1965). Likewise, mounting tensions created by nationalist and imperialist disputes since the middle of the nineteenth century led to two world wars and the dramatic reconfiguration of global economy and society after 1945. As the twenty-first century unfolds, an interested bystander may become increasingly concerned when faced with acute environmental pressures (climate change, biodiversity loss, etc.), financial crises (such as the global financial crisis of 2007–2008, still largely unresolved and re-emerged in 2020, following the coronavirus pandemic), worrying trends of widening local and global inequalities, rapidly diminishing trust in politicians and institutions (e.g. Brexit and the 2016 results of the US electoral system) and the list goes on. Will the situation be eventually stabilised or are we witnessing the development of the general crisis of the twenty-first century, which will impact and transform today's global civilisation?

Mainstream responses to these problems and to the environmental contradictions of mainstream development are typically devoted to their amelioration in an atomistic isolation from other issues, normally informed by discourses and the rationalisation of socio-economic and ecological problems as circumstantial failures of the existing politico-economic order (e.g. the concept of sustainable development as an attempt to build consensus around technological and managerial adjustments). However, a rejection of these responses appears to have taken hold in a number of countries and is spreading, most conspicuously on the political level. This is evident, for instance, in the rise of (apparent) anti-establishment (e.g. Donald Trump's victory in the US 2016 presidential election) and right-wing extremist political parties in Europe (e.g. Alternative für Deutschland, the third largest party in the Bundestag after the 2017 election in Germany), their popular support fuelled by uncertainty in the world economy and the people's declining confidence in the power of current political parties to adequately tackle socio-economic and environmental problems. In addition, there is a growing awareness among academics and social movements that current mainstream approaches to resolve the ongoing economic and environmental crises (typically along the lines of ecological modernisation and market-based environmental conservation) can only provide superficial and short-term solutions as they fail to address the wider historical, political and socio-economic roots of the problems (such as those related to patters of production and consumption, property regimes, widespread alienation and multi-scale power disputes). As a result, there has been a call from critical circles for *radical* transformation of the prevailing capitalist development system and the overcoming of its perverse repercussions at the global, national, local and personal level, and a shift towards a more radical critique may even be welcomed in the current political climate (Arnold 1990; Escobar 2007; Pepper 1993; Sachs 2010; Schweickart 2009, 2010).

The fall of the Soviet Empire led many to proclaim that this demonstrated the undeniable superiority of capitalism and the unfeasibility of other alternatives. However, at the same time, the remaining uneasiness about pending socio-ecological and existential problems have recently begun to chip away at the Western world's shattered confidence in this totalising narrative, triggering a resurgence of radical critiques and keeping the debate about the future wide open (Ingham 2008; Kaletsky 2011; Prisecaru 2012). While many maintain that the current crises are simply manageable instabilities and not terminal contradictions (Kaletsky 2011), there are a number of scholars and activists who argue that the capitalist system is primed for a radical shift, somehow analogous to the transition from feudalist to capitalist Europe in the sixteenth century (Bellamy Foster 2013; Chase-Dunn 2014; Moore 2014a, b). However, in 'a world gone cosmopolitan by default', it is difficult to locate or prefigure the 'exit' to capitalism (Harvey 2014: ix) in concrete political, economic and social practices.

Our departing point in this chapter is to claim that it is not only important but necessary to explore the meaning and routes to a post-capitalist development pathway, one which may emerge should the General Crisis of the twenty-first century truly become a reality. An anticipatory or visionary approach is also relevant to help to understand the shortcomings of present-day society. For these reasons, it has become pertinent to draw our attention to those radical scholars and social movements, which challenge capitalist relations of production and reproduction, and advocate a substantive and meaningful reconfiguration of the world system and a different interface between environment and development. The consideration of an ecosocialist alternative is an important element of those debates. The ultimate aim of this chapter is to contribute to the discussion on transformative agency in ecosocialism, and thus, to ecosocialism's philosophy of praxis (Daldal 2014). Should the crisis materialise, to whom should our global society look if it wishes to see the materialisation of an ecosocialist system?

The Basis of Ecosocialism and the Focus of This Study

Ecosocialism constitutes one such radical critique of capitalism. Founded on a synthesis of red and green theories, it asserts that the current economic and environmental crises constitute one unified, not separate, 'epochal crisis' (Bellamy Foster 2013), which has emerged as a result of the capitalist mechanisms of production and reproduction. Ecosocialist authors indict capitalism's 'grow or die' regime of accelerating accumulation for ecological destruction, the creation of exploitative working conditions, poverty and a highly individualistic society, which is addicted to commodity consumption (Ioris 2018; Kovel 2008b; Löwy 2015). It has recently been gaining prominence within development discourses, and social movements embodying some of the principles of ecosocialism are beginning to emerge worldwide (e.g. La Vía Campesina) and, despite the 'demise' of the Left in the West, many Latin American countries have had experiences with Left or Centre-Left governments, some of which have incorporated ecosocialist goals into their national constitutions (e.g. Sumak Kawsay in Ecuador) with varying levels of success and political acceptability (Adams 2009; Betto and Löwy 2010; Burkett 2006; Huan 2010; Löwy 2015; Löwy 2014; Radcliffe 2012). However, while it can be considered a persuasive political discourse for transformation, ecosocialism still faces significant epistemological, analytical and operative challenges (Huan 2010), which reflect the wider questions about the meaning and the contribution of radical political thinking in the contemporary world.

One particular gap in the work of many ecosocialist writers is the limited attention given to social actors, exactly those agents with the transformative capacity to realise the ecosocialist vision. This lacuna in the ecosocialist literature was rightly observed by Aşıcı (2012) and Aşıcı and Bünül (2011) in their review of the available literature and confirmed by the authors' own research. This oversight is disappointing given that an analysis of transformative actors constitutes an undeniably vital component of a successful radical social critique, from both a theoretical and practical standpoint (Arnold 1990; Castree 2009). This is highlighted, for instance, in the definition of transformability of socio-ecological systems (SES) provided by Walker et al. (2004) as the capacity of social actors to 'create a fundamentally new system when ecological, economic, or social (including political) conditions make the existing system untenable' (ibid.: 7). Such thinking arguably mirrors the conditions created by capitalism at this current juncture and emphasises that understanding agency is crucial to understanding large-scale transformation (Westley et al. 2013). The ecosocialist writers White et al. (2017) also hinted at the need to explore the transformative capacities of different actors, believing that the 'second stage' of ecosocialism constitutes an attempt to move beyond a critique of the status quo and towards an analysis of the conditions that will enable such a transition, including the empowerment of diverse agents.

In light of this, our research seeks to specifically contribute to the neglected topic of transformative agents in ecosocialism, from both a theoretical and empirically based standpoint. This chapter shall examine the transformative potential of key actors that are somehow associated with the ecosocialist vision, based on their outlined practices, their ability and commitment to form part of the transition to a new world order, and their capacity (power) to successfully bring about such changes. This was investigated through (1) an analysis of ecosocialist literature, including two emblematic ecosocialist manifestos (the Belém Declaration and the Réseau Ecosocialist Manifesto) and Ecuador's national development plan, selected as documents which mirror ecosocialist goals and are actively employed and followed by a group of people, and (2) a power analysis of the controversial Belo Monte Dam project in Brazil, to investigate the power of the identified transformative agents within a practical and 'real-world' context. The latter was selected as a case study based on Kovel's (2014) assertion that 'agents of transformation [will] emerge interstitially' in 'zones of emergence that appear as contradictions mature' (ibid.: 14), and the prevailing view that Southern communities will be those most immediately and effectively impacted by environmental issues, and thus see the creation of such zones (White et al. 2017).

Transformative Agents Within Ecosocialist Discourses

Within the limited ecosocialist literature on the crucial subject of the identity, practices and imaginaries of social actors, there appears to be two prevailing perspectives as to the agents of transformation who could move society into a post-capitalist, ecosocialist future reality. One view prioritises the controversies in more affluent nations and social contexts, particularly those in the Global North, while the other attributes the responsibility for the envisioned transformation to those people less estranged from nature, typically rural and indigenous groups, predominantly those in the Global South. In order to more closely examine the issue of actors within ecosocialism, the conceptualisation of transformative agency and its function within SES from Westley et al. (2013) is adopted. This approach is considered to be complementary both to the objective of this study and to the theories of ecosocialism, given that it focuses on the role of social innovation in transforming problem domains in complex systems with a view to system change (Olsson 2017).

Agency is thus examined on a broad system scale in what Westley et al. (2013) term the 'problem domain', constituted of actors, organisations and institutions concerned with or affected by a complex problem; within the problem domain, strategic action or agency is required to move the process of transformation of the SES forward. Due to the complexity of the problem domain, agency is associated with the action of a myriad of actors, classes and class segments that push the system through different stages of transformation, an effort termed 'institutional entrepreneurship'. This conceptualisation of agency is reflected in the writing of ecosocialist writers, such as Kovel (2013) who asserts that 'there is no privileged agent of ecosocialist transforming' (ibid.: 14), but the post-capitalist transformation will be realised through an abundance of individual paths, which will eventually converge, having taken place where the contradictions of capitalism manifest themselves as opportunities (Holloway 2010; Kovel 2008a, b). The theory of institutional entrepreneurs in SES allows for such a pluralistic view of transformative agents, who are facing complex problems in SES, and an understanding of their different and disaggregated activities according to their temporally and spatially specific opportunities.

Our examination employs the theory of transformative agency of institutional entrepreneurs, adopting the heuristic framework employed by Westley et al. (2013), in order to shed light on the identity of the political subjects of the ecosocialist transformation. The approach is based on a combination of Holling's 1986 adaptive cycle and Dorado's 2005 model of opportunity context, which presents the context of SES as dynamic, with different opportunities for transformative agents emerging at different intervals (Fig. 9.1). Here, opportunity context indicates whether the current situation will permit actors to identify, introduce and institutionalise new ideas and reforms. This heuristic model, combined with the theory of the institutional entrepreneur, describes how actors emerge from within certain institutions to challenge and transform the status quo, and charts the wide diversity of skills and strategies required by these agents for a successful introduction and subsequent institutionalisation of a new or modified system (in this case, ecosocialist reforms). The three different opportunity contexts (opaque, hazy and transparent) are characterised by different configurations of two essential elements: (1) the diversity of organisations forms and (2) the degree of institutionalisation. The entrepreneurs, their agency and their roles are different according to the stages of the cycle and the specificities of the opportunity context therein. Olsson (2017) asserts that this is important where large-scale transformation within complex systems is required, and thus change must occur at multiple levels, scale and sectors. Indeed, it is highly likely that any post-capitalist transformation will necessarily take a non-linear form, where the cyclical movement through opportunity contexts occurs at different spatial and temporal scales.

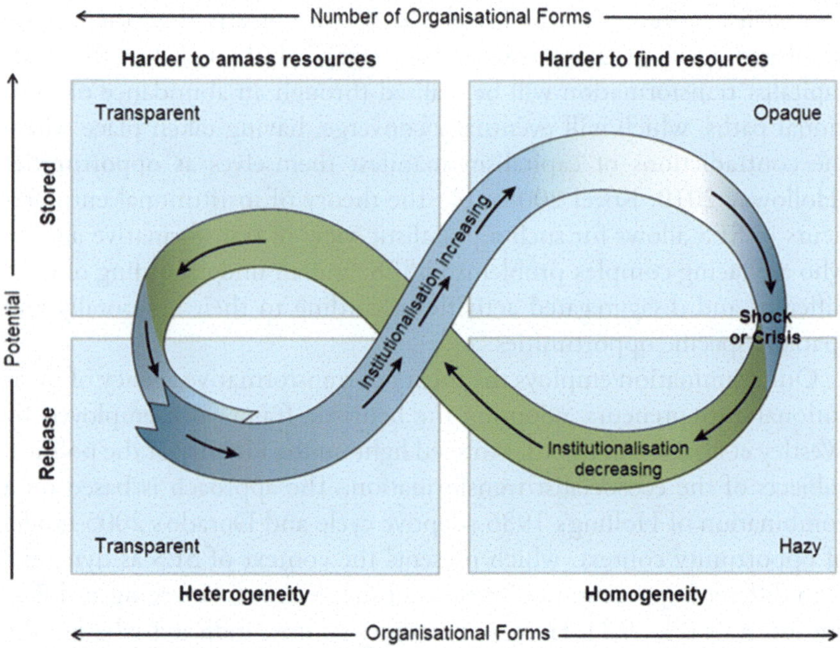

Fig. 9.1 Theory of transformative agency in socio-ecological systems. (Adapted from Westley et al. 2013)

In brief, the 'back loop' of the heuristic model constitutes the turbulent change portion of the cycle in which there exist multiple organisations and institutionalisation is low, thus allowing for novel ideas to emerge and trigger transformation of the institutional system. The new systemic configuration thus becomes increasingly stabilised, and is institutionalised as the new system of the 'front loop'; here, the degree of institutionalisation is high, and few organisations are dominant. Although Kovel believes that no specific blueprint of a transition to an ecosocialist world can be laid out in advance (Kovel 2008a; Kovel 2013), this heuristic framework adopted from the study by Westley et al. (2013) can be employed to better visualise and discuss the diversity of actors who are required to successfully bring about changes in SES and the various skills, alliances and strategies needed according to different opportunity contexts, the latter being spatially and temporally specific. We are interested in the agency that enables movement from the 'front loop'—the current problem domain—to the 'back loop' where multiple agents are mobilised, new ideas emerge and are eventually institutionalised into the new 'front loop', a reformed (ecosocialist) system. Given that the transition process to ecosocialism is theorised to take place gradually through the action of a vast array of actors, a discussion on the spectrum of transformative agents, their different capabilities and resources, is important in order to successfully take advantage of opportunities as they emerge, borne out of the multiple crises of capitalism. The agents and their actions outlined in the three ecosocialist documents listed in the previous section are examined using this heuristic framework, allowing for a discussion on whether the agents and their actions stipulated cover the full transition from the current capitalist 'front loop' to an ecosocialist future 'front loop'.

Taking the 'opaque' context as a starting point, this context is characterised by a few dominant organisations, and institutions and beliefs are well established. In order to challenge the status quo (which is identified as the ultimate cause of the complex problems society is facing), the primary strategies employed at this stage are those that introduce disturbances into the well-established system. Indigenous people and social groups, particularly those in South American countries where their power is more intact, are attributed the highest attention in ecosocialist literature and in the two ecosocialist manifestos examined, with their role primarily concerning introducing disturbances into the system. This performs the dual function of challenging established institutions and raising human consciousness. The Belém Declaration, adopted by the Ecosocialist International Network in 2009, was the result of a global mobilisation and focuses almost solely on such activities and those agents who are assigned the task of creating disturbances, largely through social activism and protests. This is an undeniably important part of the transformation process, but the document appears to operate at a high level of political activism and gives much less consideration for other strategies and other agents that will be required past the point of creating disturbances; this, therefore, raises questions as to how the ambitious goals outlined in the Belém Declaration will be met in the complex political landscape and with the high level of political demobilisation in the world today. This is a trend that tends to pervade ecosocialist literature—a gap between the agent and their actions and the theorised ecosocialist reforms that will take place.

Our call for greater attention to the complexity of social protagonism seems to be greatly justified by the need to connect local and sectoral problems with global and generalised tensions. Coordinated political disturbances, combined with internal system shocks, such as crisis internal or external to the system, may subsequently cause the opportunity context to move from an 'opaque' to 'hazy'; a change that is required, in this case, to allow for the system of capitalism to be directly and progressively challenged. In a 'hazy' context, old organisations have lost their dominance, causing institutions and beliefs to become open to interpretation and change. Recent economic and political crises across the globe combined with the growing environmental crisis have sent shockwaves through the system, but has this amounted to the creation of an 'epochal' crisis, as some have argued (Bellamy Foster 2013; Moore 2014a)? Have these shocks caused people to question the system of capitalism to the extent that its very foundations are being challenged? It can be argued that, previously, the capitalist system did not generate contradictions powerful enough to motivate the revolution predicted by Marxist scholars, but that the incorporation of the earth's ecological limits will create a demand for immediate and more radical action (Whiteside 1997). Kovel (2008a: 242), while echoing this sentiment, believes that 'none of these conditions is close to being met at present'. Indeed, an awareness of the urgency of environmental problems has existed for a number of decades, but meaningful and sustained radical action has yet to be observed; it is because of this that social actors need to be more carefully understood.

The continuing and self-reinforcing hegemony of the modern global capitalist system continues to consistently inhibit the prospects of a transition from an 'opaque' to a 'hazy' context, as it subdues challenges and threats to the system, thus restricting any movement towards a post-capitalist future. Indeed, Mankoff (1970) argues that the conditions for triggering radical social change are likely to be constantly present under capitalism, but the existing economic order is maintained due to the non-elites' acceptance of capitalism's legitimacy. Other authors also point to this false consciousness that pervades today's society, which prevents it from challenging capitalism's ideological hegemony (Kovel 2013; Pepper

1993). This evokes a more classical Gramscian view of ideological hegemony exercised not by states specifically, but through social groups and certain dominating classes operating through the state (Robinson 2005). Perhaps for this reason, Dordoy and Mellor (2000) and Johns et al. (2003) argue that capitalism can only be overcome once the crisis truly affects those affluent groups that benefit from the exploitation of people and the planet, implying that the transnational capitalist bourgeois is a pivotal change agent in the ecosocialist transition. It is thus possible that the crucial movement to a 'hazy' context, in which the foundations of capitalism can be challenged and new ideas allowed to emerge, requires action from the affluent, capitalist classes as well as social groups and indigenous communities.

In any case, anti-capitalist sentiments and thinking appear to be gaining ground in different countries to significantly different degrees, and in academia and social movements, particularly those in Latin America; therefore, it is possible that we find ourselves now in a system which is *beginning* to move into the 'back loop' (Schumpeter 2015). At this point, an essential strategy in the 'hazy' and subsequent 'transparent' context, in which institutionalised systems and beliefs are openly being challenged, is allowing new ideas and understandings to emerge (Westley et al. 2013). Here, indigenous groups and grassroots social movements are implied to play another key role, that of shaping the design of the post-capitalist future system (Löwy 2015; Kovel 2008b). These peoples' connectedness to nature assumedly forms values that reflect respect for the environment and the maintenance of sustainable practices; thus, they become a source of creativity and experimentation for the new system being designed. This notion recalls the work by Max-Neef (1992) on barefoot economics, which proposes to engage with those people who are on the periphery of, or excluded from, the capitalist economic system, in order to construct a more humble economic system that builds on their ecocentric values (Imas et al. 2012). Therefore, indigenous groups and social movements, in various associations with urban and rural working classes and other critical activisms, constitute essential transformative agents which bridge the gap between local action, wider engagement and actualised social reform, as this requires making the step from critique to construction, and from opposition to proposition (Pieterse 1997).

Supporting such activities by social groups and indigenous peoples, the Réseau Ecosocialist Manifesto, adopted by a Quebec-based ecosocialist movement, also identifies left-wing political parties as key agents, particularly throughout the 'hazy' and 'transparent' contexts, as their outlined activities include creating spaces for involvement, and accumulating resources and ideas. These initiatives are vital to the convergence and sense-making activities, which need to occur through the 'hazy-transparent', context to allow a transition to the 'front loop'. While this could possibly be achieved by indigenous groups, social movements and others acting together, it is arguably more likely to be successful when there is a designated party or organisation to lead these activities, particularly those which already have the relevant resources and experience at their disposal, such as political parties. However, caution is required here. In the so-called Western world, there has been a notable decrease in the popular support for left-wing political parties, with some political scientists using the term 'pasokification' to describe this continent-wide shift in Europe; the term was coined following the crashing support for the Greek left-wing political party Pasok from a steady 45% to 4% in 2015 (The Economist 2016).[1] Additionally, the political Left in many Western countries appears to have lost much of its identity, discourse and legitimacy, and is therefore less likely to join those groups that are directly challenging the institution of capitalism. Indeed, corporatist theory postulates a social domination model where those organisations opposed to capitalism, like left political parties and unions, are progressively integrated into an institutional structure of social control, ultimately leading to class collaboration as opposed to the more conventional class struggle predicted by Marxists (Lichten 1984; Stephens 1986). In light of this, it is perhaps unlikely that the ecosocialist transition will begin in the Global North. The situation is different in Latin America, where social programmes introduced by left-wing political parties retain popular support, although in more recent times these parties appear to be in state of flux due to the mounting contradictions of formal democracy and

[1] Although the reasons behind Pasok's downfall are considerably more complex than simple declining faith in the political left in Greece.

conventional policymaking (Bremmer 2017; The Conversation 2017; Ioris 2017, 2019; Kozloff 2016).

While the Réseau Ecosocialist Manifesto identified left-wing political parties as important transformative agents, particularly due to their available resources and leading capabilities, it rejects the potential of states and corporations to act as agents in the ecosocialist movement. In contrast to this, Ecuador's National Plan for Good Living (2013–2017) provides an interesting contribution to ecosocialist literature, as it identifies the state as the main transformative agent; although this is unsurprising, given that it is a plan issued by the Ecuadorian government (likely to be significantly amended since the recent presidential election). It is also the first ecosocialist document encountered by the authors that gives significant consideration to the important step of institutionalisation of a new systemic design—that of an ecosocialist system; this process is essential in the 'transparent' context, in order for the SES to transition to an 'opaque' context in which the new post-capitalist model has become well established. Ecuador's National Plan stipulates that the state is primarily responsible for institutionalising the changes set out in its Plan, many of which are based on the principles of *Sumak Kawsay*, the concept of a society where people live in harmony with each other and with their environment, and one which heavily mirrors the goals of ecosocialism. Indeed, Wall (2010: 138), one of the few ecosocialist scholars who acknowledges the importance of the state as a transformative agent, emphasises that, although 'an obvious starting-point is to stress the vital importance of working-class and indigenous agency', state-level power is ultimately imperative 'in order to transform the structure – legal, social, economic and even physical – necessary for the transition to an ecosocialist society' (ibid.: 143).

In addition to this, through the implementation of Ecuador's National Plan, the state aims to create an active citizenry which is critical and conscious, characteristics of key transformative agents; thus, the state is said to be carrying out its 'educative' function described by Gramsci (Daldal 2014), despite all the intrinsic difficulties associated with the political pressures that involve and pervade the state apparatus (Ioris 2014). In the Ecuadorian plan, priority is being given to investing in generating capacities and also providing platforms through which citizens can participate in and influence decision-making. In this manner, the state is not only a transformative agent but also acts in an intermediary manner as 'the main agent for collective action', that is, conducting convening and sense-making activities, crucial in the transition from a 'hazy' to a 'transparent' context (Senplades 2013: 47). Furthermore, the aim to create an active and critical society enshrined in the plan implies that the states seeks to facilitate an almost cyclical or iterative process, whereby social actors are made capable of challenging, as well as transforming, the established institution. A society with such capabilities is indisputably important, particularly given the fact that an 'ideal' system can never truly be achieved, and socio-economic institutions will always require refinement and adjustments.

Despite the potential transformative ability of the state, it tends to be heavily neglected or even rejected in ecosocialist writings, likely a result of the influence of the red-green theory foundation of ecosocialism, which views the state as primarily upholding capitalist interests (Burkett 2006; Pepper 1993). In contrast, Wall (2010: 141) emphasises that 'to get to this ecosocialist society, the state must be transformed; it cannot simply be ignored'. Indeed, the significance of this statement is enhanced when recalling that the hegemony of capitalism is maintained by the dialectical interplay of coercion and consent, and that the power of legitimate coercion is a unique attribute possessed by the state (Upchurch and Mathers 2012). And, if we apply the orthodox Marxist view of the state acting as a tool of the capitalist bourgeois to maintain capitalism's hegemony, the transformation of the state into one with an ecosocialist ontology would theoretically undermine this relationship between the former and the state. Indeed, to disregard the state would be to overlook the distinct opportunities it presents for the ecosocialist movement. The contemporary state, informed by the Hegelian political theories, adopted a 'greener' appearance, but at the cost of avoiding any fundamental change of core state commitments; it means that calls for governance and democracy that avoid tackling the distortions and ideological biases that underpin current-day environmental responses tend to reinforce and, consequently, perpetuate socio-ecological contradictions and environmental injustices (Ioris 2014). Indigenous groups and social movements are undeniably key transformative agents, most prominently in the prefiguration stage of ecosocialism; however, consideration of transformative agency beyond this stage, to that in which the radical changes demanded by ecosocialism are refined and institutionalised, is essential. In light of this, the state should perhaps be considered as another potential key actor in ecosocialism, one which plays a dual role in any large-scale transitional process, both enabling and encouraging social reform, and also institutionalising it. It is also possible that transformation of the state's ontology would significantly impinge on the ideological hegemony of capitalism. The fundamental antinomies of flexibility and legitimacy of contemporary environmental policies need to be situated in this wider politico-ecological debate about the substantial transformation of the actually existing state apparatus. The key ontological and political question is less how the state deals with the environmental policies and strategies per se, but what its ultimate commitments are and how it works to reinforce or eliminate processes of exclusion and exploitation (Ioris 2015).

In sum, the framework set out in the study by Westley et al. (2013) provides a solid basis for visualising the lengthy process of ecosocialist design, reform and institutionalisation of a post-capitalist system and the myriad of agents creating such a transformation. It allowed for gaps in certain ecosocialist documents to be identified in terms of the range of actions that will be required to bring about lasting ecosocialist reform and the transformative agents who will carry out such initiatives. For instance, the Belém Declaration only focused on activities that need to take place in an opaque context and neither the agents outlined nor their actions seek to fulfil different parts of the cycle. Moreover, there was often a notable gap between the designated actors and their actions, and the outlined ecosocialist design. The Belém Declaration focuses solely on indigenous and social groups that are charged with creating disturbances, while its proposed goals include changes such as the 'provision of an extensive free public transportation system' or 'drastic and enforceable reduction in the emission of greenhouse gases' (Angus et al. 2009: 3). Not only is it questionable how such goals can be achieved by the designated actors, but there is no consideration for the various activities that are required in order to realise these goals. In contrast, the agents outlined in the Réseau Ecosocialist Manifesto account more fully for the different activities, which need to take place throughout the cycle from the current 'front-loop' through to a new one, pertaining to those activities carried out by social and indigenous groups as well as left-wing political parties. The state, however, is viewed with mistrust and is rejected as an agent. On the other hand, the National Development Plan of Ecuador places the responsibility of instituting change on the state, in terms of both its education function and institutionalising the principles of *Sumak Kawsay* (as mentioned in other chapters of this book, please note equivalent concepts informing the political agency of indigenous groups in South America, as *Küme Mongen* for the Mapuche or *Teko Porã* and *Teko Joja* for the Guarani). Additionally, through the discussion, the possibility emerged of the transnational capitalist class, constituting a vital agent in the ecosocialist transition, particularly in the transition from an opaque to a hazy context, that is, overcoming the ideological hegemony that pervades much of society; however, it is likely that the international affluent classes will have to be significantly affected by the problems created by capitalism before they assume any transformative role in an ecosocialist transition.

It is worth noting the five transformative agents featured in this discussion based on an analysis of two ecosocialist manifestos and Ecuador's National Plan for Good Living (2013–2017): social movements, indigenous groups, the affluent capitalist class, left-wing political parties and the state. The transformation and transition to an ecosocialist system will require a wide range of actions, and thus, a full spectrum of actors to perform such actions. In addition to those mentioned here, consideration of the potential transformative agency of the working class, trade unions, non-governmental organisations (NGOs), activists and other such actors would increase the depth of discussion. Thus, there is a need for ecosocialist authors to consider a wider spectrum of transformative agents in connection with the multi-layer and multi-sectoral actions, which are required in order to bring about lasting and large-scale solutions (Carvalho 2006; Wall 2010). Thus, there is also a need for a more enhanced discussion on the arc of actions, which will be required for a successful transition beyond the mere beginning of challenging the established institution of capitalism.

Transformative Agency and Power Resources: A Case Study

Based on the available ecosocialist literature, determination of which agents of change are to play a key transformative role in ecosocialism appears to have been heavily shaped by consideration of their *perceived* social power. Power clearly plays a predictive role in the evolution of human societies, and there is an obvious relationship between transformative agency and social power (Dowding 2012; Hearn 2014). In order to examine this relationship further, particularly in the context of transformative agents in ecosocialism, a stakeholder power analysis was conducted on the case of the Belo Monte Dam (BMD), based on the method developed by Mayers (2005) and adapting the power resources framework created by Korpi (1985). This particular case study was selected because the social actors involved included the majority of those referenced in ecosocialist texts and identified in this research. It therefore provided a practical example through which to analyse the different degrees of their transformative agency based on events which spanned decades. During this time, the country of Brazil underwent two important transitions which had significant implications for the BMD project and the agents involved: a political transition from a military regime to a democracy and an economic transition from import-substituting industrialisation to the model of neoliberalisation (Saad Filho 2014).

The BMD on Brazil's Xingu River was initially proposed in 1975 under Brazil's military dictatorship. A region that makes up 54.4% of Brazilian territory and encloses 78% of the reserve of national freshwater was considered by the dictatorial government guilty by its geography and 'condemned' to go through a process of water development (a risky endeavour, taking into account the region's particularities, especially the extensive plains and complex socio-ecology). Note that the Amazon, despite the incredible amount of surface water (around a fifth of the planet's freshwater), is quite ill suited for dams, and swaths of land and forest have been flooded in exchange for relatively little electricity. The largest scheme built during this first phase was Tucuruí (inaugurated in 1984), which generates electricity especially for aluminium smelting. The project received very limited environmental consideration, as it was common for all projects advanced by the military government (although it has changed only superficially after the introduction of more comprehensive environmental regulation in the 1990s). During the filling of the Tucuruí reservoir, a large quantity of trees was left in the area and then inevitably died, leading to a massive release of methane from their decomposition. With the construction of the Tucuruí Dam, 2430 square kilometres of forest were flooded and more than 33,000 people (besides the indigenous population) had to be resettled. The scheme also inundated part of three indigenous areas (Parakanã, Pucuruí and Montanha), the effect of which was added to the impact of the transmission lines on their land; in addition, the artificial Tucuruí Lake led to the extinction of various biological species and to the proliferation of water-borne illnesses. Many of these problems remain unsolved and were again present in the next phase of dam construction.

A second phase of dam engineering was more focused on politico-economic adjustments and that coincided with the neoliberal reform of the Brazilian State in the 1990s. The contradictions of the state-led model of development—implemented in Brazil along most of the twentieth century and intensified by the military governments, making use of foreign savings through loans from multilateral banks—had resulted in growing economic inefficiencies and reliance on the continuous injection of capital by the state. Along the lines of the liberalising institutional changes introduced by the Real Plan, both generation and commercialisation of energy became then available to private national entrepreneurs, increasingly associated with international investors or energy companies. Because of the changing role of the state apparatus, increasingly focused on pro-private sector incentives, instead of direct construction and operation (which, nonetheless, continued to occur, despite the neoliberal discourse), very few hydropower schemes were built during the second period. Coordination and decision-making became significantly diffused across many agencies, without the presence of a centralised, well-equipped agency as during the military dictatorship.

More recently, since 2003, the Amazon hydropower sector has provided an emblematic demonstration of the choreography of continuities and adjustments. In the Madeira Basin, after a lengthy political dispute, the Jirau and Santo Antônio plants were licensed in July 2007, allowing the overflow of up to 529 square kilometres. The two schemes have affected several indigenous groups, such as the Karitiana, Karipuna, Urueu-Wau-Wau, Katawixi and also other isolated, unnamed tribes (some living only six miles from the Jirau dam) who were forced to migrate to new lands, where they clash with *garimpeiros* (wildcat miners) logging companies, and are exposed to diseases to which they had no immunity. The more contentious project was Belo Monte—the largest hydropower scheme entirely within the Brazilian territory—which has attracted great publicity in the international media (e.g. 29 articles posted by BBC News alone between 2010 and 2017). Belo Monte, still under construction, demonstrates that the neoliberalising platform was being cleverly adjusted to fulfil neo-developmentalist goals. It also epitomised the revival of the construction of large hydroelectric power plants in the Amazon and the encroachment of Brazilian energy demands upon neighbouring nations (because some schemes have flooded Bolivian land, and Brazilian construction companies are deeply involved in projects in Peru, Venezuela and other countries). The project was partly inaugurated in 2016, with 11,233 megawatts of installed capacity and a cost of more than US$13 billion. After its completion, Belo Monte will be the third biggest hydropower plant on the planet, just behind Itaipu (inaugurated in 1984 on the border between Brazil and Paraguay) and the Three Gorges Dam in China.

The Belo Monte project, with another name (Cararaô), was originally conceived by the military in 1975, but could not go ahead due to the collapse of the regime and its growing financial and political problems. The residents of Altamira (near the site of the dam) and the indigenous people keep up a resistance that is organised and quite influenced by the traumatic experience with Tucuruí. There was a large gathering in 1989 with more than 1000 participants, including more than 600 indigenous people, which managed to attract international attention to the dispute and led to the cancellation of a World Bank loan under negotiation (Carvalho 2006). The proposal was revisited by the neoliberal administration of President Cardoso, but it was only really implemented by the government of Presidents Lula and Dilma. A consortium of state-owned companies (called Norte Energia [more details at http://norteenergiasa.

com.br]) won the contract to build Belo Monte and manage it for 35 years. It is actually bewildering that the Lula government included many energy and environmental authorities who for a long time fiercely opposed this and other similar projects in the past. The government had to respond to the criticism raised by environmental groups and local communities (including important Indian groups) that managed to form an international coalition against Belo Monte (that included Hollywood stars and famous artists). Capacity and transmission lines were reduced, and the new design removed the sizeable reservoir, in order to minimise the negative impacts. Moreover, the controversial features of Belo Monte continue to stir protest and serious resistance (Ioris 2020).

The project was the object of a lengthy battle in the courts, which led to repeated interruptions of the construction and operation (the last interruption ordered by a judge occurred in 2017, but local groups also regularly occupy the site of the dam to protest against the unresolved problems associated with the project). The most controversial issue is related to actual viability of the project without other supporting dams upstream. Due to the long dry season and the resulting long period of low flows in the Xingu River Basin (a common feature of eastern Amazon rivers), Belo Monte has a low operation and economic performance if operating as a single dam. It means that a cascade of dams is necessary, but this would magnify the impacts on indigenous land, natural parks and farms. The national government has guaranteed, on many occasions, that Belo Monte is the only large-scale dam of the Xingu River, but the wider problem for the Brazilian society is the low legitimacy and eroded trust in public authorities. This is a global phenomenon that seriously impacts on the quality of the formal democracy of many countries, Brazil included. Similar reassurances were given in relation to the construction of other schemes in the Amazon, such as the 43 large dams to be implemented along the (even more vulnerable and controversial) Tapajós River by 2022 (Fearnside 2015). Those dams are directly benefiting from the friendly interpretation of the legislation and lax enforcement of water and environmental regulation, which means that several indigenous reserves and conservation areas will be flooded and degraded.

It is evident that the interests of the construction and business sectors for new infrastructure projects in the Amazon contrast with the persistent disregard for public participation and social demands, which reveals the contradictory and asymmetric priorities of the state. The granting of environmental and water licences, as much as public consultation required for the approval of the project, was notoriously undemocratic and strongly pushed forward by the federal government under the justification that economic growth needed additional sources of energy. It is estimated that Belo Monte displaced between 20,000 and 50,000 people (including many families left behind after the collapse of rubber production in the early 1900s), while only one-third of the compensation promised by the authorities was ever paid (Anderson 2017). For many activists and local communities, there is a distinctive feeling of betrayal and deception associated with Belo Monte, especially because of the strange new alliances formed between left-wing politicians and conservative economic and political groups (Melo 2016). The election of Lula, an important union leader, had raised great expectations, but once in office, his government (2003–2010) was allied with the most archaic and corrupted politicians and construction companies in the country. It was not coincidence that the approval of Belo Monte involved systematic diversionary operations employed to divide the opposition and, at the same time, to legitimise the controversial project. During the Dilma government (2010–2016), which succeeded Lula's, the secret service was even asked to spy indigenous leaders, anthropologists and social movements that opposed the construction of Belo Monte (Valente and Bragon 2017).

Due to the associated environmental and social consequences, the BMD project has been the focus of intense controversy since its inception, sparking substantial opposition from affected groups, extending to and including citizens, groups and NGOs at the national and international level. For the purposes of this research, the social actors involved were divided into two groups based on their response to the project: the pro-dam and anti-dam groups. Both are constituted of a wide variety of actors, ranging from local to national and international agents, but supported by contrasting types and different degrees of power resources. These are summarised in Table 9.1.

The power of the anti-dam actors stemmed primarily from their shared ideology, the belief that the environmental and social impacts of the BMD outweighed any benefits from the project. Ideological power, as a

Table 9.1 Stakeholder power analysis based on the power resources framework created by Korpi (1985)

Stakeholders	Category	Role	Primary resource(s)
Brazilian government	Primary	Pro-dam as part of its development strategy and to provide energy to the region	Physical capital, financial resources, means of violence and regulatory
Norte Energia S.A. Consortium	Primary	Consortium of public and private corporations in charge of dam construction and operations	Physical capital and financial resources
State police	Primary	Responsible for enforcing orders from the state	Means of violence
Indigenous groups	Primary	Opposition to the dam to preserve their land and their culture	Ideology
Local fishermen and small-scale farmers	Primary	Opposition to the dam to protect their livelihoods and subsistence	Ideology
International corporations (operating within Brazil)	Secondary	Corporations dependent on dam's construction for the increased energy to sustain energy-intensive industries	Physical capital and financial resources
Other nations	Secondary	Own shares in corporations involved in dam construction	Money
Active national citizens	Secondary	Opposition to dam in support of local and indigenous groups	Ideology
International community	Secondary	Opposition to dam in support of local and indigenous groups	Ideology
FUNAI (National Indian Foundation)	Secondary	Government body responsible for policies relating to indigenous peoples and is charged with preventing invasions of indigenous territories by outsiders	Regulatory

(continued)

Table 9.1 (continued)

Stakeholders	Category	Role	Primary resource(s)
UN Human Rights Council	Tertiary?	Published statements denouncing Brazil's treatment of indigenous groups	(Minimal) regulatory
International Labour Organisation	Tertiary?	Upholds the rights of indigenous people (ILO 169)	(Minimal) regulatory

power resource, is characterised by a large domain (the number of actors who are receptive to it) but a narrow scope (the range of activities of other actors that can be punished or rewarded via the resource) (Korpi 1985). Indeed, it is traditionally considered to be less powerful in the face of financial resources, physical capital and means of violence, power resources which have both a large domain and wide scope. These latter resources were embodied by the national state, Norte Energia and the World Bank, the main actors in the pro-dam camp. However, when the project was first proposed, the opposition from indigenous and local people, NGOs and the international community proved to be successful in its lobbying of the World Bank, leading to cancellation of its loan to the federal government, causing the project to be shelved (Xing 2013). While this could demonstrate the potential power of ideology when large groups of people are mobilised, Carvalho (2006) attributes the success of the anti-dam campaign not to the opposition movement itself but to the timing. The late 1980s constituted a crucial moment for the World Bank, as it was under much scrutiny for its poor environmental performance, and following the cancellation of the loan, the Brazilian government was not in a position to finance the project itself. This highlights the importance of *context* when considering the transformative potential of agents.

Following this, the design of the BMD project was amended to decrease flooding and increase capacity, and proposed again in the early 2000s. Despite the previous success of the anti-dam movement, different circumstances resulted in the eventual full licensing of the BMD (the scheme is expected to be completed by 2019). The government announced that the Brazilian National Development Bank (the main development bank, owned by the federal government) would finance 80% of the project, demonstrating the significant increase in the state's financial resources and thus its autonomy from multilateral agencies, such as the World Bank (Xing 2013). The plans for the BMD were again met with strong opposition from the anti-dam coalition, their power still primarily based on ideology, with the aim to stop construction. However, the new financial strength of the state and Norte Energia created a more pronounced power asymmetry between the two sides, allowing the state to resist the protests at all levels, including the international level, where it bypassed the temporary injunctions, resulting from the anti-dam's coalition to the United Nations (UN) Human Rights Council and the International Labour Organisation (Cato 2013; D'Elia 2012). On 7 November 2018, a delegation of the Inter-American Commission on Human Rights visited the region and, at the end, urged the Brazilian government to immediately resolve the grave violation of the basic human rights of indigenous peoples and riparian communities living close to the BMD. Unfortunately, the new Brazilian government, in power since January 2019, was elected with a strong anti-indigenous and anti-minorities manifesto, which has included the cancellation of any new recognition of indigenous lands, and also a clear policy of utility privatisation and minimum environmental regulation.

This may be demonstrative of the supremacy of financial capital as a power resource over ideology, particularly in today's capitalist economies in which accelerated economic growth is the main concern. Financial resources and physical capital certainly presented advantages over ideology as power resources in the case of the BMD. The latter has a low mobilisation and storage potential, implying that over long periods of time—the protests spanned three decades—the ability of ideology to provide significant power is reduced, particularly when engaged in an ad hoc fashion (Korpi 1985). As a result, the second wave of protests constituted a more diffuse coalition and less visible action than in 1989 (Carvalho 2006). Others also noted division among indigenous and local groups, some of which were materially enticed by the state to neutralise key activists, indicating that ideology may be compromised by other more material resources (Barrionuevo 2010; Bratman 2015; D'Elia 2012;

Watts 2014). Should ecosocialism rely exclusively on similar groups of agents, that is social movements and indigenous people (as is the prevailing view in ecosocialist literature), to further the movement's goals and eventual transition to a post-capitalist society, their power resources would have to include more than a shared ideology, in order for them to succeed, for this has shown to decrease in momentum and power over time. And ecosocialists, in contrast with orthodox Marxist scholars, favour a more gradual transition process, implying that transformative agents will require power resources with a high storage potential, these being physical and financial capital and regulatory power (Kovel 2008a, b; Holloway 2010).

In light of the dependency of agents' transformative potential on their respective power resources, the role of the state as a transformative agent becomes interesting due to its dual characteristic; it is both an agent that effects change and is also an agent that can enable or restrict other agents due to its regulatory power, that is, its role as a major distribution of power resources, and its power of legitimate coercion. The state creates and upholds the structures of society, which determine the power of agents, as they constitute the organisationally defined social means to power resources. Through defining property rights and controlling market regulations, the state can theoretically even exert influence on the economic behaviour of organisations and corporations, thereby extending its power to the economic elite, those agents that, in a capitalist society, are characterised by a high concentration of power resources (Nasra and Dacin 2010; Scott 2013). Moreover, the internationally prevailing neoliberal paradigm appears to have institutionalised a *modus operandi*, which restricts the state's power and thus further reduces the possibility that the state would take actions that run contrary to capitalist interests (Teeple 2000).

Indeed, it could be argued that the Brazilian state's support of the BMD project through its National Development Bank demonstrates its adherence to the neoliberal paradigm and to capitalist interests (emphasis on the growth of the national economy creates an ever-increasing demand for energy to support it). The emergence of the left-wing movement in Brazil in the 1980s led to the dismantling of the elitist party system that dominated national politics and began the substantial process of democratisation of decision-making and the creation of opportunities for public participation; Guedes de Oliveira (2010) terms this Brazil's 'counter-hegemony'. Despite this, however, the state's adherence to the neoliberal paradigm significantly reduced the agency of social groups through the 'restoration of class power', particularly as a result of strict property rights and the progressive neutralisation of labour unions (Harvey 2005, 2006). This has led to the concentration of financial resources and physical capital in the hands of the economic elite, the characteristic distribution of power resources in capitalist societies, as these are the agents that own and control the means of production. Therefore, while democracy may have been established at the political level in Brazil, the capacity of agents to effect the change they desired in the case of the BMD was still restrained due to the state's adherence to the neoliberal paradigm and its subsequent distribution of power resources, which allowed one group (the economic elite) to maintain its hegemony, thus undermining Brazil's 'counter-hegemony' (Saad Filho 2014; Vanhanen and Kimber 1994). This contradiction in political and economic structures recalls the arguments outlined in support for 'economic democracy', in which the structure of the economy is redesigned to engender democratisation, a condition which would be fundamental to ecosocialism (Dolack 2012; Pepper 1998).

In sum, the BMD project provided an interesting, and intriguing, case through which to examine a significant attribute of transformative agency: the power of different actors and how this pertains to their potential contribution in the ecosocialist transition. Empirically based observations drew attention to the varying levels of impact these groups can have, impacts which are dependent on the context and the power resources of the agents. Ideology, employed by social movements and indigenous people, was proven to be a persuasive force, particularly when the opposing actors faced uncertain financial resources. However, the former cannot be considered a sustainable resource and was unable to sustain its force over the four decades of the project; thus, ideology alone cannot be expected to sustain the momentum of an ecosocialist transition. The financial power of the Norte Energia and the federal state (Brazilian National Development Bank—BNDES) was seen to be a significant power resource and one which ultimately decided the fate of the

BMD project. This strengthens the argument that the state should be considered a key factor in ecosocialism, one which must be transformed as opposed to ignored.

Conclusion

This study sought to contribute, from the perspective of the nexus between environment and development, to a current lacuna in ecosocialist literature on transformative agency, that is, the actors who will bring about the theorised transition to a post-capitalist regime and their transformative potential as a function of their power resources. The former was explored against the framework created by Westley et al. (2013) of transformative agency within SES; this brought to attention the wide range of actions that must be executed in the transition from one SES to another, thus implying an equally diverse range of agents. The discussion focused on five transformative actors and their particular roles in an ecosocialist transition according to different contexts: indigenous groups, social movements, left-wing political parties, the state and the transnational bourgeois classes. The latter two are often disregarded in the limited ecosocialist literature that references the subject of transformative agents. However, given that they possess significant and effective power resources, and thus have a high transformative potential, the role of the state and affluent classes in the ecosocialist transition requires further research.

Indeed, both context and power resources were demonstrated to play a significant role in the transformative potential of different agents. Different contexts require different actors to take the leading role, depending on the action required at the time (e.g. political parties would take a leading role in the sense-making activities required during hazy and transparent contexts). In this regard, there is a need to further understand the complexity and political subjectivities of social actors in different socio-spatial circumstances. In addition, the effectiveness of certain power resources may also be dependent on contexts (e.g. the initial triumph of the anti-dam coalition on the basis of ideological power).

On a final point, the authors, here, would like to emphasise that, based on the earlier discussion, it would be pertinent for future research in this area to examine the transformative potential of agents in a *globalised* context, one which presents a number of challenges to social actions, even at the micro- and meso-scale, and has significant repercussions on the distribution of power resources. The transition which has occurred from national to global capitalism implies the need for *global* reform, and the predominant reforms advocated by ecosocialists (e.g. collective ownership over the means of production) could arguably only be implemented and sustained at a local or global level (Pieterse 1997; Kovel 2013; Löwy 2015). This rationale provides justification and substance to Kovel's (2008a: 252) proclamation that 'ecosocialism will be international or it will be nothing;.

References

Adams, W. M. (2009). *Green Development: Environment and Sustainability in a Developing World*. London/New York: Routledge.

Anderson, M. (2017). *Displaced by Brazil's Giant Belo Monte Hydroelectric Dam, 'River People' Reoccupy Reservoir*. https://news.mongabay.com/2017/03/displaced-by-brazils-giant-belo-monte-hydroelectric-dam-river-people-reoccupy-reservoir. Accessed 30 May 2019.

Angus, I., Joel, K., Löwy, M., & Follett, D. (2009). *The Belém Ecosocialist Declaration*. https://climateandcapitalism.com/2018/07/22/the-belem-ecosocialist-declaration-an-historic-document. Accessed 10 Apr 2019.

Arnold, N. S. (1990). *Marx's Radical Critique of Capitalist Society: A Reconstruction and Critical Evaluation*. Oxford: Oxford University Press.

Aşıcı, A. A. (2012). Green New Deal: A Green Way Out of the Crisis? *Environmental Policy and Governance, 22*(5), 295–306.

Aşıcı, A. A., & Bünül, Z. (2011, June 13–16). Green Recovery From the Crisis: A Comparative Analysis of Green New Deal and Ecosocialism. Paper Presented at the *9th International Conference of the European Society for Ecological Economics – Advancing Ecological Economics: Theory and Practice*. Boğaziçi University, Istanbul.

Barrionuevo, A. (2010). Bypassing Resistance, Brazil Prepares to Build a Dam. *The New York Times.* http://www.nytimes.com/2010/08/16/world/americas/16brazil.html?_r=0. Accessed 16 Apr 2019.

Bellamy Foster, J. (2013). The Epochal Crisis: The Combined Capitalist Economic and Planetary Ecological Crises. *International Journal of Socialist Renewal, 65*(5), 1–12.

Betto (Friar), & Löwy, M. (2010). Ecosocialism and Spirituality. *Capitalism Nature Socialism, 21*(2), 87–99.

Bratman, E. (2015). Passive Revolution in the Green Economy: Activism and the Belo Monte Dam. *International Environmental Agreements: Politics, Law and Economics, 15*(1), 61–77.

Bremmer, I. (2017). Ecuador, Venezuela and the Fall of the Latin American Left. *Time Magazine.* http://time.com/4719076/ecuador-venezuela-latin-america-left-wing. Accessed 12 Apr 2019.

Burkett, P. (2006). Two Stages of Ecosocialism? *International Journal of Political Economy, 35*(3), 23–45.

Carvalho, G. O. (2006). Environmental Resistance and the Politics of Energy Development in the Brazilian Amazon. *The Journal of Environment & Development, 15*(3), 245–268.

Castree, N. (2009). Crisis, Continuity and Change: Neoliberalism, the Left and the Future of Capitalism. *Antipode, 41*, 185–213.

Cato, A. (2013). Localized Movements and Socio-Environmentalism: The Belo Mote Dam Complex in Brazil. *Environmental Movements in Brazil.* http://sites.duke.edu/brazilenvironmentalmovements/topics/the-belo-monte-dam. Accessed 18 May 2019.

Chase-Dunn, C. (2014). Crisis of What: The End of Capitalism or Another Systemic Cycle of Capitalist Accumulation? *Perspectives on Global Development and Technology, 13*(1–2), 13–35.

D'Elia, A. (2012). *Belo Monte – Anúncio de Uma Guerra* (Movie). Brazil.

Daldal, A. (2014). Power and Ideology in Michel Foucault and Antonio Gramsci: A Comparative. *American Research Institute for Policy Development, 2*(2), 149–167.

Dolack, P. (2012). There Is No Democracy Without Economic Democracy. *Systemic Disorder.* https://systemicdisorder.wordpress.com/2012/09/05/there-is-no-democracy-without-economic-democracy. Accessed 15 May 2019.

Dordoy, A., & Mellor, M. (2000). Ecosocialism and Feminism: Deep Materialism and the Contradictions of Capitalism. *Capitalism Nature Socialism, 11*(3), 41–61.
Dowding, K. (2012). Why Should We Care About the Definition of Power? *Journal of Political Power, 5*(1), 119–135.
Escobar, A. (2007). 'Post-Development' as Concept and Social Practice. In A. Ziai (Ed.), *Exploring Post-Development: Theory and Practice, Problems and Perspectives* (pp. 18–32). London: Routledge.
Fearnside, P. M. (2015). Amazon Dams and Waterways: Brazil's Tapajós Basin Plans. *Ambio, 44*, 426–439.
Guedes de Oliveira, M. A. (2010). Sources of Brazil's Counter-Hegemony. *Revista Brasileira de Política Internacional, 53*(2), 125–141.
Harvey, D. (2005). *A Brief History of Neoliberalism.* New York: Oxford University Press.
Harvey, D. (2006). *Spaces of Global Capitalism: Towards a Theory of Uneven Geographical Development.* London/New York: Verso.
Harvey, D. (2014). *Seventeen Contradictions and the End of Capitalism*. New York: Oxford University Press.
Hearn, J. (2014). On the Social Evolution of Power To/Over. *Journal of Political Power, 7*(2), 175–191.
Hobsbawm, E. J. (1965). The Crisis of the Seventeenth Century. In T. Aston (Ed.), *Crisis in Europe 1560–1660* (pp. 5–58). London: Routledge.
Holloway, J. (2010). *Crack Capitalism.* Cambridge: Pluto Press.
Huan, Q. (2010). Eco-Socialism in an Era of Capitalist Globalisation: Bridging the West and the East. In Q. Huan (Ed.), *Eco-Socialism as Politics: Rebuilding the Basis of Our Modern Civilisation* (pp. 3–12). Dordrecht: Springer Science & Business Media.
Imas, J. M., Wilson, N., & Weston, A. (2012). Barefoot Entrepreneurs. *Organization, 19*(5), 563–585.
Ingham, G. (2008). *Capitalism: With a New Postscript on the Financial Crisis and Its Aftermath.* Cambridge: Polity Press.
Ioris, A. A. R. (2014). *The Political Ecology of the State: The Basis and the Evolution of Environmental Statehood.* London: Routledge.
Ioris, A. A. R. (2015). Theorizing State-Environment Relationships: Antinomies of Flexibility and Legitimacy. *Progress in Human Geography, 39*(2), 167–184.
Ioris, A. A. R. (2017). *Agribusiness and the Neoliberal Food System in Brazil: Frontiers and Fissures of Agro-neoliberalism.* London: Routledge.

Ioris, A. A. R. (2018). Amazon's Dead Ends: Frontier-making the Centre. *Political Geography, 65*, 98–106.

Ioris, A. A. R. (2019). La Politique de l'Agrobusiness et le Business de la Durabilité. In *Les Nouveaux Territoires de l'Agrobusiness* (Alternatives Sud) (Vol. 26, pp. 25–56). Louvain-la-Neuve/Paris: Centre Tricontinental & Éditions Syllepse.

Ioris, A. A. R. (2020). *Frontier Making in the Amazon: Economic, Political and Socioecological Conversion*. Cham: Springer.

Johns, D., Kovel, J., & Löwy, M. (2003). Has Ecosocialism Passed on the Tough Questions? *Capitalism Nature Socialism, 14*(2), 120–128.

Kaletsky, A. (2011). *Capitalism 4.0: The Birth of a New Economy in the Aftermath of Crisis*. New York: PublicAffairs.

Korpi, W. (1985). Power Resources Approach vs. Action and Conflict: On Causal and Intentional Explanations in the Study of Power. *Sociological Theory, 3*(2), 31–45.

Kovel, J. (2008a). Ecosocialism, Global Justice, and Climate Change. *Capitalism Nature Socialism, 19*(2), 4–14.

Kovel, J. (2008b). *The Enemy of Nature: The End of Capitalism or the End of the World* (2nd ed.). London/New York: Zed Books.

Kovel, J. (2013). Ecosocialism as a Human Phenomenon. *Ecosocialist Horizons*. http://ecosocialisthorizons.com/2013/08/ecosocialism-as-a-human-phenomenon. Accessed 20 May 2019.

Kovel, J. (2014). Ecosocialism as a Human Phenomenon. *Capitalism, Nature, Socialism, 25*(1), 10–23.

Kozloff, N. (2016). South America's Leftist Implosion: Let the Debate Begin. *Huffington Post*. www.huffingtonpost.com/nikolas-kozloff/south-americas-leftist-im_b_10174060.html. Accessed 9 May 2019.

Lichten, E. (1984). Class Struggle, Power, and Social Change in Contemporary Capitalism. *Contemporary Sociology, 13*(6), 681–683.

Löwy, M. (2014). Ecosocial Struggles of Indigenous Peoples. *Capitalism Nature Socialism, 25*(2), 14–24.

Löwy, M. (2015). *Ecosocialism: A Radical Alternative to Capitalist Catastrophe*. Chicago: Haymarket Books.

Mankoff, M. (1970). Power in Advanced Capitalist Society: A Review Essay on Recent Elitist and Marxist Criticism of Pluralist Theory. *Social Problems, 17*(3), 418–430.

Max-Neef, M. A. (1992). *From the Outside Looking in: Experiences in 'Barefoot Economics'*. Zed Books: London.

Mayers, J. (2005). Stakeholder Power Analysis. *Power Tools*. www.policy-powertools.org/Tools/Understanding/docs/stakeholder_power_tool_english.pdf. Accessed 20 May 2019.

Melo, M. A. (2016). Crisis and Integrity in Brazil. *Journal of Democracy, 27*(2), 50–65.

Moore, J. W. (2014a). The Capitalocene – Part 1: On the Nature and Origins of Our Ecological Crisis. *Journal of Peasant Studies, 44*(3), 594–630.

Moore, J. W. (2014b). The End of Cheap Nature or: How I Learned to Stop Worrying About 'the' Environment and Love the Crisis of Capitalism. In C. Suter & C. Chase-Dunn (Eds.), *Structures of the World Political Economy and the Future of Global Conflict and Cooperation* (pp. 285–314). Berlin: LIT.

Nasra, R., & Dacin, M. T. (2010). Institutional Arrangements and International Entrepreneurship: The State as Institutional Entrepreneur. *Entrepreneurship: Theory and Practice, 34*, 583–609.

Olsson, P. (2017). Synthesis: Agency and Opportunity. In F. Westley, K. McGowan, & O. Tjörnbo (Eds.), *The Evolution of Social Innovation. Building Resilience Through Transitions* (pp. 58–72). Cheltenham: Edward Elgar.

Pepper, D. (1993). *Eco-Socialism: From Deep Ecology to Social Justice*. London: Routledge.

Pepper, D. (1998). Sustainable Development and Ecological Modernisation: A Radical Homocentric Perspective. *Sustainable Development, 6*, 1–7.

Pieterse, J. N. (1997). Globalisation and Emancipation: From Local Empowerment to Global Reform. *New Political Economy, 2*(1), 79–92.

Prisecaru, P. (2012). After the Crisis: The End of Capitalism or a New and Reformed Capitalism? In D. Popescu & L. Mihaescu (Eds.), *Revista Economică: Economics of Crises versus Crisis of Economics* (pp. 680–686). Sibiu: Lucian Blaga University of Sibiu.

Radcliffe, S. (2012). Development for a Postneoliberal Era? Sumak Kawsay, Living Well and the Limits to Decolonisation in Ecuador. *Geoforum, 43*(2), 240–249.

Robinson, W. I. (2005). Gramsci and Globalisation: From Nation- State to Transnational Hegemony. *Critical Review of International Social and Political Philosophy, 8*(4), 1–16.

Saad Filho, A. (2014). Two Transitions in Brazil: Dilemmas of a Neoliberal Democracy. *Socialist Project*. www.socialistproject.ca/bullet/927.php. Accessed 6 May 2019.

Sachs, W. (2010). Environment. In W. Sachs (Ed.), *The Development Dictionary: A Guide to Knowledge as Power* (2nd ed., pp. 24–37). London/New York: Zed Books.

Schumpeter, J. (2015). Capitalism and Its Discontents. *The Economist*. www.economist.com/news/business/21669911-anti-capitalism-being-fuelled-not-just-capitalisms-vices-also-its. Accessed 22 May 2019.

Schweickart, D. (2009). Is Sustainable Capitalism an Oxymoron? *Perspectives on Global Development and Technology, 8*(2), 559–580.

Schweickart, D. (2010). Is Sustainable Capitalism Possible? *Procedia – Social and Behavioral Sciences, 2*(5), 6739–6752.

Scott, W. R. (2013). *Institutions and Organizations: Ideas, Interests, and Identities* (4th ed.). Los Angeles: Sage.

Senplades. (2013). *National Development Plan: National Plan for Good Living 2013–2017*. Quito: National Secretariat of Planning and Development.

Stephens, J. D. (1986). Review of 'The Democratic Class Struggle' by Walter Korpi. *American Journal of Sociology, 91*(4), 1018–1020.

Teeple, G. (2000). *Globalization and the Decline of Social Reform: Into the Twenty-First Century*. Aurora/New York: Garamon/Humanity Books.

The Conversation. (2017). *Why Latin America Has Not Yet Fallen Out with Left-Wing Politics*. https://theconversation.com/why-latin-america-has-not-yet-fallen-out-with-left-wing-politics-76200. Accessed 12 May 2019.

The Economist. (2016). Rose Thou Art Sick. *www.economist.com/news/briefing/21695887-centre-left-sharp-decline-across-europe-rose-thou-art-sick*. Accessed 11 May 2019.

Upchurch, M., & Mathers, A. (2012). Neoliberal Globalization and Trade Unionism: Toward Radical Political Unionism? *Critical Sociology, 38*(2), 265–280.

Valente, R., & Bragon, R. (2017). Abin Espionou Indígenas e ONGs no Governo Dilma. *Folha de São Paulo*. http://www1.folha.uol.com.br/poder/2017/05/1882257-abin-espionou-indigenas-e-ongs-no-governo-dilma.shtml. Accessed 17 May 2019.

Vanhanen, T., & Kimber, R. (1994). Predicting and Explaining Democratisation in Eastern Europe. In G. Pridham & T. Vanhanen (Eds.), *Democratisation in Eastern Europe* (pp. 63–98). London/New York: Routledge.

Walker, B., Holling, C. S., Carpenter, S. R., & Kinzig, A. (2004). Resilience, Adaptability and Transformability in Social – Ecological Systems. *Ecology and Society, 9*(2), 5. www.ecologyandsociety.org/vol9/iss2/art5.

Wall, D. (2010). *The Rise of the Green Left: Inside the Worldwide Ecosocialist Movement*. London: Pluto Press.

Watts, J. (2014). Belo Monte, Brazil: The Tribes Living in the Shadow of a Megadam. *The Guardian*. http://www.theguardian.com/environment/2014/dec/16/belo-monte-brazil-tribes-living-in-shadow-megadam. Accessed 13 May 2018.

Westley, F. R., Tjornbo, O., Schultz, L., Olsson, P., Folke, C., Crona, B., & Bodin, Ö. (2013). A Theory of Transformative Agency in Linked Social-Ecological Systems. *Ecology and Society, 18*(3). https://doi.org/10.5751/ES-05072-180327.

White, D. F., Gareau, B. J., & Rudy, A. P. (2017). Ecosocialisms, Past, Present and Future: From the Metabolic Rift to a Reconstructive, Dynamic and Hybrid Ecosocialism Ecosocialisms, Past, Present and Future: From the Metabolic Rift to a Reconstructive, Dynamic and Hybrid Ecosocialism. *Capitalism Nature Socialism, 28*(2), 22–40.

Whiteside, K. H. (1997). French Ecosocialism: From Utopia to Contract. *Environmental Politics, 6*(3), 99–124.

Xing, B. (2013). The Belo Monte Dam Nation: Analysis of the Kayapó Resistance. *Indigenous Culture in the Amazon*. http://sites.duke.edu/amazonindigenousculture/kayapo-and-the-belo-monte. Accessed 13 June 2018.

Part III

The Lived Environment and Development of the Amazon Region

10

The Indigenous Politics of Belonging: Opposing Neo-liberal Extractivism with Ethical Cosmologies

Stefano Varese

Prologue: Four Parallel Stories[1]

First Story In 1967, one of the four Mayoruna/Matsés lineages located on the Yaquerana River in the Peruvian Amazon across the border with Brazil was shelled with Napalm type of firebomb and machine-gunned by the FAP (Fuerza Aérea Peruana) (Peruvian Airforce), with the support of US Army helicopters sent from the Southern Command based in Panama. Peruvian President, Fernando Belaúnde Terry, had personally ordered the attack and requested the military assistance of the US and the International Petroleum Company based in Talara, Peru. In October of 1968, a Military Revolution, led by General Juan Velasco Alvarado sent

[1] This chapter is based on a manuscript submitted to the book project of Adrián Lerner (Yale University) and Javier Puente, PUCP-Lima (eds.) *Shaping Natural Regions: Capitalism, Environment, and the Geography of Post-Colonial Peru* (forthcoming).

S. Varese (✉)
Department of Native American Studies, University of California-Davis, Davis, CA, USA
e-mail: svarese@ucdavis.edu

A. A. R. Ioris (ed.), *Environment and Development*,
https://doi.org/10.1007/978-3-030-55416-3_10

Fernando Belaúnde Terry into exile to Argentina, expropriated the International Petroleum Company (a branch of Standard Oil of New Jersey), declared a radical Agrarian Reform by expropriating large landholding (haciendas and plantations) in the Coast and the Andes, nationalised the strategic industry of the country and initiated a reform aimed at transforming the private sector of the country into a co-owned and co-managed social sector.

The vast Amazon region of the country—more than 60% of the national territory of Peru—was left untouched until 1971 when the Revolutionary Government created a national bureau charged with carrying forward political and cultural reforms, aiming at guaranteeing the communal territorial possession of land and renewable 'resources' in the hands of each ethnic entity organised as autonomous political body. It is under this new pro-indigenous legislation that the Mayoruna-Matsés, that had been bombarded by the liberal democratic Peruvian government of Belaúnde Terry in alliance with the US Southern Command, obtained the recognition and legal communal title of close to half million hectares of primary forest. As far as I know, the Matsés of the Yaquerana region have been successful in protecting their forest and resources from loggers and other colonists during the past 50 years.

Second Story In 2014, Edwin Chota and three other Asháninka leaders of the Saweto community of the Alto Tamaya River (Ucayali) were murdered in front of their families and community members by loggers for defending their forest. In 2015, after an international scandal and the mobilisation of Peruvian and other Latin American Indigenous People organisations Asociación Interétnica de Desarrollo de la Selva Peruana (Interethnic Association for the Development of the Peruvian Rainforest) (AIDESEP), Coordinadora de las Organizaciones Indígenas de la Cuenca Amazónica (Coordinator of Indigenous Organizations of the Amazon River Basin) (COICA)), NGOs and citizen organisations, the community of Saweto was granted 80,000 hectares of forest. It took the Asháninka people of Saweto more than 40 years and four of their men assassinated by loggers and illegal colonists and 'extractivists' to obtain the rightful adjudication of their own ancestral territory. The Law of Native Communities of the Jungle protecting the lands and territories of the

Native Communities had been in the books for almost 50 years and none of the neo-liberal pseudo-democracies that ruled Peru had done anything to guarantee the autonomy and land possession of the Indigenous People.

Third Story This is the case of the Kichwaruna of Sarayaku, on the Bobonaza River of Ecuador's Amazon region. In 1996, the Ecuadorian government imposed oil concession blocks in Sarayaku territory without getting the consent of the 1200 Indigenous People who live there. The Sarayaku runa learned that their land had been opened to oil exploration only when the helicopters arrived accompanied by men with guns. The resistance and organisation of the people of Sarayaku began immediately. Initially, some of the younger members of the community thought that armed resistance could have been the answer to this neo-colonial invasion. The examples of the long ongoing guerrilla warfare of Colombia to the north and the disastrous tragic case of the civil war in Peru were enough to discourage the community from following that route. With the traditional wisdom, deep spirituality and practical knowledge, the people of Sarayaku managed to beat back oil drilling plans on their territory, expel the foreign and national oil companies and demand the Government of Ecuador for illegally licensing an oil company for doing explorations in indigenous territory without the community-informed consent. With the support of the United Nations-International Labour Organization (UN-ILO) Covenant 169, ratified by the Government of Ecuador, the new Constitution of Ecuador and a massive social movement of Indigenous People organisations in Ecuador (Confeniae, Conaie, Movimiento Pachacuti, etc.) and the civil society, in April 2012, the government of Ecuador acknowledged responsibility for illegally licensing an oil company to do business in Indian Territory. In July 2012, the Inter-American Court of Human Rights (IACHR) ruled that the government must consult with indigenous communities prior to such enterprises and pay for physical and 'moral' damages to the community.

'These victories result from the efforts of the people of Sarayaku to build a network of national and international allies and to employ international human rights instruments in their defence. They have used media to promote their struggle across the globe, and have never succumbed to intimidation, political pressure, or fatigue. The IACHR-Inter-American Court of Human Rights verdict is a major step forward in the safeguarding of indigenous rights—particularly the right to free, prior, and informed consent (FPIC) in development decisions that affect their territory—and sets a precedent for cases around the world in which the indigenous peoples struggle to defend their territories and rights'. The fight is not over for the people of Sarayaku. More than eight years after the IACHR decision, the government of Ecuador has yet to comply with part of the ruling, which requires the removal of explosives buried in community lands. In addition, Sarayaku is threatened by the Ecuadorian government's auction of 21 new oil blocks in the XI oil licensing round. Block 74, whose auction has been provisionally postponed, contains all of Sarayaku territory.[2]

Fourth Story of the Future In the month of December 2019, I became aware of a new worrisome neocolonial attack on the territorial and cultural sovereignty of the Amazonian Indigenous People and *ribereños* of Peru. Through an unnamed whistler-blower, I was informed that the current public health project in the Napo Region of Peru implemented by the Fundación Enlace Hispano de Salud (EHAS) in coordination with the Catholic University of Peru-Pontificia Universidad Católica del Perú (Pontifical Catholic University of Peru) (PUCP) is going to be expanded with a contract between The Google LOON[3] project and *Internet Para Todos-Peru* (a consortium of The Interamerican Bank of Development, the Corporación Andina de Fomento and FACEBOOK). The central idea of this *New Digital Frontier* is to reach 80,000 Peruvian rural hamlets and families in the lower jungle (Native Communities and *Ribereños*) with Internet and digital communication via a soon-to-be-installed balloon at 22 kilometres above Iquitos, with a 50-mile diameter cone of connectivity (with an Artificial Intel software support). Then Google Loon LLC project is a subsidiary working on providing Internet access to rural and remote areas. The company uses high-altitude balloons in the

[2] From website of *Amazon Watch*. https://amazonwatch.org. Accessed 10 December 2019.

[3] Loon LLC is an Alphabet Inc. subsidiary working on providing Internet access to rural and remote areas. The company uses high-altitude balloons in the stratosphere at an altitude of 18 km to 25 km tocreate an aerial wireless network.

stratosphere at an altitude ranging from 18 kilometres (11 miles) to 25 kilometres (16 miles) to create an aerial wireless network.

While the advantages of these digital technologies may be clear for the regular citizen of rural Peru, the dangers are less obvious. Since the early revelations of Edgar Snowden, the scandal of Russian interference in the US presidential elections in 2016, and the corrupted participation of Facebook in falsifying political and cultural information, the manipulation of the digital frontier has become a fundamental challenge to the democratic life of communities in the world. For Amazonian rural indigenous and mestizo people, the dangers are compounded by their vulnerability as responsible for the common goods inherited by the ancestors: the forest, the waters, the whole living and physical landscape now easily investigated and recorded by Geographic Information System (GIS) digital tools like the National Aeronautics and Space Administration's (NASA) Hyperion system that provides capabilities to detect a wide variety of minerals. Extractivist espionage may become the major threat to Indigenous People's sovereignty who in the future may not be able to ascertain how their territorial integrity and autonomy are being offered on the global financial market unbeknown to them.

Some Lessons Learned

It seems that in three of the stories, the relative success of the Amazonian indigenous communities in defending their autonomy and their territory is linked to their ability to mobilise external solidarity around the use of national and international legal instruments as effective shields of protection of their rights. What is less obvious to external observers is the fact that the strength of their argument is derived from their bio-ethical position as people who have not been seduced yet by materialism and the mirage of commoditisation of the 'world'. It is this positioning of the Amazonian Indigenous People in the world/universe/cosmos what I would like to discuss next.

Epistemological and Ethical Premise

Geological history, or the 'Long-Duration' perspective proposed years ago by French historian Fernand Braudel, may be the most appropriate epistemological and ethical tools to approach the study and understanding of Amazonia as a whole integrated living system co-created by millennia of interactive co-evolution of all its/her tangible and intangible bio-physical entities. In this sense, rather than using the neologism of Anthropocene—with its constraining reference to a late geological era defined by the onset of the Industrial Revolution and the overwhelming expansion of capitalism—I have suggested to choose a term that enlightens the joint role played by humanity and all other entities in landscaping, bio-scaping and thus ethno-scaping the tropical rainforest, the *selva*, as an anthropogenic phenomenon, a dynamic-structured production and constantly expanding reproduction of life resulting from millennia of coexistence of all the 'relatives' that inhabit the world.

It is with humbleness, then, that I must refer to the Kichwa people of Sarayaku, the large indigenous territory on the banks of the Bobonaza River in Ecuador and cite their words as the most appropriate ethical definition of what Amazonia has been and still is for thousands of lineages of Indigenous People living in harmony—in Sumak Kawsay—in this immense territory now threatened by neo-imperialist destruction. The Kichwa people of Sarayaku call their land, their territory, their 'environment', Kawsak Sacha—the Living Forest: 'Kawsak Sacha recognises that the forest is made up entirely of living selves and the communicative relations that they have with each other … these selves, from the smallest plants to the supreme beings who protect the forest, are persons – *runa* – who inhabits the waterfalls, lagoons, swamps, mountains, and rivers, and who, in turn, compose the Living Forest as a whole. These persons live together in community (*llakta*) and carry out their lives in a manner that is similar to human beings …'

The analogy that comes to mind has been pointed out by quantum physics theorist F. David Peat (2002: 31) who has asserted that Indigenous Peoples of the Americas' way of knowing is similar to the 'quantum physics picture of the material world as being the outward manifestation of patterns, forms, balances and relations of energy…likewise, (indigenous peoples) speak of relationship among the powers and spirits that surround them…. These relationships carry with them obligations and the necessity of carrying out periodic ceremonies of renewal …'. The indigenous conception and treatment of the world—in our case, the Amazonian forest—as a living and willing web of inter-relationships rather than a fixed set or structure of inanimate categories without intentionality implies that for Indigenous People knowledge-science occurs always in a context where 'nothing is abstract since all things happens within a landscape of and by virtue of a web of inter-relationships' (Peat 2002: 234).

These critical and comparative approaches to ways of knowing has been accompanied by profound revisions of Western modern science even questioning the ability of Indo-European languages and logics to operate in a cosmology, where objects can be simultaneously members of two exclusive categories: evil and good, live and dead, material and spiritual, 'waves and particles'. This epistemological—at the same time, ethical and ontological—revolution has been driven by the fundamental insights of quantum scientists such as Bohr, Heisenberg, Einstein and others who forced modernity to re-introduce metaphysics into our world view and accept 'ontological uncertainty', 'complementary descriptions' rather than single descriptions that exhaust the phenomenon in question and 'mutually contradictory accounts' (Apffel-Marglin 2011). But most of all, the quantum revolution has confronted us with the existence of mysteries and the ancient philosophical quandary of how to coexist with them in peace and harmony in what the Sarayaku people call the *Sumak Kawsay*.

How Many Amazons in Peru?

In Peru, *La Selva* or *Montaña* or *La Amazonía* is a vast region of more than half of the country that does not have a single official definition. Since the enactment in 1974 of the Law of Native Communities (Decree Law 20653) during the Military Revolution of Velasco Alvarado, and a series of other legislative measures in the following administrations, the

Amazon region has been treated more as a cultural ideological construct than as a precise geo-ecological region with its specific policy requirements. The now defunct *Instituto Nacional de Recursos Naturales, Ministerio de Agricultura* (INRENA) identified 77,528,030 hectares, the 60.32% of the national territory as *Amazonía Peruana*. This extension of national territory is covered by tropical rain forest that begins on the Eastern slopes of the Andes at an altitude of approximately 3800 metres. The actual forested area—the original ecosystem of the tropical rain forest—of the whole *Selva* has been assessed in few opportunities during various administrations with notable distortions due to political opportunism. Until 1974, 4.5 mm hectares had been deforested; in 1990, the number had grown to 6.9 mm hectares (Dourojeanni et al. 2010). The current situation of deforestation is practically unknown since the various administrations have no interest in revealing the level of destruction caused by the total lack of regulations and control of private and state-run extractive activities.

A simple projection of the data provided by a study performed by Dourojeanni et al. for the year 2000 reveals that an average of yearly deforestation of 150,000 hectares per year would add another 2,550,000 hectares of deforestation for the current year of 2017. These numbers contradict the official position of the Peruvian government that the level of deforestation of the Amazon is a little more than 9%, while studies of the United Nations in the mid-1980s were already showing that the entire South American forested area had been denuded of 40% of its original cover, and the two major Amazonian countries—Brazil and Peru—were responsible, respectively, of 20% and almost 10% of the loss (Dourojeanni et al. 2010: 44).

This flimsy and possibly manipulated quantitative information about the reality of the Amazon indicates, however, how dangerously deceitful is the neo-liberal ideology, claiming the emptiness and untouched wealth of this part of Peru that needs to be occupied, 'colonised' and developed for the good of the rest of the country. Since the times of President Ramón Castilla in 1848 and through the conservative and liberal governments of the country all the way to the first administration of Fernando Belaúnde Terry in the 1970s and the following neo-liberal administrations, the political 'imaginary' imposed on the country has been that the Amazon,

La Selva, is where lies the country's future of wealth. This political geography of deception has sustained for more than 150 years the fiction of a *Terra Nullius* rich of untapped resources that can be occupied directly as an internal colony or better yet given in concession to foreign enterprises. Former President Fernando Belaúnde Terry in the early 1960s expressed this oligarchic colonialist mythology in the title of his first political platform and book 'The Conquest of Peru by Peruvians' [*La conquista del Perú por los Peruanos*], letting unsolved the questions of who are the Peruvians and who are the conquered.

From Geography of Deception to Ecological Realism

In terms of geo-historical perspectives, the Eastern slopes and foothills of the Andes of Colombia, Ecuador, Peru and Bolivia, between an altitude of 1500 metres and 3800 metres above the sea level, constitute a complex mountainous and hydrological system of extreme biological diversity that announces the vast lowland of Amazonia, the largest tropical rainforest of the world. In each country this boundary of cloud forest receives a different vernacular name: *montaña* in Peru and Ecuador, *yunga* in Bolivia, *ceja de selva* ('brow of the jungle') in most of these countries or simply *selva alta* (high jungle) and *El Oriente*. The geo-cultural area loosely referred to by the term *montaña* extends from the headwaters of the Caquetá, in southern Colombia, to the headwaters of the Mamoré in southern Bolivia (Taylor 1996).[4] The extreme humidity of this piedmont region, caused by the ascending warm air of the Amazon crushing against the cold and dry Andean mountain, creates a blanket of fog—the cloud—that is burnt by the sun only for a few hours every day. The eastern Andean piedmont is also the ecological frontier of the major Amazonian cultivar: the *yuca* root (*Manihot esculenta)*. Beyond the 1500–2800 metres of altitude, the basic staple of most Amazonian Indigenous People does not grow well. Bananas, plantains and others neo-tropical fruit trees are still viable cultivars and so

[4]Taylor's *Bibliographic Essay* at the end of the chapter is possibly one of the most complete lists of ethnohistorical references on the *montaña* available today.

is corn (*Zea maiz*), a plant of astounding adaptability and *coca* (*Eritroxylon coca*), one of the earliest plants to be domesticated by the ancient Indigenous People of the Coastal region of Peru thousands of years ago.

This ecological divide between the Andes and Amazonia has constituted historically a barrier and at the same time a gate for the movement of peoples, cultivars and cultural practices of production, circulation and consumption. Since the early archaeological studies of Andean societies in the 1940s, scholars such as A. Posnansky, Julio C. Tello and A. Lipshutz in Bolivia, Peru and Chile, respectively, and J. Imbelloni in Argentina had carved a niche to study and understand the archaeological remnants of 'Ancient' cultures and their relations to other regions of the Americas and across the Pacific and the Atlantic Oceans. Peruvian Quechua-speaking scholar J. C. Tello put forward the argument that some of ancient pre-Inca societies of the Central and Northern Andean region of Peru showed their clear Amazonian origin. J. C. Tello's hypothesis would have to wait more than 30 years to be proven by new archaeological studies of the Upper Amazon region of Peru by Lathrap (1970, see also Lathrap 1968), as well as ethnological studies of indigenous communities living in the upper jungle of Ecuador and Peru (Varese 2016).

Eastward, as the Andes drops to an altitude of approximately 600 metres, their formation changes into softer hills and increasingly into plains and wetlands criss-crossed by slow-moving rivers carrying the rich sediments of millennia of human agricultural activity and geological Andean existence. Here begin the other Amazonia, The Lower Jungle [*La Selva Baja*], the immense extension of tropical rainforest and slow-flowing rivers rich in Andean silt that is deposited year after year on the riverbanks or *varzeas* and the flooded forest and lagoons; in this case, the Andean-Amazonian territory would constitute about 74.4% of the total national territory (Dourojeanni et al. 2010).

Ethics of Time and Place

One of the central principles of the Andean-Amazonian cosmologies is the notion of time-space warp—*pachakuti*. Time is written in space-place, space marks time. Humans, non-humans and other-than-humans all coexist in a time-space warp where past-present-future are not separate entities but a coexistent phenomenon so that the spirit protector of the animal, the spirit of the tree, the spirit of the water-river (*yaku)* is existing in the primordial original time, the present and the future at the same time. When I cut a tree to make a canoe, I am cutting the tree of the original time, the current tree and the future expression of the tree. That is why I have to ask permission, pay an offering, mourning the loss-death and celebrate the continuity of life. The ritual, even in its bear simplicity, is what enacts the world by acknowledging the origin, continuity and future of each entity of the forest with whom (not which) I interact. In modern Euro-American terminology, this indigenous cosmology is defined by the categories of History (times-space warp and 'long duration'); epistemology, way of knowing; and ethics: consequences of choices, decisions, treating the web of inter-relationships as 'all my relatives'.

The long historical presence and continuity of these indigenous ethical and cosmological praxes in the Peruvian Amazon region have been impacted drastically by the Euro-American invasion, colonial establishment, modernity and late expansion of unconstrained capitalist market economy and its extremely destructive extractivist version. A few hard-core, deeply rooted principles of the indigenous cosmologies have survived in clandestinity or as ideal practices of past generations and ancestors not anymore viable except as symbolic re-enactment in ceremonies, performative activities or oral narratives. In all the indigenous nationalities of Peru's Amazonia, the strength of native ethics tends to re-emerge when the community is threatened by external aggression. At least since the early seventeenth century, a series of punctual, local, regional or pan-ethnic movements of resistance and opposition to the aggressive penetration and expansion of the early and late capitalist economy in all its different manifestations have dotted the history of Amazonia.

From opposition to early colonial *encomiendas, haciendas* and missions, to the modern slavery of the rubber boom and plantations and the increasingly violent dispossession of territory lead by oil, mining, logging industries and agribusiness corporations, as well as roads, hydroelectric and waterways infrastructures initiatives with the full cooperation of the liberal state, indigenous/native communities have expressed organised responses to all these acts of blatant colonialism. A future brief history of colonial/neo-imperial aggression will certainly have to pay special attention to the decade of 1980–1990 that witnessed the horrors of a civil war waged against the Indigenous People by idiotic ideologues disguised as Maoists on one side and Neo-liberal chicanery on the other, both armed to their teeth by arm dealers and corrupt politicians, both determined to establish their version of hell on earth. The Campa-Ashánika people of the Selva Central lost more than 5000 women, men and children during this tragic period of their millenarian history. Thousands of other people of the Amazon region were displaced and lost their lands becoming refugees in their own country. No just peace for the people was ever achieved in central Amazonia that was rather transformed in the blossoming marketplace of drugs, weapons, illegal logging and ecological destruction.

Reshaping the Ethical Landscape of Amazonia

In approximately 200 years—since the nominal Independence of Peru in the early nineteenth century—the landscape and the ethnoscape of the Upper and Lower Amazon have been drastically modified by the imposition of different incarnations of the capitalist market economy: from the slavery and indentured peonage forced to gather precious hardwoods, animals and rubber tapping of the late nineteenth and early twentieth centuries, to the exorbitant usury lending to subsistence *ribereños* farmers, to the massive illegal and government-approved expropriation of communal territories for looting, logging, and mineral and oil exploitation, to the plain robbery of private and communal indigenous lands, all types of unbridled capitalist trickery have been tried and have shaped the social and natural landscape through what Harvey (2003) has called the 'accumulation by dispossession'. The complex web of anthropogenic ecosystems that for millennia were co-developed and constantly co-managed and renewed by Native/Indigenous Peoples together with 'All-Their-Other-than-Human-Relatives' had been turned into the hunting grounds for the 'aimless, senseless chaos of private interests' without any state government regulations or any ethical restriction.

A closer analysis of the socio-economic and environmental history of the past two centuries of Peru's Amazon region shows a classic example of Karl Marx's illustration of primitive accumulation (Bottomore 1983; Harvey 2003; Polanyi 1964) with all its range of processes: commodification and privatisation of land; forceful expulsion of Native/Indigenous People and Andean and Coastal colonist farmers *chacareros* and Amazonian *ribereños*; conversion of different forms of collective, communal and 'tribal' property rights into exclusive private property rights; suppression of the notion and rights of the commons (lands, forests, water, surface minerals, animals); commodification of labour and the suppression of alternative indigenous forms of production, circulation, consumption; internal colonial processes of appropriation of natural 'resources'; monetisation of exchange and elimination of the principles of reciprocity; commoditisation of land; normalisation of usury; and, lastly, a credit system that traps the impoverished borrowers into a cycle of permanent dispossession.

Looking back at the last 60 years of Peru's Amazonian ecological history, it seems clear that of all the victims of this violence exercised on the Amazon region by the capital in collusion with the state, the Indigenous People—the Native Communities—have been able to endure the aggression and through their political and cultural struggle have brought to the centre of national and international attention the ethical and environmental principles that should shape the relation of the rest of the country with the Amazon. The most enlightened participants of the Pan-Amazonian-Andean indigenous movement have been proposing what some scholar are calling a *biocentric perspective* on the rights of nature and peoples (Gudynas 2016), and I prefer to call a *cosmocentric perspective*, in order to emphasise the deep spiritual and ethical dimension of this ontological revolution (Varese 2013). The Amazonian Indigenous Peoples concept of 'socio-nature', and even the ontological shift to the notion of *kawsak sacha-living jungle* proposed recently by the *Sarayakuruna* intellectuals José and Patricia Gualinga, was already contained in the philosophical and ethical platform of the Kichwa people of Sarayaku along the Bobonaza River of Ecuardor announced more than 15 years ago as the indigenous response to the 'state-enabled immoral ecologies on economic activities – frontier-colonial agribusiness, oil exploitation, large-scale mining and commercial logging … extracted from the (indigenous lands) without (permission and without) reciprocally giving back (anything as just compensation)'.[5]

Today, in the midst of the apocalyptic scenario of global warming that threatens the survival of humanity and hundreds of thousands of species, the moral treatment of Amazonia and the reverence that all its tangible and intangible beings deserve become the fundamental challenge of our times, not only for the peoples living in Amazonian countries but for all the humanity of good will and ethical common sense. The Amazonian Indigenous Peoples—as all the other thousands of Native Peoples of the Americas—are telling us the *Original Instructions,* the *First Teachings,* the stories that makes the world an expression of place-based spiritual responsibility and cosmic cognitive pluralism. This is the sentient that minimises materialist science and its devastating commoditisation of reality, while shifting our focus towards a post-materialist science-knowledge where mind and consciousness become the centre of life.[6]

Sixty years ago, Theilhard de Chardin, the heterodox Catholic mystic palaeontologist, reminded us that 'consciousness is not a by-product of evolution; it is the purpose of evolution…as human, who occupy the pinnacle of the tree of life, we have a sacred obligation to participate responsibly in evolution … (and thus) the age of nations is past … (and) the task before us now, if we would not perish, is to build the earth' (paraphrased from Pruett 2015: passim 315–319). Today, as I write these final words of my chapter, I am reminded by the Kichwa people of Sarayaku in Ecuador of similar ethical considerations when they assert that '…the tropical rain forest of their Amazonian territory the "Kawsak Sacha is a living sentient being, with consciousness, constituted by all the entities of the jungle, from the smallest ones to the largest and the supreme ones. The Kawsak Sacha comprehends all the beings of the animal, vegetal, mineral, spiritual and cosmic world that are in constant communication

[5] Free paraphrased quote from an unpublished manuscript by Chris Ebdon, Doctoral Candidate at Yale University, Newhaven, CT.

[6] Manifesto for a Post-Materialist Science (7–9 February 2014, Tucson, AZ), *International Summit on Post-Materialist Science: Summary Report.*

among themselves and with us – the humans – offering to each other and to us the needed elements to renew our life, our thoughts, our spirits in a permanent flow of life energy that keeps the Native People in harmony with the Universe".'[7]

References

Apffel-Marglin, F. (2011). *Subversive Spiritualities. How Rituals Enact the World.* Oxford/New York: Oxford University Press.

Bottomore, T. (Ed.). (1983). *A Dictionary of Marxist Thought.* Cambridge, MA: Harvard University Press.

Dourojeanni, M., Barandiarán, A., & Dourojeanni, D. (2010). *Amazonía Peruana en 2021.* Lima: Sociedad Peruana de Derecho Ambiental.

Gudynas, E. (2016). *Deep Ecologies in the Highland and Rainforests – Finding Naess in the Neotropics.* CLAES (Arne Naess Chair in Global Justice and the Environment). University of Oslo.

Harvey, D. (2003). *The New Imperialism.* Oxford/New York: Oxford University Press.

Lathrap, D. W. (1968). The Hunting Economies of the Tropical Forest of South America: An Attempt at Historical Perspective. In R. Lee & I. De Vore (Eds.), *Man the Hunter.* Chicago: Aldine Publishing.

Lathrap, D. W. (1970). *The Upper Amazon: Ancient Peoples and Place.* London: Thames and Hudson.

Peat, F. D. (2002). *Blackfoot Physics. A Journey into the Native American Universe.* Boston/York Beach: WeiserBooks.

Polanyi, K. (1964 [1944]). *The Great Transformation. The Political and Economic Origins of Our Times.* Boston: Beacon Press.

Pruett, D. (2015). *Reason and Wonder. A Copernican Revolution in Science and Spirit.* Harrisonburg: Alba Enterprises LLC.

Taylor, A. C. (1996). The Western Margins of Amazonia from the Early Sixteenth to the Early Nineteenth Century. In S. Frank & S. B. Schwartz (Eds.), *The Cambridge History of the Native Peoples of the Americas, Volume III, South America.* Cambridge: Cambridge University Press.

[7] *Pueblo* Originario Kichwa de Sarayaaku, *Declaración. Kawsak Sacha- Selva Viviente, Ser Vivo y Consciente, Sujeto de Derewchos. Pueblo Originario Kichwa de Sarayaku,* Puyo y Sarayaku, June 2018.

Varese, S. (2013). La Ética Cosmocéntrica de los Pueblos Indígenas de la Amazonía: Elementos para una Crítica de la Civilización. In S. Varese, F. Apffel-Marglin, & R. Rumrrill (Eds.), *Selva Vida. De la Destrucción de la Amazonía al Paradigma de la Regeneración*. Copenhagen/México/La Habana: IWGIA/UNAM/Casa de las Américas.

Varese, S. (2016). Relations Between the Andes and the Upper Amazon. *Oxford Research Encyclopedia of Latin American History*. http://latinamericanhistory.oxfordre.com/view. Online Publication June 2016.

11

Voiceless Development, Toxic Injustice, Criminal Resistance: A Study of Peruvian Natural Resource Extraction Through the Political Ecology of Voice

Adrian Gonzalez

In his first 100 days in office, Brazil's new president, Jair Bolsonaro, moved with impressive speed to dismantle the country's environmental protection framework. Only hours after taking office in January 2019, the president enacted an executive order, which transferred the regulation and creation of new indigenous reserves to the agriculture ministry (Phillips 2019). According to the environment minister, Ricardo Salles, further measures have been implemented to counteract the previous administrations 'poor management' of the environment (Kaiser 2019). These include reducing the authority of the Brazilian Institute of Environment and Renewable Natural Resources (IBAMA), Brazil's environmental agency, and the Chico Mendes Institute for Biodiversity Conservation (ICMBio), which manages conservation areas, and eliminating senior official positions (IBAMA), Brazil's environmental agency, and the Chico Mendes Institute for Biodiversity Conservation (ICMBio), which manages conservation areas, and eliminating senior official

A. Gonzalez (✉)
School of Geography and Planning, Cardiff University, Cardiff, UK
e-mail: GonzalezA1@cardiff.ac.uk; adrian.gonzalez.2013@alumni.rhul.ac.uk

A. A. R. Ioris (ed.), *Environment and Development*,
https://doi.org/10.1007/978-3-030-55416-3_11

positions and replacing staff with military personnel (Brandford and Borges 2019; Kaiser 2019). The socio-environmental implications of developing 'unproductive' Amazon regions (Rocha 2019) have been vast. According to data from the government's satellite monitoring agency, deforestation in May 2019 (the first of three months in which forestry clearance and logging surge following the region's rainy season) totalled 739 square kilometres (285 square miles), equivalent to two football pitches every minute (Watts 2019; also TerraBrasilis n.d.). This was an increase from 550 square kilometres in May 2018 and more than double the deforestation rate detected two years earlier (Boadle 2019). This ongoing Amazonian destruction has led to rubber tappers and indigenous tribes uniting in opposition (Benassatto 2020).

From a wider regional political perspective, the rise of a far-right candidate like Bolsonaro has only occurred in Colombia (Goldstein 2019). Yet, Brazil's ongoing neoliberalisation of nature (McCarthy and Prudham 2004) through agribusiness and natural resource development, is evident throughout Latin America (Bebbington and Bury 2013; Deonandan and Dougherty 2016). Neoliberalism's focus on natural resource extraction, known as the 'extractive imperative' (Arsel et al. 2016), has caused significant environmental impacts (Johnston et al. 2019) and contestation, particularly in Global South contexts (Goyes et al. 2017). Conflict is apparent not only in terms of physical resistance to resource extraction (Nepomuceno et al. 2019) but also in the way that these struggles are disseminated or 'narrated' by national state and corporate actors (Dietz 2019). In Bolivia, for example, former president, Evo Morales, until his recent coup-d'état in November 2019, was swept to presidential power in 2005 on the back of 'indigenous nationalism' (Postero 2010) and an anti-neoliberal social protest movement (Singer 2007). However, the ongoing controversy surrounding the Isibore Sécure National Park and Indigenous Territory (TIPNIS) road development project, highlights the resilience of neoliberalism in a post-neoliberal age (Gautreau and Bruslé 2019). The former Morales government asserted that societal opposition is caused by the influence of international non-governmental organisation (NGO) 'pseudo-environmentalist' ideologies (Hope 2016) and developed country 'colonial environmentalism' (Collyns 2017). This is rather than recognising the tension between hydrocarbon extractivism on the one hand, and alternative development and environmentalism on the other (Hope 2016).

This chapter explores the contestation caused by this extractive imperative by focusing on the ability and willingness of citizens to hold oil companies accountable for pollution, an issue which particularly affects Global South extractive countries such as Ecuador, Nigeria and Peru (Adunbi 2015; Bebbington and Bury 2013; Sawyer 2004). Drawing upon recently published PhD research conducted in Peru's Loreto Region, it outlines the findings analysed through the political ecology of voice (PEV) theoretical framework.

The Political Ecology of Voice

To understand the causes of environmental pollution, one must grapple with wider environment-society relations, particularly natural resource management processes and underlying influences such as power, space and scale. There is therefore a need for a theoretical framework such as political ecology that can provide an integrative, holistic foundation for analysis. Political ecology, first emerging as a field of study in the 1970s, integrated political economy with ecology, with the aim of critically examining power relations and access and control of resources (Blaikie and Brookfield 1987; Blaser and Escobar 2016). Early political ecology work has given way to more recent conceptualisations. It is now influenced by areas such as feminism (Elmhirst 2011) and urban studies (Heynen 2013), which have sought to internationalise and decolonise the field (Loftus 2019), whilst there has been further academic debate on ontology (Schulz 2017). Its focus, however, remains on understanding how political, cultural and economic issues can cause injustice, and the ways this can be mitigated (Bryant 2015). Despite the absence of a single definition, there are certain core aspects, which provide a degree of cohesion to the field, while maintaining room for ongoing exploration.

Power plays a fundamental role within the framework (Ahlborg and Nightingale 2018) via a focus on control of land or resources by different actors at multiple scales and in various spaces (Bryant 2015). Power is intrinsically linked to scale and space. Political ecologists highlight how struggles of, and negotiations for, power, occur across context-situated, multi-level governance mechanisms and interactions (Perreault et al. 2015). Meanwhile, the framework emphasises how specific localities such as frontiers (Willow and Wylie 2014) or urban areas, can act as spaces of exclusion (Adams et al. 2018) or contestation (Behrsin and Benner 2017) that reinforce hegemonic power interactions and processes. Lastly, political ecology studies reflect the temporality of environment-society relations, recognising the need to not only study past (historical) actors, processes or events to understand the present (Simon and Peterson 2019) but also that 'present' studies are situated within a temporally dynamic, fluid environment. For example, Hvalkof (2006) shows how the indigenous Peruvian Asháninka people of Gran Pajonal transformed from exploited colonial labourers to active political agents. These four facets of political ecology (power, scale, space and time) are crucial for a study focusing on the ability and willingness of local people to hold oil companies accountable for pollution by reporting or 'voicing' their concerns. This idea of voice has been explored by one academic, Albert Hirschman, and is the second important element of the PEV framework.

Albert Hirschman, one of the twentieth-century founding fathers of political economy, explored public choice through his 1970 book *Exit, Voice and Loyalty: Responses to Decline in Firms, Organisations and States*. In it, Hirschman, set out the actions which consumers and citizens take when faced with unacceptable situations caused by businesses and states (Hirschman 1970). The first action is exit, in which consumers stop purchasing a firm's products, members leave an organisation or people choose to emigrate (Hirschman 1970, 1995). Generally, it is a neat (exit or non-exit), impersonal (no physical interaction between industry and consumer), indirect (impact of exit from company felt through loss of profit) silent, private decision and activity (ibid.). Voice, on the other hand, is radically different. It is an individual or collective action, conducted in a public, vocal manner and is messy, due to the different gradients that encompass it, from violent 'participation explosions' (Hirschman 1980: 432) down to faint grumbling (Hirschman 1970). Principally, it is an active action, the most notable being radical, participatory explosive acts, but it can also occur passively, for example, via rumours. These vocal acts occur in different situations, such as speaking with one's peers (horizontal voice), and conversations with higher-level actors like politicians (vertical voice) (O'Donnell 1986). This makes dialogue partly influenced by political freedoms and the opportunity for this action (Brenkert 2006; Goodwin and Jasper 2011).

In scenarios where public happiness is affected, such as an automobile safety problem, food hazard (Hirschman 1981) or, as PEV contends, an environmental issue (Gonzalez 2015), voice is more relevant than exit. This is because people have a strong motivation to defend their quality of life. Strong group loyalty, for example, to your local community (Dowding et al. 2000; Hirschman 1970; Zuindeau 2009) and particular livelihood attachments, such as employment (Groves 2015), prevent exit and activate voice (Hirschman 1970) due to time horizons, in which it is unlikely for citizens to immediately exit without first vocalising their concerns. Exit becomes an action of last resort or one pursued by the wealthy (Hirschman 1978) or those with weaker loyalty and livelihood attachments.

Voice is therefore an active expression of protestation against disagreeable issues (Gonzalez 2018a) and can broadly occur in several ways. There is *individual* voice in which an individual's action is to enact the desired change solely through this act (Dowding et al. 2000). However, in the context of environmental issues, vocalisation is predominantly *collective*, not only because of the variable number of people affected but also because solutions occur through vertical voice dialogue that is best achieved through collective solidarity in the face of hostility from state and wider society forces (Barrett 2014; Davenport 2000). Individuals vocalising within a vertical voice situation, and their lack of power in comparison to state and business elites, will not be able to 'achieve powerful speech, a legitimate, known, recognised, authorised and authoritative voice' (Bourdieu 1986: 301), without collectively working together. Radical voice actions are an extreme form of collective vocalisation.

One must also consider *who* is vocalising; within natural resource management and environmental contamination scenarios, there are multiple voices or stakeholders (Stoll-Kleemann and Welp 2006) present that can be amalgamated into three broad categories. Firstly, there is *individual*, that is, citizen voice articulation. There is also *institutional* voice, that is, the voice of the government or resource extraction industry (REI). The third is *collective*, including community-based organisations (CBOs), trade unions, student movements and NGOs. Therefore, while each actor has a different voice, they are also *interconnected* due to institutional and collective voice actors' ability to influence citizen voice, how and by whom it is heard (Gonzalez 2015) and whether it is recognised or listened to. PEV investigations involve 'the study of a specific temporal, economic, political, social and geographical environment in which various stakeholders (e.g. citizens…CBOs and…NGOs) utilise their voice over an environmental issue' (Gonzalez 2015: 467; see Fig. 11.1). Drawing upon political ecology and its core tenants—power, scale, space and time—PEV studies explore the ability of people, communities and other actors to vocalise environmental concerns and the influences and challenges in doing so. This incorporates investigating the political environment and the 'freedom of voice', space, scale, accessibility and development, non-state actor interactions and impact and, lastly, socio-economic and ethnic status. In turn, all these interconnections occur at different scales and spaces, operating in a temporally dynamic environment.

Black Gold Amidst the Green: Peru's Loreto Region

Peru's Amazon forms part of the wider western Amazonia region, one of the most biodiverse areas of the world (Pitman et al. 2002). The Food and Agricultural Organisation's most recent report for Peru found that 58 per cent of its landmass was covered by rainforest, giving it the second largest area after Brazil (Food and Agricultural Organisation of the United Nations, FAO 2016). Several Peruvian regional political territories

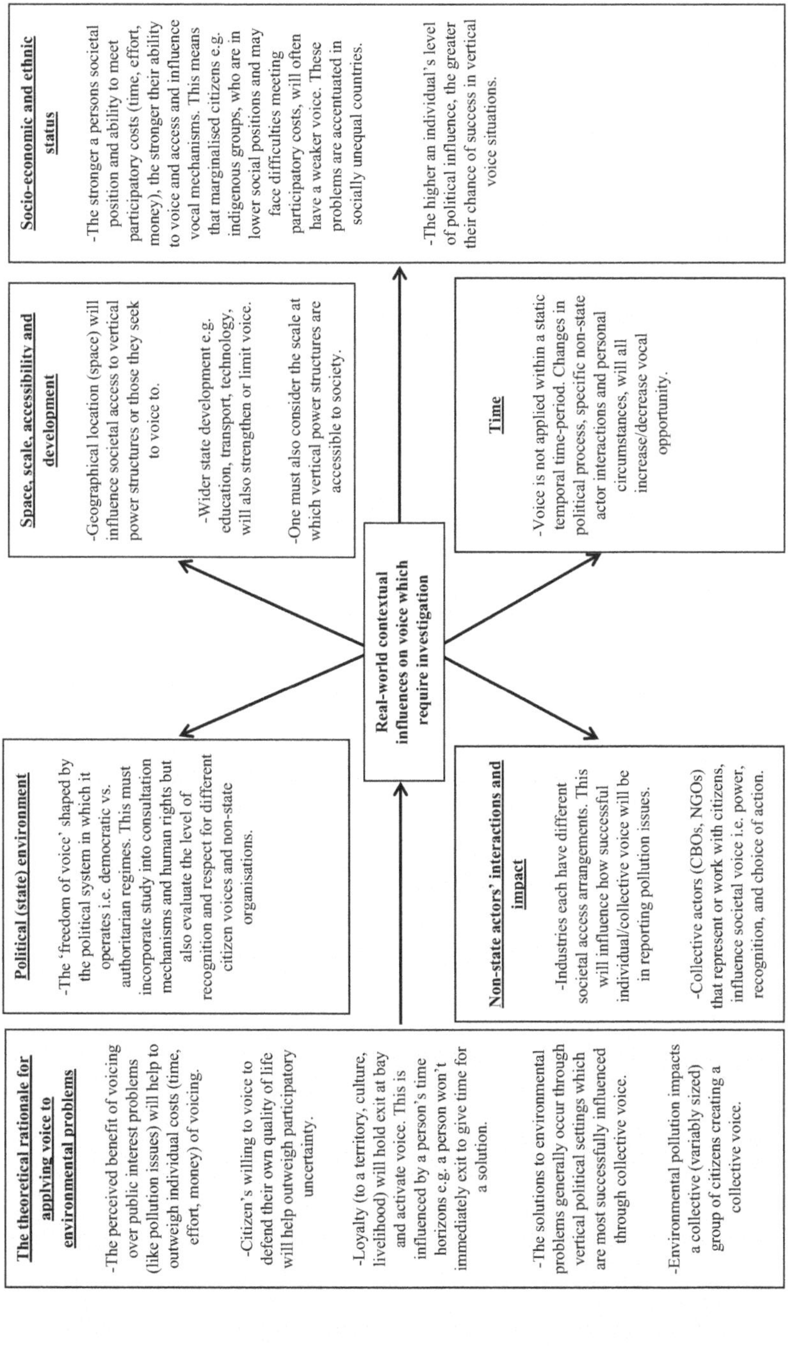

Fig. 11.1 The political ecology of voice (PEV). (modified from Gonzalez 2015: 479)

encompass this western Amazonia region, but the one encompassing the greatest proportion is Loreto. This region, 98 per cent of which is covered in rainforest (Brierley et al. 2014), covers nearly 369,000 square kilometres, making it bigger than Germany (Finer et al. 2013). It is home to approximately one million inhabitants, 45 per cent of whom live in the regional capital Iquitos, whilst the remainder reside in over 2000 smaller, remote river-edge communities (Brierley et al. 2014). Significantly, for this study into the extractive imperative, it is also one of Peru's most important onshore oil production sites.

Peru has a long history of oil production. After Edwin Drake's successful drilling of the first oil well in Titusville, Pennsylvania in 1859, initial discoveries in Peru were made on the Pacific Northwest coast and include the site of the western hemisphere's second oil well in an area just north of La Brea y Pariñas on 2 November 1863 (Pinelo 1973). Northwest Peru became the site of the country's early successes, centred on three key oil fields (La Brea, Lobitos and Zorritos). However, a decline in production in the 1930s necessitated exploration in unconventional rainforest or *selva* sites (Thorp and Bertram 1978). It was not until the early 1970s that oil was first discovered in the Loreto Region by the state-run company Petroperu (Haller et al. 2007), culminating in two key oil blocks that remain the longest running in the region. The first was Block 8, leased in 1969 and now operated by Argentinian firm Pluspetrol, located along the Corrientes River and within the Pacaya Samiria National Park. The second, Block 1AB, now Block 192, leased in 1971 and currently operated by Canada's Frontera Energy, is situated near the Ecuadorian border along the Corrientes, Pastaza and Tigre rivers. In 2014, these Blocks accounted for just under 40 per cent of Peru's total national oil production (Orta-Martinez et al. 2018). A further three Blocks have been leased for exploration. The crude oil is transported from these oil wells to the refinery operations on the Pacific coast by the North Peruvian pipeline and its northern branch, which were built in the mid-1970s (Philip 1984) (Fig. 11.2).

These oil Blocks and pipeline infrastructure have been a source of significant environmental pollution for several decades which has had an adverse impact on local indigenous community health and livelihoods (Rosell-Melé et al. 2018; Yusta-Garcia et al. 2017). Contamination

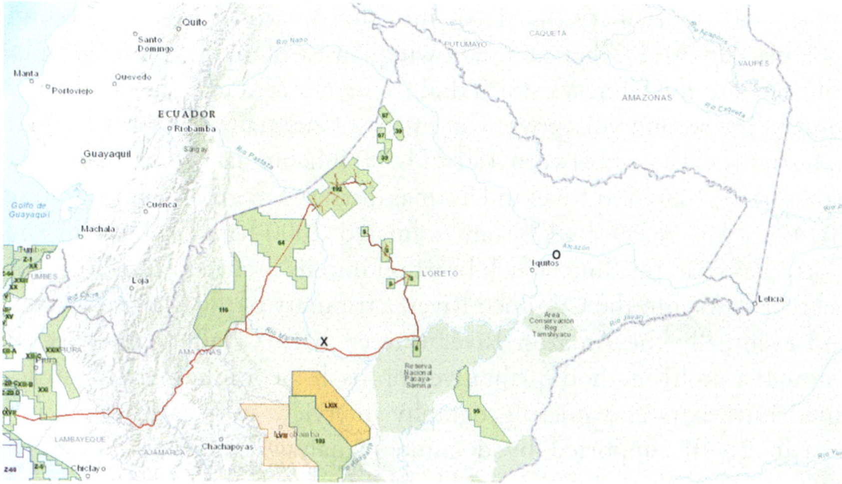

Key:
O = Approximate location of Barrio Flordio
X = Approximate location of Cuninico

Fig. 11.2 Location of Loreto's exploratory and production blocks and case study villages. (Perupetro n.d.)

remains a constant issue. In 2016, seven ruptures from the pipeline occurred, spilling an estimated 10,000 barrels of oil (Law in Action n.d.; Peru Support Group 2016), while a further 8000 barrels were lost in November 2018 (Cespedes 2018). This environmental pollution and local community health impacts, the chronic *selva* underdevelopment and the absence of free, prior and informed consent from local people, have led to ongoing indigenous protests, including river and pipeline blockades (The Chaikuni Institute and ORPIO 2018), oil installation seizures (Reuters 2019) and alleged pipeline vandalism (Cespedes 2018).

My doctoral research had focused on the REI Petroperu, that operates Loreto's pipeline infrastructure, and the experiences of two oil communities affected by the company's contamination. The first village was Barrio Florido, a non-indigenous community comprising roughly 800 people on the banks of the Amazon River, adjacent to the Iquitos Petroperu refinery in Puchana district, Maynas province. Villagers described two small spillages. The first was in 2009 when heavy rain caused four oil barrels to spill into the Ramirez River, a small creek situated between the village and the refinery that feeds into the Amazon River. The second took place in 2011, when an unknown quantity of oil from storage tanks connected to the refinery's storm drain system leaked into the same water source. The second village was Cuninico, a Cocama (also spelt Kokáma) indigenous village of between 450 and 500 inhabitants who reside on the banks of the Marañón River in Urarinas district, Loreto province. In June 2014, they were affected by an estimated 2000-barrel spill from the North Peruvian pipeline, which burst oil into its dredged flotation channel that flows into the Cuninico River, a tributary of the Marañón River, and eventually the Amazon River (Fraser 2016). Data were collected through a multi-method qualitative approach incorporating semi-structured interviews and micro-geography interview analysis (Elwood and Martin 2000) supported by document analysis. The semi-structured interviews, conducted with a paid local interpreter, were audio recorded with participant consent and anonymised. An interview guide was used, tested through pilot interviews and refined through contact summary sheets. Each participant was provided with a coded category made up of their broad professional occupation, ethnicity, geographical location and interviewee and organisation (if applicable) number (*n*). It was supported by secondary source analysis of relevant grey literature including government and industry documentation alongside academic research.

Pre-fieldwork contact was made with potential community gatekeepers, found through a Google Internet search. These were three NGOs, E-Tech International (NGO1R1), The Peru Mission (NGO2R2) and Alianza Arkana (NGO7R1), and an indigenous environmental monitoring CBO, Red Ambiental Loretana (CBO1R1), all, barring E-Tech International, based in Iquitos. A theoretical sampling approach was used for data collection, assisted through the snowballing method (Taylor and Bogdan 2016). Altogether, 110 semi-structured interviews were conducted with 105 interviewees from the different stakeholder categories and other citizens of interest. Several UK translators created the interview transcripts, which were analysed through QSR International Nvivio software, in which data were coded through different typologies via a weak form of inductive content analysis (ibid.). Inductive analysis was integrated with predominantly analyst-based typologies alongside indigenous or emic typologies (Bazeley 2013) so that interviewees were 'heard' and utilised in a way that did not alter their voice. Emotive phrases have been written in Spanish with a bracketed English translation to connect with the fieldwork context and participant voice. The coding process was underpinned by development of a loose research storyline based upon the PEV theoretical framework.

'A Beggar Sitting on a Bench of Gold': Peru's Political Environment and Freedom of Voice

As the PEV theoretical framework diagram indicates, the political environment has a significant influence on voice. Broadly, one can distinguish between authoritarian and democratic contexts. To gain power and stability, extreme authoritarian regimes will try and limit participatory encounters (Stiefel and Wolfe 1994) by collapsing public and private society through the suppression of voice. Meanwhile, in democracies, freedom of speech and equality are central components of democratic participation, most visible in regular free and fair elections and societal right to defend one's own interests without fear of reprisal or external interference. As Hirschman (1982: 4–5) summarises, 'it is easier to participate in a public protest when one only loses time through the act of participation than when one thereby risks losing one's head'. How can we evaluate Peru's political environment and what impact does it have on citizens vis-à-vis the extractive imperative and reporting environmental pollution?

Peru has a chequered political history. Since independence in 1821, the country has witnessed 13 constitutions and only 9 of 19 elected governments completing their terms in office (Taft-Morales 2013). The military was at the forefront of shaping early Peruvian post-independence politics in the nineteenth and twentieth centuries, which helped to stifle democracy until the twenty-first century, when successive free and fair elections have been held since 2001. Problems in Peru of course remain, including police abuse, women's and girl's rights and freedom of expression for journalists (Human Rights Watch 2020). Whilst contemporary Peru continues to consolidate democratic norms and principles, its political environment affords clear civil rights and responsibilities that appear to enable a predominantly open freedom of voice for stakeholders to voice (Gonzalez 2018a).

The nation's 1993 constitution sets out various rights under Title 1: The Person and Society Chapter I, Article 2, which provide civil rights and liberties surrounding freedom of expression (Constitute n.d.). These rights are protected and enforced through the Office of the Ombudsman. At a regional and international level, Peru is a member of the Organization of American States (OAS)[1] and is also a signatory of relevant treaties including the International Covenant on Civil and Political Rights (ICCPR) (ratified 1978), Convention on the Rights of the Child (ratified 1990) and the International Labour Organization (ILO's) Indigenous and Tribal Peoples' Convention 1989 (ILO Convention 169) (ratified 1994). Though Peru suffers from weak state institutions and regulations that undermine the implementation and oversight of these democratic norms (Gustafsson and Scurrah 2019; Wright 2015), one would still expect this political environment to provide citizens with a strong freedom of voice. However, focusing on the political environment surrounding hydrocarbon extraction highlights a far more suppressive and dangerous freedom of voice. The development of onshore hydrocarbon sites in Peru necessitates a legal obligation to consult with local people. This occurs through two mechanisms, the law of prior consultation, a seven-stage consultation process based upon ILO Convention 169 (Peru Support Group n.d.), and environmental impact assessments and their informative workshops and public hearings (Stetson 2012). Analysis indicates that citizens are denied the opportunity to provide free, prior and informed consent to development projects (Gonzalez 2018a), a situation summarised by one NGO interviewee as "*a salute to the bandido* (bandit)" (NGO2R1c).

One of the primary causes of this consultative suppression is Peru's advancement of a highly neoliberal extractivist agenda, termed a '*selva*

[1] Key Organization of American States (OAS)-affiliated mechanisms are the Inter-American Commission on Human Rights and Inter-American Court of Human Rights. The key conventions are the American Declaration on the Rights and Duties of Man (signed 1948) and American Convention on Human Rights 'Pact of San Jose, Costa Rica' (ratified 1978) and its additional protocols.

hydrocarbon and development vision (SHDV)' (Gonzalez 2018a). The emphasis on extractivism and large-scale development projects[2] is implemented through the deliberate dispossession of indigenous citizens, not only of their opinions surrounding these projects but also their land. A symbolic phrase in Peru is that it is 'a beggar sitting on a bench of gold', unable to exploit its extractive oil wealth due to indigenous people who are hindering the nation's progress. This racialised development thinking (Drinot 2006) and silent racism (de la Cadena 2001) can be traced back to indigenous people's exploitation in the Amazonian rubber boom during the nineteenth century (Hvalkof 2000). Peru has also sought to violently stifle social protests surrounding these hydrocarbon projects through new or modified legislation. For example, a 2008 legal amendment removed the requirement for the government to declare a state of emergency before army deployment (Supreme Decree 012-2008-DE-CFFA) (Latin American Andean Group Report 2010), whilst armed forces are permitted to use lethal force against 'hostile groups' (Law No. 1095/2010) (Front Line Defenders 2014). Meanwhile, SHDV opponents, notably environmental and human rights defenders, are targeted by the state through smear campaigns to delegitimise their work or judicial harassment, weakening societal trust of these actors and increasing the likelihood of violence against them (Global Witness 2018).

Silencing Through Fear: Petroperu—Community Engagement

This political repression is also evident in the relationship that Petroperu has with Barrio Florido and Cuninico. Research has found that REIs are able to wield a huge degree of power and influence over local extractive communities, particularly in remote Global South localities (Maconachie and Hilson 2013). They can control or withhold citizen access to

[2] These include the Amazon Water Project, which will deepen and widen the Amazon and its major tributaries to improve navigability for larger vessels (Giardino 2018) and the Moyobamba-Iquitos electricity concession, a project to construct a 600 kilometre 220 kilovolt transmission line across the *selva* to Iquitos (Lo Lau 2018).

governance mechanisms (Himley 2008), whilst their investment in local Corporate Social Responsibility (CSR) development projects (Gilberthorpe and Banks 2012) can often lead them to effectively becoming 'the state' through basic service provision, such as healthcare or energy (Gonzalez 2016). Efforts to improve REI community engagement into more purposeful interactions have recently occurred. The Organisation for Economic Co-operation and Development (OECD) has developed the concept of meaningful engagement, which is defined as ongoing, two-way engagement, conducted in good faith and responsive to community concerns, for example, remedial action to environmental pollution (OECD 2015). However, a PEV study (Gonzalez 2018b) indicates that this is not the reality for either case study community.

Cuninico residents confirmed that Petroperu had never visited the village prior to the 2014 pipeline spill, which left local people without any idea of who operated the pipeline or how to contact them. This is surprising, given that Petroperu has staff located only three hours away at the North Peruvian pipeline pumping station 1. Villagers were only able to alert Petroperu's coastal refinery through a phone number on the back of a school textbook sponsored by the company and distributed by the Ministry of Education. In Barrio Florido, their geographical proximity means there is more interaction, but it is controlled by Petroperu. The refinery manager holds infrequent meetings with the village's locally elected volunteer authorities, principally about the cultivation work for which residents are paid to complete by Petroperu.[3] Wider village petitions to Petroperu can only be made through these local elites. Unlike Cuninico, there is also wider Petroperu CSR support such as the supply of water for the village water tank, the provision of groceries during the rainy season and electricity. The provision of this energy alongside low-skilled manual labour jobs, which almost all villagers relied on, represents the most important aspect of this community's engagement with Petroperu (ibid.).

One can see that meaningful engagement is either non-existent or strictly controlled by Petroperu, that has sought to decouple itself,

[3]This involved cutting away the vegetation surrounding the refinery pipeline, which runs to Iquitos.

whenever possible, from *selva* inhabitants. Crucially, this power enables them to establish a 'climate of fear' in both villages that seeks to suppress the willingness of citizens to report and hold the company accountable for pollution issues. Interviews with Cuninico residents assert that 150 predominantly male villagers were offered remediation jobs in exchange for their silence about the spill occurring. They were given the most difficult and dangerous task of locating the pipeline break without any training, protective clothing or heavy lifting equipment, highlighting Peru's entrenched discrimination. Despite immediate health problems, their poverty and job loss warnings from Petroperu staff initially prevented them from speaking out. Similarly, in Barrio Florido, their dependency on the refinery for electricity and jobs meant that there was great concern about criticising Petroperu. Rumours about the possible loss of electricity or the termination of short-term employment contracts were rife in the village, helping stifle vocal discontent about company malpractice. Moreover, the absence of impartial volunteer authorities, almost all of whom worked in the refinery, damaged community trust in their local elites who many believed were bribed or threatened with job losses to keep them quiet (ibid.).

Despite this power inequality and climate of fear, both villages were able to vocalise their environmental struggles for two key reasons. Firstly, strong resolute citizens were key. In Barrio Florido, two residents, who did not work for the refinery and therefore were unaffected by the rumours of possible job losses, chose to alert the state about the pollution events. Meanwhile, Cuninico's *Apu* (community leader) was viewed by residents as a principled, honourable man who would fight for the fair treatment of the village. He was convinced by another villager, with strong ties to the Catholic Church, to alert Catholic priests to the spill, thereby indicating the importance that external trusted non-state actors can have in these situations (ibid.).

Inaccessible and Ignored: Shadow Environmental Citizenship in Peru's Loreto Region

So far, Peru's political context and Petroperu's community relationships makes it challenging for local people to hold oil companies accountable for pollution. The situation is exacerbated by PEV research (Gonzalez 2019), indicating the peripheral access that rural, predominantly indigenous, communities have to environmental justice through the state. Environmental justice has evolved from an initial focus on the distribution of environmental pollution to a three-way concept (Schlosberg 2007) that also acknowledges the importance of participation (how and who participates) and recognition (accommodating all visions and values) (Fraser 2000; Schlosberg 2004). The participatory element of environmental justice indicates the significance of access, a concept that Ribot and Peluso (2003) have reflected on in their influential theory of access paper. Several key aspects are important to mention. These are access to authority, which is intrinsically linked to social and economic selectivity (Robbins 2000). Middle- and high-income earners can more easily bear the participatory costs, for example, time, effort and money, than low-income households (Laurian 2004), especially those in rural areas, whose access in relation to human mobility and state capacity is more limited (Kuklina and Holland 2018). Access also often occurs through social identity or community membership, for example, gender and ethnicity (Ribot and Peluso 2003), whilst NGOs can determine access to environmental justice along identity lines (Buckingham and Kulcur 2009). PEV acknowledges these different access features by exploring space, scale, accessibility and development and social, economic and ethnic impacts.

Centralisation (CBO1R1; IUA3; NGO2R1b) was the key word used by interviewees to describe the state in Loreto, which was predominantly focused on Iquitos and other provincial capitals (Gonzalez 2019). In rural localities, the state presence, evident in the form of basic infrastructure such as educational, health, legal, transport and energy provision, alongside active political representation was *absent* (CBO4R3; NSI4R1; RGR3), demonstrating the high levels of chronic poverty found within

Loreto's districts (World Bank Group 2017). Both case-study villages highlighted this situation, with neither having legal or secondary school educational provision nor limited access to medical facilities, clean water and transport links. Cuninico residents were reliant on generators for electricity, whilst Barrio Florido had electrical provision from Petroperu's refinery rather than the state (Gonzalez 2018b). This underdevelopment significantly weakened citizen access to the state and therefore environmental justice. The absence of Information and Communications Technology (ICT) connectivity means that an online environmental justice mechanism, the National Information System on Environmental Complaints (SINADA), administered by the Agency for Environmental Assessment and Enforcement (OEFA), is inaccessible for rural Loreton communities (OEFA n.d.). Though complaints can be made in person at OEFA's decentralised offices located in each regional capital (in this case, Iquitos), the infrequent and expensive nature of the Loreton non-public water transport system further impedes citizen access to state-based environmental justice.[4] This is especially true for rural indigenous communities like Cuninico whose increased distance and poverty as subsistence farmers and fisherman make trips to Iquitos all but impossible (Gonzalez 2019).[5]

Access to environmental justice is further constrained by the negligible village interaction with elected political representatives who often only visit during electoral campaigns. Aside from discrimination, a key driver is socio-spatial inequality, in which there is an explicit effort by urban Peruvian residents to detach themselves from a perceived inferior space (the rural) and culture (indigenous) (Paredes 2008). This situation leads to low levels of societal trust in the government (Gonzalez 2019) but also forces communities, like Barrio Florido and Cuninico, to live beyond formal state frameworks via their own informal governance systems.

[4] Transport vessels can be divided into two broad categories; large, slow, cargo and passenger-carrying *motonaves fluviales* (or *lancha*) and smaller *botes fluviales* (or *bote*) such as canoes or speed boats (*delizadores* or *rápido*) (Vuori 2009). Barrio Florido had four or five 30-minute weekly *bote* services, costing 12 *soles* (equivalent to US$4 in January 2020). Cuninico had only one weekly 11-hour *lancha* or *rápido* service to Nauta, the port city 98 kilometres from Iquitos costing 90 *soles* by *lancha* or 180 *soles* by *bote* (equivalent to US$29 and US$54, respectively).

[5] For rurally remote indigenous Peruvian inhabitants, earnings are predominantly below US$1 per day (Brierley et al. 2014).

However, these can exacerbate rather than improve state access. Unsurprisingly, citizens in both communities were left with limited options to vocalise and access state-based environmental justice. Two Barrio Floridon residents were forced to ring a local radio station to alert authorities to both oil spills, an example of an unorthodox informal voice action, in which vertical or horizontal voicing occurs through an irregular method. Meanwhile, Cuninico alerted the authorities through difficult but ultimately successful communication with Petroperu and the two Catholic Church priests (Gonzalez 2018b).

However, the state's absence of support for these oil spills highlights the inequality of environmental justice in Loreto. Despite different state agencies such as the Ministry of Environment of Peru and Office of the Ombudsman visiting both villages, residents asserted that environmental quality results were not distributed. Other practical support and recognition for these contamination events were poor. In Barrio Florido, the Office of the Ombudsman failed to confirm to citizens whether Petroperu had a legal case to answer. In Cuninico, citizens were given only limited food and water provisions for four months after the spill, leaving ongoing clean water access a significant challenge (Gonzalez 2019). Through PEV, one can conclude that these citizens have been left with shadow environmental citizenship (Gonzalez 2019, 2020). This concept, building on the idea of environmental citizenship advocated by Dobson and Bell (2006), asserts that rural, predominantly indigenous, communities cannot access or claim their environmental rights from the state, which fails to adequately support or recognise these citizenship grievances. Shadow environmental citizenship forces communities to utilise unorthodox or radical voice actions, for example, oil pipeline sabotage or rely on their links to non-state actors such as NGOs, to gain access to, and recognition for, environmental injustice (Gonzalez 2019, 2020).

The use of radical and unorthodox vocal acts and reliance on non-state actors both present challenges for local people seeking environmental justice. Radical, contentious acts can appear to achieve success. For example, in 2006, non-violent indigenous Achuar mobilisations in Loreto helped lead to The Dorrissa Accords, which stipulated the reinjection of all produced water by Pluspetrol in Blocks 8 and 1AB/192 and investment in local development schemes (Orta-Martinez et al. 2018). However, the failure by the state to implement the developmental aspects of the agreement indicates a clear strategic response by the state to placate societal concerns through tokenistic dialogue. Moreover, the resumption of ongoing contentious radical acts (Taj and Bell 2019) enables the state to depict indigenous people as 'savages' and environmental and human rights defenders as 'troublemakers', helping to weaken wider societal support and protect the SHDV. At the same time, environmental injustice issues remain ignored and unaddressed. Environmental justice through non-state actors is also very complex. Globally, one finds that non-state actors have gained a significant but challenging role in administering community development schemes and fostering societal empowerment in a wide array of areas. These roles are amplified in environments like rural Loreto, where state underdevelopment and shadow environmental citizenship operate. In relation to environmental justice, NGOs can help to scale up local place-based justice struggles (Perreault 2003) and offer citizens new spatial opportunities of interaction. Consequently, whilst voice remains an active expression of protest against a disagreeable issue (Gonzalez 2018a), in the context of environmental justice, it can also be viewed as an attempt to make injustice problems *visible* and *heard* by the wider world, the latter apparent through the redressing of specific injustice issues (Gonzalez 2020).

The difficulty is that each non-state organisation will have their own ideas of what justice looks like and how it can be achieved. In Loreto, it was apparent that non-state actor supported justice was sought through three broad methods—support for radical action, for example, Red Ambiental Loretana (CBO1R1) facilitating the 2006 Achuar indigenous protests, private negotiated financial agreements between REIs, communities and the respective CBO federations and lastly, legal action. Whilst the potential pitfalls with radical action were described earlier, some interviewees (NGO4R2b; NSI2R1) also expressed concerns with negotiated agreements, which were seen to stoke community tensions and not achieve a valid form of legal justice. Yet, indigenous impoverishment, ongoing underdevelopment and environmental degradation highlight why these communities do deserve financial compensation for these issues (ibid.). In the case of Cuninico, their links to the Catholic Church, and the affiliated Catholic Human Rights Commission, enabled the local injustice claim to become visible by securing access to a transnational legal mechanism (the Inter-American Commission on Human Rights) who recognised and listened to their grievances and have sought to remediate them (OAS 2017). Legal successes like this are not guaranteed as the historic 1994 Aguinda v. Texaco court case in Ecuador indicates (see BBC News 2018). Nevertheless, if successful, it can help reform illegal or poor operational practices, which can have wider beneficial impacts for other communities suffering injustice. For example, last year, the Awajún and Wampins peoples, in collaboration with the non-state actor Legal Defence Institute (IDL), won a historic legal victory as a Peruvian court ordered the suspension of all hydrocarbon activities in Block 116 for REI failure to consult and secure their consent (Forest Peoples Programme 2018).

One can see that non-state actors help create an array of environmental justice visions in Loreto that does little to equalise environmental justice or combat shadow environmental citizenship. This is not a criticism per se; rather, it is a reflection that these actors are not best placed to implement environmental justice, as they do not operate everywhere in Loreto (for example, Barrio Florido), and cannot be expected to provide core state services such as legal provision or education. Whilst Cuninico's voice has been heard and their injustice issue listened to, other community expressions and concerns remain unheard and unaddressed (Gonzalez 2020).

Conclusion

Overall, these PEV studies reveal how the extractive imperative in Loreto makes it difficult for rural citizens to hold oil companies accountable for pollution. Peru's extractive imperative, evident from the SHDV, is preventing citizens from having an open freedom of voice. Nationally, there is an absence of free, prior, informed consent, whilst the state seeks to delegitimise and attack environmental and human rights defenders. In Loreto, Petroperu disconnect themselves from meaningful societal participation, instead establishing a climate of fear to suppress critical voices, a situation accentuated by state absence and the dependency that communities have on their CSR development initiatives. Chronic underdevelopment, poverty and state disinterest in rural localities further reinforce community isolation and consequently help lead to shadow environmental citizenship. Like Peru's development views, one can see the absence of accessible and equitable (environmental) citizenship rights for rural, predominantly indigenous communities borne out of discrimination and racism. It is therefore unsurprising to find citizens utilising radical or unorthodox vocal acts to be heard or seek environmental justice, as the increasing levels of social unrest, particularly surrounding hydrocarbon development, indicate (Defensoria del Pueblo 2017, 2018). The state responses of suppression and placation maintain the cyclical nature of Peru's state-society relations in which environmental and developmental concerns remain ignored. Meanwhile, rural societal reliance on non-state actors does not equalise environmental justice or guarantee the accountability of REIs.

Success like Cuninico's is possible, but these PEV Loreton studies indicate multifaceted and connected power, space and scale challenges in ongoing temporal realities that serve to disrupt and limit the ability of rural communities to be heard and listened to. One must also consider this study in relation to the wider Latin American context. Loreto's challenging PEV environment operates in a country with a broadly moderate political context in comparison to the present situation unfolding in Bolivia, Brazil, Colombia and Venezuela. Research shows that in these extremist political circumstances, respect for the rule of law and participatory, inclusive dialogue is usurped in favour of wholesale repression and brutality (Human Rights Watch 2018a, b, c, d). For local communities and non-governmental actors, the hostile threat posed by the state is exacerbated, a situation requiring the widespread use of radical, contentious acts to try and be heard. Whilst the PEV studies in Peru indicate that the extractive imperative cuts across the political spectrum, the challenges of voicing are more keenly felt in these populist, authoritarian settings, in which (environmental) citizenship rights are ignored.

References

Adams, E. A., Jurah, L., & Ajibade, I. (2018). 'Spaces of Exclusion' in Community Water Governance: A Feminist Political Ecology of Gender and Participation in Malawi's Urban Water Use Associations. *Geoforum, 95*, 133–142.

Adunbi, O. (2015). *Oil Wealth and Insurgency in Nigeria*. Bloomington: Indiana University Press.

Agency for Environmental Assessment and Enforcement (OEFA). (n.d.). *Environmental Complaints Service*. www.oefa.gob.pe/en/sinada. Accessed 23 Oct 2019.

Ahlborg, H., & Nightingale, A. J. (2018). Theorising Power in Political Ecology: The Where of Power in Resource Governance Projects. *Journal of Political Ecology, 25*, 381. https://journals.uair.arizona.edu/index.php/JPE/article/view/22804. Accessed 24 June 2019.

Arsel, M., Hogenboom, B., & Pellegrini, L. (2016). The Extractive Imperative in Latin America. *The Extractive Industries and Society, 3*, 880–887.

Barrett, P. M. (2014). *Law of the Jungle: The $19 Billion Legal Battle Over Oil in the Rain Forest and the Lawyer Who'd Stop at Nothing to Win*. New York: Penguin Random House.

Bazeley, P. (2013). *Qualitative Data Analysis: Practical Strategies*. London: Sage.

BBC News. (2018, September 8). Chevron Wins Ecuador Rainforest 'Oil Dumping' Case. www.bbc.co.uk/news/world-latin-america-45455984. Accessed 10 Apr 2019.

Bebbington, A., & Bury, J. (2013). *Subterranean Struggles: New Dynamics of Mining, Oil, and Gas in Latin America*. Austin: University of Texas Press.

Behrsin, I., & Benner, C. (2017). Contested Spaces and Subjectivities of Transit: Political Ecology of a Bus Rapid Transit Development in Oakland, California. *Journal of Transport Geography, 61*, 95–103.

Benassatto, L. (2020, January 16). Brazilian Tappers and Forest Tribes Unite Against Bolsonaro. *Reuters*. https://uk.reuters.com/article/uk-brazil-indigenous/brazilian-tribes-and-forest-tappers-unite-against-bolsonaro-idUKKBN1ZF08M. Accessed 16 Jan 2020.

Blaikie, P., & Brookfield, H. (Eds.). (1987). *Land Degradation and Society*. London: Routledge.

Blaser, M., & Escobar, A. (2016). Political Ecology. In J. Adamson, W. Gleason, & D. Pellow (Eds.), *Keywords in the Study of Environment and Culture* (pp. 164–167). New York: New York University Press.

Boadle, A. (2019, June 4). Satellite Data Shows Amazon Deforestation Rising Under Brazil's Bolsonaro. *Reuters*. www.reuters.com/article/us-brazil-environment-deforestation/satellite-data-shows-amazon-deforestation-rising-under-brazils-bolsonaro-idUSKCN1T52OQ. Accessed 10 June 2019.

Bourdieu, P. (1986). The Antinomies of Collective Voice. In A. Foxley, M. S. McPherson, G. A. O'Donnell, & A. O. Hirschman (Eds.), *Development, Democracy, and the Art of Trespassing: Essays in Honour of Albert. O. Hirschman* (pp. 301–302). Notre Dame: University of Notre Dame Press.

Brandford, S., & Borges, T. (2019, May 8). Dismantling of Brazilian Environmental Protections Gains Pace. *Mongabay*. https://news.mongabay.com/2019/05/dismantling-of-brazilian-environmental-protections-gains-pace. Accessed 10 June 2019.

Brenkert, G. G. (2006). *Political Freedom*. London: Routledge.

Brierley, C. K., Suarez, N., Arora, G., & Graham, D. (2014). Healthcare Access and Health Beliefs of the Indigenous in Remote Amazonian Peru. *The American Journal of Tropical Medicine and Hygiene, 90*(1), 180–183.

Bryant, R. L. (Ed.). (2015). *The International Handbook of Political Ecology*. Cheltenham: Edward Elgar Publishing Limited.

Buckingham, S., & Kulcur, R. (2009). Gendered Geographies of Environmental Injustice. *Antipode, 44*(4), 659–683.

Cespedes, T. (2018, November 28). Vandals Causes 8,000 Barrels of Oil to Spill in the Peruvian Amazon. *Reuters*. www.reuters.com/article/us-peru-oil-spill/vandals-cause-8000-barrels-of-oil-to-spill-in-peruvian-amazon-idUSKCN1NX2LK. Accessed 1 July 2019.

Collyns, D. (2017, August 15). Bolivia Approves Highway Through Amazon Biodiversity Hotspot. *The Guardian*. www.theguardian.com/environment/2017/aug/15/bolivia-approves-highway-in-amazon-biodiversity-hotspot-as-big-as-jamaica. Accessed 21 June 2019.

Constitute. (n.d.). Peru's Constitution of 1993 with Amendments Through 2009. *Constituteprogect.org*. www.constituteproject.org/constitution/Peru_2009.pdf?lang=en. Accessed 18 Oct 2019.

Davenport, C. (Ed.). (2000). *Paths to State Repression: Human Rights Violations and Contentious Politics*. Lanham: Rowman & Littlefield Publishers.

De la Cadena, M. (2001, September 3–5). The Racial Politics of Culture and Silent Racism in Peru. Paper Presented at the United Nations Research Institute for Sustainable Development (UNRISD). *Racism and Public Policy Conference*, Durban, South Africa. www.unrisd.org/80256B3C005BCCF9/

(httpAuxPages)/EE7EB1E30A96C11F80256B6D00578643/$file/dcadena.pdf. Accessed 15 Jan 2020.

Defensoria del Pueblo. (2017). Reporte de conflictos sociales No. 166. www.defensoria.gob.pe/wp-content/uploads/2018/07/Reporte-Mensual-de-Conflictos-Sociales-N-166-Diciembre-2017.pdf. Accessed 10 Apr 2019.

Defensoria del Pueblo. (2018). Reporte de Conflictos Sociales No. 174. www.defensoria.gob.pe/wp-content/uploads/2018/09/Conflictos-Sociales-N%C2%B0-174-Agosto-2018.pdf. Accessed 10 Apr 2019.

Deonandan, K., & Dougherty, M. L. (Eds.). (2016). *Mining in Latin America: Critical Approaches to the New Extraction*. Abingdon: Routledge.

Dietz, K. (2019). Contesting Claims for Democracy: The Role of Narratives in Conflicts Over Resource Extraction. *The Extractive Industries and Society, 6*(2), 510–518.

Dobson, A., & Bell, D. (Eds.). (2006). *Environmental Citizenship*. Cambridge, MA: Massachusetts Institute of Technology.

Dowding, K., John, P., Mergoupis, T., & Van Vugt, M. (2000). Exit, voice and loyalty: analytic and empiric development. *European Journal of Political Research, 37*(4): 469–495.

Drinot, P. (2006). Nation-building, Racism and Inequality: Institutional Development in Peru in Historical Perspective. In J. Crabtree (Ed.), *Making Institutions Work in Peru: Democracy, Development and Inequality Since 1980* (pp. 5–23). London: Institute for the Study of the Americas.

Elmhirst, R. (2011). Introducing New Feminist Political Ecologies. *Geoforum, 42*(2), 129–132.

Elwood, S. A., & Martin, D. G. (2000). 'Placing' Interviews: Location and Scales of Power in Qualitative Research. *The Professional Geographer, 52*(4), 649–657.

Finer, M., Jenkins, C. N., & Powers, B. (2013). Potential of Best Practice to Reduce Impacts from Oil and Gas Projects in the Amazon. *PLoS One, 8*(4), e63022. www.ncbi.nlm.nih.gov/pmc/articles/PMC3641117. Accessed 1 July 2019.

Food and Agriculture Organisation of the United Nations. (2016). *Global Forest Resources Assessment* 2015. How Are the World's Forests Changing? Rome: FAO.

Forest Peoples Programme. (2018, August 22). *Historic Legal Victory for Awajún and Wampis People*. www.forestpeoples.org/en/node/50275. Accessed 29 Aug 2019.

Fraser, N. (2000). Rethinking Recognition. *New Left Review, 3*(3), 107–20.

Fraser, B. (2016). Oil in the Forest: A Rash of Spills in the Amazon Has Scientists and Villagers Scrambling to Understand Oil's Effects on the Ecosystem. *Science, 353*(6300), 641–643.

Front Line Defenders (FLD). (2014). *Environmental Rights Defenders at Risk in Peru*. www.frontlinedefenders.org/en/statement-report/environmental-rights-defenders-risk-peru. Accessed 18 Oct 2019.

Gautreau, P., & Buslé, L. P. (2019). Forest Management in Bolivia Under Evo Morales: The Challenges of Post-neoliberalism. *Political Geography, 68*, 110–121.

Giardino, N. (2018, October 23). Peru's Natives Say Amazon Waterway Project Threatens Food Sources. *Al Jazeera*. www.aljazeera.com/indepth/features/peru-natives-amazon-waterway-project-threatens-food-sources-181022164802018.html. Accessed 18 Oct 2019.

Gilberthorpe, E., & Banks, G. (2012). Development on Whose Terms? CSR Discourse and Social Realities in Papua New Guinea's Extractive Industries Sector. *Resources Policy, 37*(2), 185–193.

Global Witness. (2018). At What Cost? Irresponsible Business and the Murder of Land and Environmental Defenders in 2017. https://www.globalwitness.org/en/campaigns/environmental-activists/at-what-cost. Accessed 27 Mar 2019.

Goldstein, A. A. (2019). The New Far-right in Brazil and the Construction of a Right-wing Order. *Latin American Perspectives, 46*(4), 245–262.

Gonzalez, A. (2015). The Political Ecology of Voice (PEV): An Innovative Approach to Examining Environmental Pollution and the Accountability of Economic Actors. *Journal of Political Ecology, 22*(1), 466.

Gonzalez, A. (2016). Poverty, Oil and Corruption: The Need for a Quad-Sector Development Partnership (QSDP) in Nigeria's Niger Delta. *Development Policy Review, 34*(4), 509–538.

Gonzalez, A. (2018a). "A Salute to the Bandido": A Political Ecology of Voice (PEV) Exploration into Peru's Freedom of Voice Surrounding Hydrocarbon Development. *Society & Natural Resources, 31*(6), 649–665.

Gonzalez, A. (2018b). Control, Fear and Suppression: A Political Ecology of Voice Study into Resource Extraction Industry-community Engagement in Peru's Loreto Region. *The Extractive Industries & Society, 5*(3), 319–329.

Gonzalez, A. (2019). Making "A Racket" But Does Anybody Care? A Study of Environmental Justice Access and Recognition Through the Political Ecology of Voice. *Geoforum, 102*, 142–156.

Gonzalez, A. (2020). What Justice and for Whom? A Political Ecology of Voice Study into 'Senses of Justice' in Peru's Loreto Region. *Environment and Planning E: Nature and Space* (Under review).

Goodwin, J., & Jasper, J. M. (Eds.). (2011). *Contention in Context: Political Opportunities and the Emergence of Protest*. Stanford: Stanford University Press.

Goyes, D., Mol, H., Brisman, A., & South, N. (Eds.). (2017). *Environmental Crime in Latin America: The Theft of Nature and the Poisoning of the Land*. London: Palgrave Macmillan.

Groves, C. (2015). The Bomb in my Backyard, the Serpent in my House: Environmental Justice, Risk, and the Colonialisation of Attachment. *Environmental Politics, 24*(6), 853–873.

Gustafsson, M.-T., & Scurrah, M. (2019). Strengthening Subnational Institutions for Sustainable Development in Resource-rich States: Decentralised Land-use Planning in Brazil. *World Development, 119*, 133–144.

Haller, T. M., Blöchlinger, A., John, M., Marthaler, E., & Siegler, S. (Eds.). (2007). *Fossil Fuels, Oil Companies, and Indigenous Peoples: Strategies of Multinational Oil Companies, States, and Ethnic Minorities*. Zurich: LIT Verlag.

Heynen, N. (2013). Urban Political Ecology I: The Urban Century. *Progress in Human Geography, 38*(4), 598–604.

Himley, M. (2008). Geographies of Environmental Governance: The Nexus of Nature and Neoliberalism. *Geography Compass, 2*(2), 433–451.

Hirschman, A. O. (1970). *Exit, Voice, and Loyalty: Responses to Decline in Firms, Organisations and States*. Cambridge, MA: Harvard University Press.

Hirschman, A. O. (1978). Exit, Voice, and the State. *World Politics, 31*(1), 90–107.

Hirschman, A. O. (1980). "Exit, Voice and Loyalty": Further Reflections and a Survey of Recent Contributions. The Milbank Memorial Fund Quarterly. *Health and Society, 58*(3), 430–453.

Hirschman, A. O. (1981). *Essays in Trespassing: Economics to Politics and Beyond*. Cambridge: Cambridge University Press.

Hirschman, A. O. (1982). *Shifting Involvements: Private Interest and Public Action*. Oxford: Martin Robertson.

Hirschman, A. O. (1995). *A Propensity to Self-subversion*. Cambridge, MA: Harvard University Press.

Hope, J. (2016). Losing Ground? Extractive-led Development Versus Environmentalism in the Isiboro Secure Indigenous Territory and National Park (TIPNIS), Bolivia. *The Extractive Industries and Society, 3*, 922–929.

Human Rights Watch. (2018a). Bolivia: Events of 2018. www.hrw.org/world-report/2019/country-chapters/bolivia#981f12. Accessed 7 Nov 2019.

Human Rights Watch. (2018b). Brazil: Events of 2018. www.hrw.org/world-report/2019/country-chapters/brazil. Accessed 7 Nov 2019.

Human Rights Watch. (2018c). Colombia: Events of 2018. www.hrw.org/world-report/2019/country-chapters/colombia. Accessed 7 Nov 2019.

Human Rights Watch. (2018d). Venezuela: Events of 2018. www.hrw.org/world-report/2019/country-chapters/venezuela#981f12. Accessed 7 Nov 2019.

Human Rights Watch. (2020). *Peru: Events of 2019*. https://www.hrw.org/world-report/2020/country-chapters/peru. Accessed 14 September 2020.

Hvalkof, S. (2000). Outrage in Rubber and Oil: Extractivism, Indigenous Peoples, and Justice in the Upper Amazon. In C. Zerner (Ed.), *People, Plants, & Justice: The Politics of Nature Conservation* (pp. 83–116). New York: Columbia University Press.

Hvalkof, S. (2006). Progress of the Victims: Political Ecology in the Peruvian Amazon. In P. Blaikie & H. Brookfield (Eds.), *Land Degradation and Society* (pp. 195–232). London: Routledge.

Johnston, J. E., Lim, E., & Roh, H. (2019). Impact of Upstream Oil Extraction and Environmental Public Health: A Review of Evidence. *Science of the Total Environment, 657*, 187–189.

Kaiser, A. J. (2019, May 9). 'Exterminator of the Future': Brazil's Bolsonaro Denounced for Environmental Assault. *The Guardian*. www.theguardian.com/world/2019/may/09/jair-bolsonaro-brazil-amazon-rainforest-environment. Accessed 10 June 2019.

Kuklina, V., & Holland, E. C. (2018). The Roads of the Sayan Mountains: Theorising Remoteness in Eastern Siberia. *Geoforum, 88*, 36–44.

Latin American Andean Group Report. (2010). *Latin American Regional Report Andean Group*. http://latinnews.com/media/k2/pdf/CVC2.pdf. Accessed 18 Oct 2019.

Laurian, L. (2004). Public Participation in Environmental Decision Making: Findings from Communities Facing Toxic Waste Cleanup. *Journal of the American Planning Association, 70*(1), 53–65.

Law in Action. (n.d., October 2). *How Many Oil Barrels Have Spilled in the Amazon*. www.alphabetics.info/international/how-many-oil-barrels-have-spilled-in-the-amazon. Accessed 1 July 2019.

Lo Lau, J. (2018). Peru: The River That Will Not Flow. *Mongabay*, November 22, 2017. https://news.mongabay.com/2017/11/peru-the-river-that-will-not-flow. Accessed 18 Oct 2019.

Loftus, A. (2019). Political Ecology I: Where is Political Ecology? *Progress in Human Geography, 43*(1), 172–182.

Maconachie, R., & Hilson, G. (2013). Editorial Introduction: The Extractive Industries, Community Development and Livelihood Change in Developing Countries. *Community Development Journal, 48*(3), 347–359.

McCarthy, J., & Prudham, S. (2004). Neoliberal Nature and the Nature of Neoliberalism. *Geoforum, 34*, 275–283.

Nepomuceno, Í., Affonso, H., Fraser, J. A., & Torres, M. (2019). Counter-conducts and the Green Grab: Forest Peoples' Resistance to Resource Extraction in the Saracá-Taquera National Forest, Brazilian Amazonia. *Global Environmental Change, 56*, 124–133.

O'Donnell, G. (1986). On the Fruitful Convergences of Hirschman's Exit, Voice, and Loyalty and Shifting Involvements: Reflections from the Recent Argentine Experience. In A. Foxley, M. S. McPherson, G. A. O'Donnell, & A. O. Hirschman (Eds.), *Development, Democracy, and the Art of Trespassing: Essays in Honour of Albert O. Hirschman* (pp. 249–268). Notre Dame: University of Notre Dame Press.

Organisation for Economic Co-operation and Development (OECD). (2015). *Due Diligence Guidance for Meaningful Stakeholder Engagement in the Extractives Sector*. www.oecd.org/daf/inv/mne/OECD-Guidance-Extractives-Sector-Stakeholder-Engagement.pdf. Accessed 24 Oct 2019.

Organisation of American States (OAS). (2017). *IACHR Grants Precautionary Measure for the Communities of Cuninico and San Pedro in Loreto, Peru*. www.oas.org/en/iachr/media_center/PReleases/2017/205.asp. Accessed 24 Oct 2019.

Orta-Martinez, M., Pellegrini, L., & Arsel, M. (2018). "The Squeaky Wheel Gets the Grease"? The Conflict Imperative and the Slow Fight Against Environmental Injustice in Northern Peruvian Amazon. *Ecology and Society, 23*(7), 7. https://doi.org/10.5751/ES-10098-230307.

Paredes, M. (2008). *Weak Indigenous Politics in Peru*. Centre for Research on Inequality, Human Security and Ethnicity (CRISE), CRISE Working Paper No. 33. https://assets.publishing.service.gov.uk/media/57a08bc0e5274a27b2000d25/wp33.pdf. Accessed 24 Oct 2019.

Perreault, T. (2003). Changing Places: Transnational Networks, Ethnic Politics, and Community Development in the Ecuadorian Amazon. *Political Geography, 22*(1), 61–88.

Perreault, T., Bridge, G., & McCarthy, J. (Eds.). (2015). *The Routledge Handbook of Political Ecology*. Abingdon: Routledge.

Peru Support Group. (2016). *Communities Blockade River Marañón in Protest at Oil Spills*. http://www.perusupportgroup.org.uk/article-1208.html. Accessed 1 July 2019.

Peru Support Group. (n.d.). *Law on Prior Consultation of Indigenous or Native Peoples as Recognised by International Labour Organisation Convention No. 169*. https://landportal.org/library/resources/law-right-prior-consultation-indigenous-or-native-peoples-recognized-convention. Accessed 18 Oct 2019.

Perupetro. (n.d.). *Mapa de Lotes de Contrato*. https://perupetro.maps.arcgis.com/apps/webappviewer/index.html?id=6a830a470b934f0687c8ed84c2bacacc. Accessed 6 Aug 2019.

Philip, G. (1984). *Bonanza Development? The Selva Oil Industry in Peru 1968–1982*. London: Institute of Latin American Studies, University of London.

Phillips, D. (2019, January 2). Jair Bolsonaro Launches Assault on Amazon Rainforest Protections. *The Guardian*. www.theguardian.com/world/2019/jan/02/brazil-jair-bolsonaro-amazon-rainforest-protections. Accessed 10 June 2019.

Pinelo, A. J. (1973). *The Multinational Corporation as a Force in Latin American Politics: A Case Study of the International Petroleum Company in Peru*. New York: Praeger Publishers.

Pitman, C. A., Terborgh, J. W., Silman, M. R., Nunez, P., Neill, D. A., Ceron, C. E., Palacios, W. A., & Aulestia, M. (2002). A Comparison of Tree Species Diversity in Two Upper Amazonian Forests. *Ecology, 83*(11), 3210–3224.

Postero, N. (2010). Morales's MAS Government: Building Indigenous Popular Hegemony in Bolivia. *Latin American Perspectives, 37*(3), 18–34.

Reuters. (2019, March 26). Amazonian Residents Seize PetroTal's Oil Installations in Peru. www.reuters.com/article/us-peru-oil-petrotal/amazonian-residents-seize-petrotals-oil-installations-in-peru-idUSKCN1R701Q. Accessed 6 Aug 2019.

Ribot, J. C., & Peluso, N. L. (2003). A Theory of Access. *Rural Sociology, 68*(2), 153–181.

Robbins, P. (2000). The Rotten Institution: Corruption in Natural Resource Management. *Political Geography, 19*(4), 423–443.

Rocha, J. (2019, January 29). Bolsonaro Government Reveals Plan to Develop 'Unproductive Amazon. *Mongabay*. https://news.mongabay.com/2019/01/bolsonaro-government-reveals-plan-to-develop-the-unproductive-amazon. Accessed 10 June 2019.

Rosell-Melé, A., Moraleda-Cibrián, N., Cartró-Sabaté, M., Colmer-Ventura, F., Mayor, P., & Orta-Martinez, M. (2018). Oil Pollution in Soils and Sediments from the Northern Peruvian Amazon. *Science of the Total Environment, 610–611*(1), 1010–1019.

Sawyer, S. (2004). *Crude Chronicles: Indigenous Politics, Multinational Oil, and Neoliberalism in Ecuador*. Durham: Duke University Press.

Schlosberg, D. (2004). Reconceiving Environmental Justice: Global Movements and Political Theories. *Environmental Politics, 13*(3), 517–540.

Schlosberg, D. (2007). *Defining Environmental Justice: Theories, Movements, and Nature*. Oxford: Oxford University Press.

Schulz, K. A. (2017). Decolonising Political Ecology: Ontology, Technology and 'Critical' Enchantment. *Journal of Political Ecology, 24*(1), 125.

Simon, G. L., & Peterson, C. (2019). Disingenuous Forests: A Historical Political Ecology of Fuelwood Collection in South India. *Journal of Historical Geography, 63*, 34–47.

Singer, M. M. (2007). The Presidential and Parliamentary Elections in Bolivia, December 2005. *Electoral Studies, 26*(1), 200–205.

Stetson, G. (2012). Oil Politics and Indigenous Resistance in the Peruvian Amazon: The Rhetoric of Modernity Against the Reality of Coloniality. *The Journal of Environmental Development, 21*(1), 76–97.

Stiefel, M., & Wolfe, M. (1994). *A Voice for the Excluded: Popular Participation in Development: Utopia or Necessity?* London: United Nations Research Institute for Social Development.

Stoll-Kleemann, S., & Welp, M. (Eds.). (2006). *Stakeholder Dialogues in Natural Resource Management: Theory and Practice*. Berlin: Springer-Verlag.

Taft-Morales, M. (2013). *Peru in Brief: Political and Economic Conditions and Relations with the United States*. Congressional Research Service. https://fas.org/sgp/crs/row/R42523.pdf. Accessed 18 Oct 2019.

Taj, M., & Bell, A. (2019, September 10). Indigenous Protests in Peru Halt Output at Frontera Oil Block. *Reuters*. www.reuters.com/article/us-peru-frontera-energy-oil/indigenous-protests-in-peru-halt-output-at-frontera-oil-block-idUSKCN1VV2GO. Accessed 24 Oct 2019.

Taylor, S. J., & Bogdan, R. (2016). *Introduction to Qualitative Research Methods: A Guidebook and Resource*. Hoboken: Wiley.

TerraBrasilis. (n.d.). *Analyses Legal Amazon*. http://terrabrasilis.dpi.inpe.br/app/dashboard/alerts/legal/amazon/aggregated. Accessed 10 June 2019.

The Chaikuni Institute and ORIPO. (2018). *The Black Snake of Peru's Amazon: The North Peruvian Pipeline*. https://chaikuni.org/wp-content/uploads/2018/10/The-Black-Snake-of-Perus-Amazon-Chaikuni-Institute.pdf. Accessed 6 Aug 2019.

Thorp, R., & Bertram, G. (1978). *Peru 1890–1977: Growth and Policy in an Open Economy*. London: The Macmillan Press Limited.

Vuori, M. (2009). *Accessibility as a Determinant of Opportunities. A Case Study from Peruvian Amazonia*. MSc Dissertation, University of Helsinki.

Watts, J. (2019, June 4). Deforestation of Brazilian Amazon Surges to Record High. *The Guardian*. www.theguardian.com/world/2019/jun/04/deforestation-of-brazilian-amazon-surges-to-record-high-bolsonaro. Accessed 10 June 2019.

Willow, A. J., & Wylie, S. (2014). Politics, Ecology, and the New Anthropology of Energy: Exploring the Emerging Frontiers of Hydraulic Fracking. *Journal of Political Ecology, 21*(1), 222.

World Bank Group. (2017). *Peru: Systematic Country Diagnostic*. http://documents.worldbank.org/curated/en/919181490109288624/pdf/Peru-SCD-final-3-16-17-03162017.pdf. Accessed 24 Oct 2019.

Wright, C. (2015). *Emergency Politics in the Third Wave of Democracy: A Study of Regimes of Exemption in Bolivia, Ecuador, and Peru*. Lanham: The Rowman & Littlefield Publishing Group.

Yusta-Garcia, R., Orta-Martinez, M., Mayor, P., González-Crespo, C., & Rosell-Melé, A. (2017). Water Contamination from Oil Extraction Activities in Northern Peruvian Amazonian Rivers. *Environmental Pollution, 225*, 370–380.

Zuindeau, B. (2009). Responding to Environmental Risks: What Can Albert Hirschman Contribute? *Ecological Economics, 69*(1), 155–165.

12

Ethnogenesis and Environmentalism in Contemporary Brazilian Amazonia: A Study in Comparative Frontier History

Paul E. Little

Until the mid-twentieth century, most frontier history was written from the perspective of dominant national societies, in which the forces of civilisation triumphed over those of primitivism and barbarism. A more recent cohort of critical studies of frontier dynamics focused on the traumatic consequences of genocide—the physical extermination of a social group—and ethnocide—the deliberate eradication of the culture or way of life of a people, often with a focus on the conquest and defeat of indigenous peoples. Hemming's monumental histories chronicle the 'Conquest of the Incas' (1970), the 'Conquest of the Brazilian Indians' (1978) and the 'Defeat of the Brazilian Indians' (1987); Todorov analyzes the symbolic encounters of interpersonal communication between Cortez and Moctezuma to reveal how the Spaniards achieved 'The Conquest of America' (1984), while Brown's (1970) 'Bury my Heart at Wounded Knee', despite its poetic title, still focuses on the conquest and defeat of indigenous people on the US Western frontier. Ethnohistorian Wright

---

P. E. Little (✉)
University of Brasília (UnB), Brasília, Brazil

A. A. R. Ioris (ed.), *Environment and Development*,
https://doi.org/10.1007/978-3-030-55416-3_12

(1990) provides an apt rejoinder to these studies when he affirms that indigenous groups have been 'conquered but not defeated'.

Social scientists and historians have generally given much less attention to the subsequent revitalising processes of ethnogenesis: the emergence of new ethnic and political identities amongst subjugated social groups. A new set of ethnohistorical studies, however, move beyond narratives of conquest and extinction by providing analyses of the historical agency of all social actors involved in colonial, imperial and national interactions and are generally more ethnically variegated as multiple points of view and interests are presented. The inclusion of ethnic narratives and practices of resistance to the powerful external forces of colonialism and capitalist expansion broadens our understanding of the complex intercultural interactions on frontiers (Comaroff and Comaroff 1997).

Frontier Waves and Cycles of Ethnogenesis

By viewing frontier dynamics over the past five centuries, in which the ebb and flow of frontier waves is taken into account, recurring patterns in the processes of sociocultural change can be detected. Many frontiers operate on the boom and bust cycles of global commodities such that they can open, expand, retract and close over time as the primary exploited resource becomes depleted or loses its market value. Each global commodity installs its own mode of extraction or production in the frontier region, thereby upending existing interethnic relations, provoking cases of ethnocide and establishing new intercultural power dynamics. Once the economic impulse of a frontier wave has been expended and moves into its bust cycle, it leaves a powerful wake of disrupted social and economic structures operating within a national context of relative neglect, since the frontier region is no longer in the forefront of economic expansion.

In the decades after a frontier bust, in what are pejoratively referred to as the 'backwaters of history', much restructuring of ethnic and political entities occurs. In this way, frontier conjunctures of structural breakdown tend to be fertile moments for engendering processes of ethnogenesis by providing the social cauldron from which new ethnic identities can emerge. As such, ethnocide and ethnogenesis can be understood as 'complementary aspects of the same social process of social and cultural change' (Whitehead 1996: 34) that 'exist in the most tenuous balance' (Whitten 1976: 285). Since each frontier wave produces a unique set of circumstances which facilitate the realignment of ethnic, social and political identities, we can posit the existence of distinct 'cycles of ethnogenesis' that follow processes of ethnocide.

In its most basic meaning, ethnogenesis is a process whereby an ethnic group comes into existence. Powers (1995: 9) offers a succinct definition of it as 'a continuous cultural process that is simultaneously reproductive and transformative'. As an analytical tool, the concept of ethnogenesis highlights 'peoples' simultaneously cultural and political struggles to create enduring identities in general contexts of radical change and discontinuity' (Hill 1996: 1). Ethnogenic phenomena challenge the long-held notion of cultures as discrete entities that clash into each other much like billiard balls. At the end of his study of the economic and sociocultural impacts of global capitalist expansion, Wolf (1982: 387) argues against a model of 'fixed, unitary, and bounded culture' in favour of a concept of culture as 'a series of processes that construct, reconstruct, and dismantle cultural materials, in response to identifiable determinants'.

In developing an analytic framework for understanding ethnogenic processes, I start with Sahlin's (1985: xiv) concept of 'structure of the conjuncture' which he defines as 'the practical realisation of the cultural categories in a specific historical context, as expressed in the interested action of the historic agents' and adds that 'this notion of praxis as a situational sociology of meaning can be applied to the general understanding of cultural change'. Li (2014: 193) modifies this concept by giving it far greater cultural flexibility, thus removing what she considers to be its undue insistence on an 'enduring cultural essence'. I posit that frontier history requires an additional modification of placing ethnohistorical analysis within distinct conjunctures of structural breakdown in which the actions of surviving members of the group restructure their social organisations and ethnic identities to meet the new realities of an economic frontier in decline.

The unleashing of the destructive forces of ethnocide in a frontier region can provoke a wide array of responses ranging from the fission of existing groups into separate entities to cooperative encounters between ethnic groups. In facing external pressures, these groups often establish new trading relationships, engage in intermarriage and form political alliances. The formation of new social groups can result from ethnic mergers (Albers 1996), from polyethnic federations and the tracing of common descent from original ancestors (Bilby 1996). Social identity is an important dimension of the ethnogenic process that has strong political implications. Whitehead (1996: 21) affirms that 'new group *identities* were created and old ones fell into disuse, not necessarily that persons were themselves destroyed or born'. The capacity of these groups to resist and respond to external pressures allows them to reconstitute themselves as distinct ethnopolitical entities and become legitimate political actors within the frontier public sphere as they seek to gain official recognition of their new ethnic identities. This political emergence gives national visibility to formerly 'invisible' groups and places their territorial claims on the public policy agenda.

Historical Processes of Ethnogenesis in the Americas

The Americas have been a key site of ethnogenesis for centuries, based on its unprecedented history of global migratory flows and incessant frontier expansions marked by European conquests, the African slave trade and indentured servitude. The arrival of Europeans over five centuries ago led to a demographic collapse of indigenous populations due to disease, warfare, slavery and malnutrition, which has been referred to in the academic literature as the 'Great Dying' (Wolf 1982: 133–135), but also characterised as the 'American Indian Holocaust' (Thornton 1990), in which numerous indigenous societies were driven to extinction. Denevan (1992) calculates that there were approximately 54 million people living in the Americas at the time of the European contact in 1492. The population decline in the indigenous population was in some regions as high

90% after the first century of European contact (see Mann 2005). Ribeiro (1957), in a mid-twentieth century study of Brazilian indigenous people, estimated that at that time, there were only 100,000 'Indians in more or less autonomous tribal life'. Despite these precipitous population declines, however, thousands of indigenous societies survived all across the Americas and were forced to adapt to the structures of colonial control.

The civilisational encounters between indigenous empires and Spanish invaders in Mesoamerica and the Andes produced radically altered ethnic landscapes. Wolf (1959), in a sweeping history of Mexico and Guatemala, chronicles how a new ruling structure was established, fuelled by a commodity boom in silver, that replaced that of the collapsed Aztec ruling class during the tumultuous post-conquest century (1519–1600). With the global depression of the seventeenth century, Mesoamerican society experienced a bifurcation with, on one hand, the expansion of the *haciendas*—large, privately owned farms and ranches—and the consolidation of a colonial elite and, on the other hand, a retreat to the countryside of indigenous populations who had been severed of ties to their pre-Hispanic past and its social structures. These groups were organised into relatively autonomous Indian communes which, three centuries later, still form the structural backbone of present-day indigenous communities. Here is an example of ethnogenesis on a civilisational scale in which the rupture caused by violent conquest and the collapse of an empire generated, among its surviving descendants, new indigenous groups and identities with decentralised organisational structures within a national context of racial subordination.

The sixteenth and seventeenth centuries also produced major ethnic restructuring in the Andean highlands. The disease catastrophes of the 1520s and 1530s and the subsequent Spanish conquest and disarticulation of the debilitated Inca State caused a sharp decline of the indigenous population throughout most of the sixteenth century. Starting in the 1590s, however, northern Andean society witnessed a demographic explosion amongst indigenous peoples whereby, 'the work of cultural reproduction was carried out by Amerindian commoners who reconstructed Andean social organisation inside the sites to which they migrated – cities, haciendas, *obrajes*, etc'. (Powers 1995: 10). Through an intense, ongoing process of revaluing pre-Hispanic categories, including that of the *cacicazgo* (chiefdom), these northern Andean indigenous peoples 'not only resisted, adapted, and survived, but reinvented themselves as a culture' (ibid.: 12).

The Great Plains of North America—encompassing what is now northern Mexico, the US Midwest and southern Canada—were another important site of indigenous ethnogenesis from the sixteenth through the nineteenth century. In the southern Plains and its periphery (northern Mexico, New Mexico and Texas), Spanish invasion, warfare and demographic pressures inaugurated a two-century long process (1580–1780) of different cycles of ethnogenesis in which smaller bands forged cultural and political unity with other groups to meet the demands of the new order. The successive emergence of Jumano, Coahuiltecan and Apache peoples culminated in the rise of the powerful Comanche nation by the end of the eighteenth century (Anderson 1999). In the northeastern Plains (Manitoba, Saskatchewan, Montana, North Dakota), the Plains Assiniboin, Cree and Ojibwa faced mounting pressures by the late eighteenth century from land invasions by immigrants, major changes in the fur market and access to European goods and the ravages of repeated smallpox epidemics (1780–82; 1837–38; 1850), which decimated their population. These groups initially formed a 'polyethnic alliance formation' in response to these constraints and, by the mid-nineteenth century, they had merged through intermarriage, co-residency and joint collaboration in subsistence, ceremonial and military activities (Albers 1996).

In the forested tropical lowlands of South America, the European occupation of the continent unleashed complex, often complementary, processes of ethnocide and ethnogenesis. In what is now Surinam, beginning with first contact in 1499, intercolonial rivalries of the Dutch, English and French over trade with native peoples led to a reconfiguration of interethnic alliances and the establishment of 'relatively monolithic native power blocs under the control of coastal Arawak and Carib leaders', which produced a 'general reorientation of their political and economic systems toward the coast and away from the uplands and their contacts with the Amazon Basin' (Whitehead 1996: 21). Numerous inland indigenous groups, who suffered from territorial invasions, slaving expeditions and disease, were unable to maintain separate ethnic identities but melded into either Carib or Arawak identities and allegiances, two expanding ethnic formations with adversarial identities that form the basis the contemporary ethnic configuration of the Guyanas, but which 'obscured the heterogeneity of the ethnic origins of their members' (ibid.: 32).

In the Upper Amazon, repeated epidemics of smallpox and measles over a 120-year period (1642–1762) decimated the large indigenous groups who inhabited riverside areas and controlled trade with indigenous groups who lived along tributaries. During this time, the vast headwaters region of the Napo, Aguarico, Curaray and Pastaza rivers (Ecuador) served as refuge zone for individuals of Quijos, Záparo, Jívaro and Andean indigenous groups who were fleeing indentured servitude, the mission priests and disease (Naranjo 1977). Quichua, a language of the Andean highlands that was used by Jesuit missionaries in the Upper Amazon region in their evangelisation of disperse indigenous groups, became the lingua franca for interethnic trade and served as a common language in uniting remnants of small indigenous societies, in a process that Hudleson (1981) calls 'Quichuaisation'. Today, the Quijos Quichua (Napo and Aguarico River basins) and the Canelos Quichua (Curaray and Pastaza River basins) form the largest indigenous group of the Ecuadorian Amazon with a population of over 60,000.

A quite different space of ethnogenesis emerged from the forced African diaspora of the transatlantic slave trade that brought over 10 million Africans to the Americas (and provoked the deaths of millions of others who did not survive the horrific conditions of the Middle Passage), with nearly 90% of them arriving in Brazil and the West Indies (Behrendt 2010). The colonial slave system that was installed throughout these countries separated families, mixed ethnic groups and repressed cultural expressions of African languages and traditions. Throughout this period, numerous slave rebellions gave witness to cases of open resistance to the slave system.

Another type of resistance to the slave system was flight or *marronage*, whereby groups of slaves fled to the forested hinterlands and established and defended semi-autonomous communities. These maroon societies were prevalent in Brazil, the French Caribbean, the Spanish Americas, Jamaica, the Guianas and, to a lesser extent, the United States (Price 1979a). Each maroon society set up its own system of concealment and defence from the plantation system and the colonial powers that propagated it, which included tactics such as guerrilla warfare and alliances with indigenous peoples. A wide variety of organisational styles characterised these societies, ranging from centralised states, to shifting federations and isolated bands (Price 1979b: 16). In addition, each society developed its own mechanisms of economic adaptation to new environments, usually founded upon swidden horticulture, hunting and fishing and, in many cases, adopting technologies of their indigenous neighbours. What is notable in almost all the cases of the development of maroon societies is that they represent a process of creating cultural dynamics in which 'Africans in the New World, who at first often shared little more than a common continental origin and the experience of enslavement, developed distinctively Afro-American ways of dealing with life from the very beginning' (ibid.: 27).

The Garifuna ethnic group of the Caribbean coast of Central America, also known as 'Black Caribs', present a unique case of ethnogenesis involving multiple inter-island migrations. The Caribbean island of St. Vincent, which served as a place of refuge for slaves escaping from other Caribbean islands throughout the eighteenth century, facilitated an intense, sometimes conflictive, process of biological and cultural intermixing with the Island Carib indigenous peoples. Upon the English takeover of the island from the French, the Black Caribs were transferred, in 1797, to Roatan Island, which was under control of the Spaniards, who subsequently settled them on the mainland in Trujillo, Honduras. Throughout the first half of the nineteenth century, this group spread out along the Caribbean coast of what is now Nicaragua, Honduras, Guatemala and Belize, where over the next century and a half they developed a subsistence economy based in horticulture and fishing and a unique set of cultural expressions (Gonzalez 1969). The Garifuna language, for example, which is dually rooted in the Arawak and Carib indigenous languages, incorporates vocabulary from French, English, Spanish and words of African origin, and has been designated by UNESCO as a Masterpiece of the Oral and Intangible Heritage of Humanity, along with Garifuna music and dance (Griffin 2005).

Ethnogenesis and Environmentalism in the Brazilian Amazonia (1980–2020)

The analysis now shifts from the Americas-wide, macro-historical perspective to a short-term historical perspective (1980–2020) within the geographic scope of Brazilian Amazonia. Within this timeframe, a current cycle of ethnogenesis is identified, one that emerged in the aftermath of the developmentalist frontier wave of the Brazilian military dictatorship of the 1960s and 1970s. One distinguishing aspect of this cycle is the prominent role played by the environmental movement, most notably socioenvironmentalism, in its embrace of Amazonian traditional peoples and the increased political power and social legitimacy that this has provided for them in their struggle for national recognition of their territorial rights.

A Powerful Developmentalist Frontier Wave (1964–1985)

The developmentalist frontier wave, implemented in Amazonia by the Brazilian military dictatorship starting in the mid-1960s, provoked unprecedented changes in the sociocultural and political landscape of the region. When the Brazilian armed forces installed a dictatorship in Brazil in 1964 and ruled the country for the succeeding 21 years, they looked to the country's vast Amazon region as an important site for national development. It was seen as a 'solution' to acrimonious land reform issues in other parts of the country by, in the words of ruling General Emílio Médici, sending 'people without land to a land without people', a phrase which disregarded the hundreds of indigenous societies that were living in Amazon and controlled vast territories there. On the economic front, the dictatorship sought to forcibly incorporate Brazilian Amazonia into the national economy, turning it into a resource frontier for the expansion of capitalist relations (Becker 1982). The government justified these efforts using the argument that if Brazil did not 'integrate' this region now, it risked 'losing' it to others [*integrar para não entregar*].

In the 1970s, successive military administrations launched a series of large-scale development and infrastructure projects, such as the construction of interregional roads (e.g. the Transamazon and Cuibá-Santarém motorways) and large-scale hydroelectric dams (e.g. Samuel; Balbina; Tucuruí), the installation of industrial-scale mines (e.g. Grande Carajás, which included the construction of a railroad and ocean port) and agricultural colonisation projects (e.g. Polonoroeste; Transamazon), all of which produced grave impacts on hundreds of indigenous lands, *quilombos*, and peasant and riverside communities across the Amazon region. An article in a major national Brazilian magazine heralded that these programmes would allow colonists to 'truly occupy a world that has up until now has been closed to them' by 'tearing down the forest fence and breaking out of the prison of its rivers' (Manchete 1973). By the early 1980s, this development model was in crisis, both politically for lack of support at home and economically due to the global crisis of overextended debt to developing countries. Nonetheless, this developmentalist wave left in its wake major economic, demographic and sociocultural transformations that allowed subjugated groups some ethnopolitical 'breathing room' in order to restructure themselves to meet the needs of this new sociopolitical conjuncture.

The Emergence of Brazilian Socioenvironmentalism (1985–2012)

The decade of the 1980s in Brazil witnessed a rapid increase in the formation and mobilisation of social movements—indigenous peoples, Afro-descendent populations, rubber tappers, landless peasants, women's groups, Christian base communities, environmentalists, among many others—clamouring for the re-democratisation of Brazilian society and vying for influence in the National Constitutional Convention. After the drafting and passage of a progressive Federal Constitution in 1988, they turned their attention to exercising their newly gained constitutional rights as Brazilian citizens within a framework of respect for cultural and ecological diversity.

Brazilian socioenvironmentalism was forged in this decade through strategic alliances between national social movements and the international environmental movement, which had become alarmed at the devastation caused by accelerating deforestation of the Amazon rainforest. Brazil's hosting of the United Nations Conference on Environment and Development in Rio de Janeiro in 1992, at the time the largest global UN Conference ever held, was a potent stimulus for these multiple social actors to coordinate actions and present alternatives to existing development models (Svirsky and Capobianco 1997). Born under the sign of 'sustainable development', a newly minted concept of the 1987 Brundtland Commission Report Our Common Future (WCED 1987), this movement affirmed that environmental policy for the Amazon would only be effective and durable if it involved local communities in land use decisions and distributed the benefits derived from natural resource use in an equitable and just manner (Santilli 2005). The Pilot Program to Conserve the Brazilian Rain Forest, financed by the Group of Seven Industrialised Countries and implemented by the Brazilian government, was launched in the aftermath of the Rio Conference and quickly became a leading site for the formulation of socioenvironmental policies and their institutionalisation in Brazilian federal and state governments.

The consolidation of socioenvironmental policies in the 1990s was facilitated, in part, to the waning of the developmentalist frontier wave that had been promoted by the dictatorship and provided a new political space for the diverse, widely dispersed groups who had been marginalised by frontier expansion. Amazonian peoples began to reorganise and re-identify themselves by using the opportunities opened up by global environmental concerns and, in the process, initiated a new cycle of ethnogenesis in Brazilian Amazonia. In this contemporary case, ethnogenesis was not the result of genocide, demographic decline or interethnic and interracial mixing, as had occurred in previous cycles, but rather through a process by which social groups who had been living in the Amazon for decades in marginal conditions and who had been largely invisible to national leaders were now asserting their identities with less fear of facing discrimination and prejudice.

As they sought to participate as political actors in the debates over the future of the Amazon, they simultaneously reconstructed their ethnic and racial identities in a positive frame. These groups were aided in this endeavour by other changes in Brazilian society, such as a renewed emphasis on cultural diversity in which the existence of numerous indigenous societies was now heralded as a beneficial attribute of society. Social and natural scientists were demonstrating that regions of high biodiversity were directly correlated with those that had a high degree of sociodiversity (Bensusan 2005), thus providing an important linkage between the social movements of indigenous peoples and those of environmentalists (Conklin and Graham 1995). Groups that were formerly considered to be anachronistic were now being recast within the global discourse of sustainable development as the vanguard of a revitalised Amazonian economy buttressed by the use of traditional ecological knowledge and the sustainable use of natural resources.

Indigenous Peoples

Indigenous peoples offer the most forceful example of ethnopolitical resurgence in contemporary Brazilian Amazonia. During the military dictatorship, indigenous peoples were subjected to an aggressive assimilationist agenda, as mandated by the 1973 Statute of the Indian, and the repression of their incipient attempts to organise themselves politically. Meanwhile, the progressive wing of Catholic Church, through its Indigenist Missionary Council (CIMI), began convening assemblies of indigenous leaders from different ethnic groups where they were able to learn about other groups and share their common concerns. This led to the formation of the Union of Indigenous Nations (UNI) in 1980, the first national-level indigenous organisation in the country (Ramos 1998). Throughout the decade of the 1980s, Brazil witnessed a flowering of regional and local Indigenous peoples' organisations, leading to the creation, in April of 1989, of the Coordinating Body of Indigenous Organisations of the Brazilian Amazon (COIAB) at a regional meeting of indigenous leaders, which would become the largest indigenous organisation in Brazil, encompassing 75 member organisations from the nine states of the Brazilian Amazon.[1]

The promulgation of the Federal Constitution of 1988 and the guarantees it provided gave indigenous peoples a strong basis from which to defend their rights. For the first time in the country's history, the Brazilian government recognised the rights of indigenous peoples to the lands that they traditionally occupy and to use their language and practice their cultural traditions (Articles 231 and 232). The historic meeting in Altamira, Pará, in February of 1989, capped off a decade of advances in the political and symbolic power of indigenous peoples. At this meeting, the mobilisation of 600 Kayapó warriors, in alliance with the international environmental movement, led to the cancelling, by the World Bank, of its financing for the Kararaô hydroelectric dam on the Xingu River, which would have flooded thousands of acres of indigenous territories.[2] In this context of ethnopolitical resurgence, the stigma associated with being Indian, along with the racial discrimination that accompanied it, gradually receded and allowed long-suppressed indigenous identities to re-emerge. Oliveira (1998), in a path breaking study of the so-called mixed-race Indians of northeastern Brazil, traces the process of ethnogenesis of indigenous peoples, with the number of indigenous groups in the region rising from 10 to 23 in the span of just 40 years through the emergence of new social identities and the reinvention of existing ethnicities.

The Brazilian Amazon harbours numerous recent cases of ethnogenesis in this sociopolitical context. Starting in 1998, three riverside communities—Taquara, Marituba and Bragança—in the Tapajós River Basin initiated an internally led process of re-elaborating their indigenous identities, reconstructing their immemorial cultural traditions and restructuring their modes of political and cultural organisation. They claimed they had always belonged to the Munduruku ethnic group and called on the National Indian Foundation (FUNAI) to formally recognise their indigenous lands (Ioris 2014). During this same epoch, members of the Baré ethnic group who lived in poor neighbourhoods in the Amazon city of

[1] Acre, Amapá, Amazonas, Maranhão, Mato Grosso, Pará, Rondônia, Roraima and Tocantins.

[2] Two decades later, the Belo Monte dam, a massive hydroelectric complex that uses run-of-the-river technology, was constructed at this site.

Manaus began reinventing their belief system and associated rituals and recovering the use of their indigenous language. They then started openly asserting their reconstituted Baré indigenous identity within an urban context (Melo 2006). Along the Solimões (Middle Amazon) River, communities that a generation earlier had migrated downstream from the Peruvian Amazon started reconstituting their indigenous past and affirming, in their new Brazilian context, a Kokama indigenous identity (Ramos 2004).

Quilombos

The establishment of *quilombos* over the past three centuries, whether by active escape by slaves to the hinterland or by remaining in place after the decline of the plantation system, made them nearly invisible to Brazilian national elites which ran the country from the urban coastal regions. Throughout the twentieth century, Brazilian national discourse rarely acknowledged the existence of *quilombos* and generally considered them to be throwbacks to an earlier age, rather than contemporary Brazilian communities that were constantly adapting to changing realities. With the expansion of frontier economies in Brazil during this period, such as the March to the West of the 1930s and 1940s and the subsequent national expansions into Amazonia in the 1960s and 1970, the invasion of the territories of these *quilombola*[3] communities increased and they were no longer isolated from the forces of national society. Starting in the 1980s, *quilombola* communities experienced a political resurgence as they revitalised their identities by reaffirming their centuries-long presence on the land and demanding formal recognition of their collectively held territories. This was accompanied by revalorisation of their history of resistance, now seen as a testament of their cultural and political strength. This movement was part of the nationwide democratisation process, though it harboured a different set of conditions due to the specific history of slavery in Brazil and the virulent forms of racism that plague Afro-descendent peoples in the country.

[3] The individual inhabitants of a *quilombo* are known as *quilombolas*.

The national remembrance, in 1995, of the 300th anniversary of the death of Zumbi dos Palmares, the leader of the most famous colonial-era *quilombo* which flourished for over a century in the hinterlands of Brazil, brought thousands of people to the capital city of Brasília and showcased the dramatic revitalisation of *quilombo* identity. This event coincided with the I National Gathering of Rural Black Communities in Brasilia and was followed, in 1996, by the First and Second Meetings of the National Commission of *Quilombola* Communities in the states of Bahia and Maranhão. The subsequent establishment of the Palmares Foundation within the Federal government served to bring official recognition to the nation's Afro-descendant population and, with its location within the Ministry of Culture, highlighted the cultural contribution of their presence in Brazilian society as a whole.

Article 68 of the transitory dispositions of the 1988 Federal Constitution granted 'definitive property rights' to the 'remnants of communities of *quilombos*' and required the federal government to provide them with legal titles. Subsequent mapping of Brazil's *quilombos* documented that there were 2847 *quilombola* communities in Brazil, with 442 of them located in the Brazilian Amazon, most of which were in the Amazonian state of Pará (Anjos 2006). In 2003, 15 years after the Constitution mandated that quilombos be titled within a 5-year period, Presidential Decree 4.887 established guidelines for the identification, recognition, delimitation, demarcation and titling of *quilombola* communities and authorised the National Institute of Colonisation and Agrarian Reform to implement the Decree. However, active resistance by landed elites has consistently blocked these efforts and by 2019, only 182, or 6%, of *quilombos* had gained effective title to their lands (Souza 2019), indicating that this 30-year long ethnogenic process has not yet been consolidated.

Seringueiros, Ribeirinhos and Other Traditional Peoples

Another ethnogenic front of this cycle opened up in the Brazilian Amazon during this period amongst diverse Amazonian peasant communities, often referred to generically as *caboclos* (see Adams et al. 2006). They began reframing their social identities around their traditional ecological knowledge and their sustainable use of Amazonian ecosystems in alliance with the environmental movement, giving rise to a host of 'traditional peoples' who have socioenvironmental attributes that are now highly valued by society at large. While the 1988 Constitution recognised the territorial rights of indigenous peoples and *quilombos*, other traditional peoples had no such guarantees and were forced to gain recognition of their collective rights to their territories by other means (Almeida 2006). The alliance with the environmental movement played an important role in changing the broader societal perception of traditional peoples from that of being vestiges of a pre-modern past, towards one that placed them in the forefront of the global search for models of sustainable development. In Brazil, the primary institutional space of traditional peoples is the Ministry of the Environment (and not the Ministry of Culture) and is directly linked to policies of sustainable development.

The social and political emergence of the Brazilian *seringueiros* (rubber tappers) in the 1980s is one of the most visible cases of ethnogenesis of a traditional people. Starting in the late 1970s, loggers and cattle ranchers started clear-cutting forests in the state of Acre as part of an uncontrolled frontier expansion on the lands where rubber tappers had their homes and used the standing forest to extract latex that was made into rubber. The rural workers' union that represented the rubber tappers resisted these encroachments on their traditionally held lands and were met with counterattacks such as the assassination of Wilson Pinheiro in 1980, a labour leader of the rubber tappers, and one of hundreds of rural Amazonian leaders to be killed over the ensuing four decades (Martins et al. 2018).

The *seringueiros* redoubled their actions in defence of the forests and, under the inspired leadership of Chico Mendes, established a political alliance with the national and international environmental movement, which was in ascent during this decade. Mendes, who became an international figure in the fight to defend the Amazon from destruction, was instrumental in organising the rubber tappers into the National Council of Rubber Tappers (CNS) which pressured the federal government to protect their lands. His assassination in 1988 by local ranchers created an international outcry and pressure which served to strengthen the rubber tappers movement. In 1989, the Brazilian federal government established the territorial category of Extractive Reserve, which granted collective title to lands that rubber tappers and other agro-extractivist communities used to gain their livelihoods. When Lula da Silva became president in 2003, he named Marina Silva, a former rubber tapper from Acre, as Minster of Environment and she began a process within the federal government of formally recognising traditional peoples of Brazil. In 2004, the National Commission for the Sustainable Development of Traditional Communities was established and subsequently organised the First National Meeting of Traditional Peoples, held in Luziânia, state of Goiás, in 2005 with representatives from 15 categories of traditional peoples.[4] These representatives forged a document, outlining their social, political and territorial claims. In 2007, the federal government established the National Policy for the Sustainable Development of Traditional Peoples and Communities.

Ribeirinhos (riverside fishing communities) are another traditional group that has experienced an ethnopolitical resurgence tied to environmentalism. These communities, which are widely dispersed throughout the floodplains of the Amazon River and its tributaries, are descendants of Indigenous peoples and many of their adaptive strategies, technologies, place names, myths and folklore have been derived or adapted from prior indigenous practices (Galvão 1976; Slater 1994). Yet, due to migration, a high degree of miscegenation with European and African peoples over the centuries, and social conflict, including the devastating Cabanagem rebellion (1835–1840) when tens of thousands of rural Amazonians were killed, it has been difficult for these riverside dwellers to claim a specific ethnic identity. Given this unique past, however, the *ribeirinho* population is generally placed within the category of traditional peoples, which distinguishes them from recent extra-Amazonian arrivals such as colonists, goldminers and ranchers.

[4] The 15 groups, with their designation in Portuguese, were: *Seringueiros*; *Quebradeiras de Coco Babaçu*; *Pescadores Artesanais*; *Quilombos*; *Geraizeiros*; *Caiçaras*; *Comunidades Fundos de Pasto*; *Ciganos*; *Faxinais*; *Agroextractivistas*; *Pantaneiros*; *Pomeranos*; *Sertanejos*; *Comunidades de Terreiro*; and *Povos Indígenas*. It is difficult to translate these designations since many of them refer to specific geographic areas or to local socioeconomic structures.

The importance of this environmental connection can be seen in the new value given to the social category of *ribeirinho*. In contemporary Amazonia, the term *ribeirinho* tends to have the positive connotation of a person who uses traditional, environmentally sound practices, particularly when compared with the category of *pescador* (fisherman), which has the negative connotations of being a predator of natural resources. In this case, the reframing of a group's socioecological identity provided *ribeirinhos* with a new source of political power as they sought to protect their aquatic use zones (e.g. lakes, rivers, flooded forests) from encroachment by industrial fishing fleets which were rapidly depleting fish stocks across wide swaths of the floodplain (Esterci 2002). By the late 1990s, local associations and organisations emerged all along the Amazon River floodplain in the defence of traditional rights of riverside communities and they began involving local and state governments in a host of unprecedented initiatives geared towards community management of fisheries. The principal instrument used by these groups was the development of fishing agreements, whereby specific floodplain regions were zoned and restrictions agreed upon at local assemblies and enforced by voluntary environmental agents.

These efforts were strengthened with the convening of the First Amazonian Community Fishing Management Meeting held in Manaus in May of 2002 and a second meeting held in Belém in July of 2003. One of the most important results of the first meeting was the development of a set of guidelines for the formal recognition of fishing agreements, which was subsequently adopted as an environmental policy instrument (Guideline 29) by the Federal Environmental Protection Agency (IBAMA) on 31 December 2002. These two regionwide meetings represented a milestone in the political articulation of floodplain communities that had historically been highly fragmented and rarely occupied an important space in the regional political structure. They also promoted an exchange of experiences between fishermen and scientists and were a forum for the launching of actions geared to consolidate the community management of fisheries.

The Neodevelopmentalist Frontier Wave (2000–2020)

By the beginning of the twenty-first century, a neodevelopmentalist variant of the Brazilian economic development agenda had emerged, fuelled by a 15-year-long commodity boom (2000–2014) and supported by strong state financing from the National Economic and Social Development Bank (BNDES) and by Chinese development and export/import banks (Little 2014). This frontier wave included a strategic effort to expand the international reach of Brazilian construction companies through the building of large-scale infrastructure projects throughout Latin America and Africa, a system which prospered throughout the boom, only to enter into crisis with a decline in commodity prices, the fallout from several international corruption scandals involving bribes from Brazilian companies and the impeachment and removal from office of President Dilma Rousseff of the Worker's Party. Meanwhile, the rural landowners lobby [*ruralistas*] organised a concerted campaign to roll back environmental policies that they considered to be a hindrance to the continued expansion of large-scale agriculture and ranching. After winning a legislative battle in 2005 that allowed the planting of genetically modified organisms (over the objections of then Environmental Minister Marina Silva), they were able to reverse numerous policies for the protection of forests through the passing of a new Forest Code law in 2012, which represented a significant defeat for the socioenvironmental movement.

With the demise of the neodevelopmentalist frontier wave and the rise of former army captain Jair Bolsonaro to the presidency in 2019, a blatant anti-indigenous, anti-environmentalist stance took control of the federal government. Bolsonaro's attitudes towards Brazil's Indigenous peoples are similar to those of the military dictatorship that ruled the country from 1964 to 1985, when indigenous peoples suffered numerous assaults and many other were driven from their lands to make way for large infrastructure projects and cattle ranches. For the first time in decades, an openly anti-Indigenous position based upon racist stereotypes is now being embraced by political leaders.

Illegal and illicit activities, which previously experienced some degree of basic enforcement, have been unmoored from their legal constraints and are not only able to advance without fear of reprisals but, in many cases, are now encouraged (and conducted) by the political elite in power. Wildcat goldminers, contraband loggers and large landowners are increasingly invading the territories of indigenous peoples and other traditional peoples throughout the Brazilian Amazon. Armed bands of land grabbers, known as *grileiros*, have been staging attacks on indigenous peoples' communities and assassinations of indigenous leaders have increased in the face of apparent impunity. The agribusiness lobby continues to exercise immense power in the National Congress, wields great influence with the president, and is now lobbying to open up indigenous lands to large-scale agricultural expansion and industrial mining. During the 30 years after the passage of the 1988 Federal Constitution, there was continual progress towards consolidating a legal, territorial and regulatory framework that was in alignment with the rights enshrined in the Magna Carta. Much of that framework is now under attack and the possibility of a significant rollback of rights looms large. The mobilisation of social movements and civil society has become one of the few bulwarks against these trends, coupled with strong international pressure for continued protection of the Amazon rainforest and its peoples. If these countermeasures are unsuccessful, it may mean that this particular cycle of ethnogenesis may be coming to an end and the onslaught of a new frontier wave may be in the offing.

Conceptual Coda

Comparative frontier history in the Americas showcases the main mechanisms of sociocultural change within a long-term perspective. The historical analysis of frontiers reveals highly complex, asymmetrical relations between social groups that cannot be subsumed by the concepts of conquest, subjugation and extinction, even though all of these elements may be present. Though ethnocide is clearly a traumatic experience for any social group, it does not necessarily mean the end of their cultural existence. The inability to see, understand or accept the processes of hybridity, syncretism, merger and bricolage has made many of these groups invisible within the national assimilationist discourse, which has dominated national societies throughout the Americas.

A fuller understanding of ethnogenesis lays bare the omissions of much social change theory by focusing on the social and cultural history of subjugated peoples, their sustained resistance to external pressures and the purposeful assertion of their ethnic rights. Subjugated groups are viewed not only as victims of larger historical forces, but also as fulfilling a protagonist role in recreating modes of social organisation and ethnic identity. This concept allows social scientists and historians to move beyond the extinction paradigm, which did not admit the agency of subjugated groups, and demands a more nuanced, historical rendering of the sociocultural change process. The repeated historical presence of ethnogenic phenomena is a testament to the cultural resilience of human societies. Far from being a peripheral sociocultural phenomenon, ethnogenesis is a defining characteristic of the continual process of culture-making and re-making.

The case of the Brazilian Amazon examined here shows some of the novel ways that ethnogenesis occurred within the sociopolitical context of the demise of the dictatorship and the return of a democratic political regime. It also highlights the central role that global environmentalism played, particularly regarding the Amazon rainforest, in strengthening the relative position of these new social groups within Brazilian society. In the aftermath of the impacts of the first Amazonian developmentalist frontier, indigenous peoples, *quilombos*, *seringueiros* and *ribeirinhos*, among many other groups, came out of the shadows and asserted identities that emphasised their environmental and sociocultural contributions. These groups are now bracing for a yet another frontier wave that will reshuffle interethnic relations and create new structural fissures but, if frontier history is any indication, will surely be met with creative sociocultural responses.

References

Adams, C., Murieta, R., & Neves, W. (2006). *Sociedades Caboclas Amazônicas: Modernidade e Invisibilidade*. São Paulo: AnnaBlume.

Albers, P. C. (1996). Changing Patterns of Ethnicity in the Northeastern Plains, 1780–1870. In J. D. Hill (Ed.), *History, Power, and Identity: Ethnogenesis in the Americas, 1492–1992* (pp. 90–118). Iowa City: University of Iowa Press.

Almeida, A. W. B. (2006). *Terras de Quilombo, Terras Indígenas, "Babaçuais Livres", "Castanhais do Povo", Faxinais e Fundos de Pasto: Terras Tradicionalmente Ocupadas*. Manaus: PPGSCA-UFAM.

Anderson, G. C. (1999). *The Indian Southwest, 1580–1830: Ethnogenesis and Reinvention*. Norman: University of Oklahoma Press.

Anjos, R. S. (2006). *Quilombos: Tradição e Cultura da Resistência*. São Paulo: Aori Comunicações.

Becker, B. K. (1982). *Geopolítica da Amazônia: A Nova Fronteira de Recursos*. Rio de Janeiro: Zahar.

Behrendt, S. D. (2010). The Transatlantic Slave Trade. In R. L. Paquette & M. M. Smith (Eds.), *The Oxford Handbook of Slavery in the Americas* (pp. 251–274). Oxford: Oxford University Press.

Bensusan, N. (2005). Biodiversidade, Recursos Genéticos e Outros Bichos Esquisitos. In A. Rios & C. Irigaray (Eds.), *O Direito e o Desenvolvimento Sustentável: Curso de Direito Ambiental* (pp. 31–69). São Paulo: Peirópolis.

Bilby, K. (1996). Ethnogenesis in the Guianas and Jamaica: Two Maroon Cases. In J. D. Hill (Ed.), *History, Power, and Identity: Ethnogenesis in the Americas, 1492–1992* (pp. 119–141). Iowa City: University of Iowa Press.

Brown, D. (1970). *Bury My Heart at Wounded Knee: An Indian History of the American West*. New York: Owl Books.

Comaroff, J. L., & Comaroff, J. (1997). *Of Revelation and Revolution. Volume 2: The Dialectics of Modernity on a South African Frontier*. Chicago: University of Chicago Press.

Conklin, B. A., & Graham, L. R. (1995). The Shifting Middle Ground: Amazonian Indians and Eco-politics. *American Anthropologist, 97*(4), 695–710.

Denevan, W. M. (1992). Native American Populations in 1492: Recent Research and a Revised Hemispheric Estimate. In W. M. Denevan (Ed.), *The Native Population of the Americas in 1492* (2nd ed., pp. xvii–xxxvii). Madison: University of Wisconsin Press.

Esterci, N. (2002). Conflitos Ambientais e Processos Classificatórios na Amazônia Brasileira. *Boletim Rede Amazônia, 1*(1), 51–62.

Galvão, E. (1976 [1955]). *Santos e Visagens: Um Estudo da Vida Religiosa de Itá, Baixo Amazonas*. São Paulo: Companhia Editora Nacional.

Gonzalez, Nancie L. S. (1969). Black Carib Household Structure: A Study of Migration and Modernization. Seattle: University of Washington Press.

Griffin, W. (2005). *Los Garífunas de Honduras: Cultura, Lucha y Derechos bajo el Convenio 169 da la OIT*. San Pedro Sula: Central Impresora.

Hemming, J. (1970). *The Conquest of the Incas*. London: Macmillan.

Hemming, J. (1978). *Red Gold: The Conquest of the Brazilian Indians*. Cambridge, MA: Harvard University Press.

Hemming, J. (1987). *Amazon Frontier: The Defeat of the Brazilian Indians*. London: Macmillan.

Hill, J. D. (1996). Introduction: Ethnogenesis in the Americas, 1492–1992. In J. D. Hill (Ed.), *History, Power, and Identity: Ethnogenesis in the Americas, 1492–1992* (pp. 1–19). Iowa City: University of Iowa Press.

Hudleson, J. E. (1981). *The Expansion and Development of Quichua Transitional Culture in the Upper Amazon Basin*. PhD thesis, Columbia University.

Ioris, E. (2014). *Uma Floresta de Disputas: Conflitos sobre Espaços, Recursos e Identidades Sociais na Amazônia*. Florianópolis: UFSC.

Li, T. M. (2014). *Land's End: Capitalist Relations on an Indigenous Frontier*. Durham: Duke University Press.

Little, P. E. (2014). *Mega-development Projects in Amazonia: A Geopolitical and Socioenvironmental Primer*. Lima: Red Jurídica Amazónica/ARA/DAR.

Manchete. (1973). *Amazônia: Um Novo Brasil*. Rio de Janeiro: Edição Especial, February.

Mann, C. C. (2005). *1491: New Revelations of the Americas Before Columbus*. New York: Vintage.

Martins, E., Rodrigues, G., Sautchuk, J., Neves, M. V., & Weiss, Z. (2018). Mártires da Luta. In *Chico Mendes: Herói do Brasil* (pp. 20–25). Sindicato dos Trabalhadores e das Trabalhadoras Rurais de Xapuri; SINPRO/DF.

Melo, J. (2006). *Baré (Aruak) reinventados em Manaus e Manaus Reinventada pelos Baré: Um Estudo de Etnogênese Urbana*. PhD thesis, Universidade de Brasília.

Naranjo, M. F. (1977). Zonas de Refugio y Adaptación Étnica en el Oriente: Siglos XVI, XVII, XVIII. In M. Naranjo et al. (Eds.), *Temas Sobre la Continuidad y Adaptación Cultural Ecuatoriana* (pp. 105–168). Quito: Ediciones de la Universidad Católica.

Oliveira, J. P. (1998). Uma Etnologia dos 'Índios Misturados'? Situação Colonial, Territorialização e Fluxos Culturais. *Mana: Estudos de Antropologia Social, 4*(1), 47–78.

Powers, K. V. (1995). *Andean Journeys: Migration, Ethnogenesis, and the State in Colonial Quito*. Albuquerque: University of New Mexico Press.

Price, R. (Ed.). (1979a). *Maroon Societies: Rebel Slave Communities in the Americas* (2nd ed.). Baltimore: Johns Hopkins Press.

Price, R. (1979b). Introduction: Maroons and Their Communities. In R. Price (Ed.), *Maroon Societies: Rebel Slave Communities in the Americas* (2nd ed., pp. 1–30). Baltimore: Johns Hopkins Press.

Ramos, A. R. (1998). The Specter of Nations within the Nation. In *Indigenism: Ethnic Politics in Brazil* (pp. 168–194). Madison: University of Wisconsin Press.

Ramos, L. (2004). *As Formas Kokama de Estar na História: Etnicidade, Política e Narrativa*. MSc dissertation, Universidade de Brasília.

Ribeiro, D. (1957). Culturas e Línguas Indígenas do Brasil. *Educação e Ciências Sociais, 2*, 6.

Sahlins, M. (1985). *Islands of History*. Chicago: University of Chicago Press.

Santilli, J. (2005). *Socioambientalismo e Novos Direitos: Proteção Jurídica à Diversidade Biológica e Cultural*. São Paulo: Fundação Peirópolis.

Slater, C. (1994). *Dance of the Dolphin*. Chicago: University of Chicago Press.

Souza, M. (2019). Menos de 7% dos Territórios Quilombolas Reconhecidos Têm Títulos de Propriedade. *Jornal da USP*. https://jornal.usp.br/ciencias/ciencias-humanas/menos-de-7-territorios-quilombolas-reconhecidos-tem-titulo-de-propriedade. Accessed 10 Jan 2020.

Svirsky, E., & Capobianco, J. P. (Eds.). (1997). *O Ambientalismo no Brasil: Passado, Presente e Futuro*. São Paulo: Instituto Socioambiental.

Thornton, R. (1990). *American Indian Holocaust and Survival: A Population History since 1492*. Norman: University of Oklahoma Press.

Todorov, T. (1984 [1982]). *The Conquest of America*. Trans. R. Howard. New York: Harper Colophon.

WCED (World Commission on Environment and Development). (1987). *Our Common Future*. Oxford: Oxford University Press.

Whitehead, N. L. (1996). Ethnogenesis and Ethnocide in the European Occupation of Native Surinam, 1499–1681. In J. D. Hill (Ed.), *History, Power, and Identity: Ethnogenesis in the Americas, 1492–1992* (pp. 20–35). Iowa City: University of Iowa Press.

Whitten, N. E., Jr. (1976). *Sacha Runa: Ethnicity and Adaptation of Ecuadorian Jungle Quichua*. Urbana: University of Illinois Press.

Wolf, E. (1959). *Sons of the Shaking Earth*. Chicago: University of Chicago Press.
Wolf, E. (1982). *Europe and the People without History*. Berkeley: University of California Press.
Wright, R. (1990). Conquistados, mas não Vencidos. *Anuário Antropológico, 87*, 249–257.

13

Brazilian National Integration Policies and the Amazon: Discourses of Modernisation Between the Past and the Present

Vitale Joanoni Neto
and Regina Beatriz Guimarães Neto

Preliminary Considerations

Even at the beginning of the twenty-first century, exploration and conservation of the Amazon remains a controversial issue. There are simplifications, anachronisms and common-sense analyses that naturalise economic exploitation and, above all, conceive of this territory as backward, inhospitable or savage. Existing documentary sources about the Brazilian Amazon allow for the production of knowledge about its nature, biodiversity, space and, especially, the different human groups living there. Since the seventeenth century, this knowledge has been reinterpreted and reused in various contexts, such as to justify Portuguese possession of the

V. Joanoni Neto (✉)
Federal University of Mato Grosso (UFMT), Cuiabá, Brazil

R. B. Guimarães Neto
Federal University of Pernambuco (UFPE), Recife, Brazil

A. A. R. Ioris (ed.), *Environment and Development*,
https://doi.org/10.1007/978-3-030-55416-3_13

territory, in conflicts between *sertanistas* (hinterland people) and the church over the 'right' to control indigenous populations and in the creation of an identity linked to the hinterland, as a way of constructing a dichotomy between this uncivilised space and more civilised coastal areas or of framing it as the birthplace of a 'true' Brazilian national identity. In addition, it has also been used to justify the idea that the Amazon's economic potential was not being properly exploited and that its 'empty areas'—a phrase that disregards the various indigenous territorialities in the area—were undermining national security.

The advancement of research in all areas of knowledge has made this significant South American territory the object of naturalisations and conceptions that disregard its social, cultural and ethnic complexity, in addition to its biodiversity. In dealing specifically with Brazil, it is useful to analyse how the dichotomies present in many narratives and public discourses have no basis in reality. The Brazilian Amazon, which corresponds to approximately 60% of the national territory, was systematically colonised by Europeans from the sixteenth century onwards, at a high social cost to the indigenous population and causing irreparable damage to biodiversity. Military and religious groups, in addition to *sertanistas*, headed west from the mouth of the Amazon River, in search of Indians and gold, largely organising their advances inland from the Piratininga Plateau (today, the city of São Paulo). The narratives resulting from these incursions gave rise to an image of the Amazon that lasted until the twentieth century. It is worth remembering that the recognition of this territory as part of Brazil was guaranteed by successive international treaties, and that its integration into national economic, social and political dynamics occurred only gradually. (Raminelli, 1996; 2013)

While conflicts and territorial disputes before the twentieth century had various motivations or justifications (e.g. territorial expansion or consolidation, conquest of indigenous people or the commercialisation of spices from the forest), during the last century, an emphasis on the power of capitalist investments in the Amazon territory was more prominent. This was due to several factors, including interest in the area's extractive potential, as in the case of the exploitation of *Hevea brasiliensis* (rubber latex tree), large investments in mineral exploration (from iron ore and bauxite to niobium, tantalite and petroleum) and the advance of agricultural production areas (with a particular emphasis on agribusiness). This expansion and domination led to a great concentration of land and the loss of thousands of hectares of forest coverage. This in particular has raised concerns and has been the subject of criticism from several international environmental organisations, based on meaningful data:

> The meteoric rise in soy production in the state of Mato Grosso is eating up rainforest and savannah at a staggering rate, with 1.2 million hectares under production in 1991; 6.2 million hectares in 2010; and 9.4 million hectares by 2016. (Brandford and Torres 2017)

Between the past and the present, the economic picture presented today leads us to critically question how the Amazon has participated in systemic cycles of capitalist accumulation (Arrighi 1996). European domination and the successive advances of commercial and financial capital in the Amazon following Brazilian independence can be read as precursors to the mechanisms and strategies of globalisation, associated with the international division of labour. However, it is fundamental to reflect on the mechanisms and political devices that have historically sustained and been embedded in the various forms of economic production and capital investment. One of these political mechanisms is the use of various instruments by governments to advance the interests of capital.

It is our purpose in this chapter to reflect on the projects of successive Brazilian State governments, especially after the military coup of 1964, articulated with important civil sectors, in which the Amazon has become the strategic locus for the development model adopted in the country. This analysis is based on an assessment of the general power strategy of the Brazilian dictatorship, which seeks to elucidate what the Amazon has meant in political and economic terms in the context of experience and governmental rationality, as manifested in a range of public policies. From a historical perspective, this discussion consists of approaching the past by examining the present. In this context, it is important to study governmental discourses and practices in terms of the ways they have supported the redefinition of power relations in Brazil, in particular by focusing on the various political mechanisms embodied in laws, decrees, plans and institutional programmes and in the State's own administrative reform. This reform, which was carried out during the dictatorship and was not debated in the Brazilian National Congress, was supported by the powers of exception of the Second Institutional Act of December 1966 (Bercovici 2010). Business logic prevailed as an administrative practice, and this has persisted in Brazil's political-administrative structure since the end of the military regime.

Particular attention will be paid to the National Integration Programme, a series of projects and policies implemented by successive Brazilian governments from the nineteenth century onwards during the National State period and continuing under the Vargas dictatorship in the New State [*Estado Novo*, 1937–1945]. The New State presented the country with a proposal for the integration of 'several *Brazils*', an 'internal imperialism', which would ensure economic development with the March towards the West (Ricardo 1970). The policy of occupation of the Midwest encompassed more than this region: it anticipated the reoccupation of the Amazon, with a clear disregard for indigenous territorialities. (Freitag, Stanczyk, 2010; Souza 2000). It is worth remembering that the Brazilian Institute of Geography and Statistics (IBGE), created in 1937, structured the regional division of Brazil (1942) as part of Vargas's policy of breaking with 'negative regionalism', producing another regionalism 'born of the very centralisation of the authoritarian regime' (Gomes 2013: 67). The aim of this policy was to promote the 'occupation of demographic vacuums' by absorbing the 'population surpluses' which, according to official discourse, were putting pressure on the Centre-South of the country, and sending these people to produce raw materials and foodstuff at low costs to subsidise the implementation of industrialisation in the Southeast. In the government's view, these measures would solve regional imbalances through the implantation of a 'demographic policy' that would encourage 'internal migration'.

The state of Mato Grosso was one of the main targets for this project. Later, the creation of the Legal Amazon through Law 1806 of 06 January 1953, and the installation of the Superintendence of the Plan for Economical Valorisation of the Amazon (SPVEA), incorporated three states into the Brazilian Amazon: Maranhão (located west of the 44th meridian), Goiás (located north of the 13th parallel south, now the state of Tocantins) and Mato Grosso (located north of the 16th parallel south).

This was the result of a political concept aimed at planning and promoting the development of the region, not a geographical imperative. The national integration policy was reaffirmed by the military governments after 1964 as a development strategy set out a plan for economic investments across the Amazonian territory in association with the Northeast region of Brazil. The National Integration Programme (NIP) was introduced in 1970 and, in addition to incorporating the capitalist ideology of business expansion, ensuring fiscal incentives and other financial benefits for entrepreneurship, was an axis of policies that aimed to articulate basic regional intervention strategies. Designated as 'strategic sectoral goals' by the Ministry of Planning (Brazil 1970), these policies directly affected the Brazilian Amazon (redefined according to Law 5173 of 27 October 1966).

In this scenario, a series of practices and discourses can be identified that aimed to justify governmental actions and policies for the Amazon, directing public resources for business investments or even for State institutions such as the Amazonian Development Authority (SUDAM) and the National Institute for Colonisation and Agrarian Reform (INCRA). These policies were implemented along with strong official publicity for the land offer policy, which redirected displaced populations from various regions of Brazil to the Amazon (Guimarães Neto 2002; Joanoni Neto 2007). In addition, other interventions, including the use of police and military force, were used to promote inter-regional displacements of ethnic groups (such as indigenous communities) that traditionally lived in parts of the Amazon with the potential to be exploited for capital, especially by private companies. In this way, a new political and economic configuration was produced with the support of the institutionally organised State according to the logic of the military that commanded it and structured it on a legal and police basis that used its own legality devices and used a complex and sophisticated information and repression structure (Figueiredo 2015). It is possible to observe, in this context of land privatisation and intense concentration of financial investments in the expansion of livestock and agribusiness, how the State enterprise apparatus increasingly interfered with the preservation of local biodiversity and socio-diversity.

It is necessary to clarify that the policy of national integration under the military governments was not the result of internal demand, as has historically been claimed by various people living in the Amazonian territory, or even a result of local populations recognising a need for better public services. It must be understood as resulting from a political position of the State, within a framework of national and international political changes that began at the end of the Second World War, with further advances during the Cold War. In this context, the Amazon came to be seen as vital to Brazilian national security. The military, in tune with various business sectors and large parts of the political class, tried to align Brazil with the Western capitalist bloc in the global political framework, taking the Soviet bloc as its main antagonist, along with the Left in Latin America. In military discourse, such as communications from the Escola Superior de Guerra (ESG—Institute for Advanced Policy, Strategy and Defence Studies), particularly the writings of General Golbery do Couto e Silva, the effective economic integration of the Amazon as the most dynamic productive centre in the country was presented as an imperative for national security policy.

After the military coup in March 1964, the salvationist discourse of a country under threat from communism re-emerged in a new form. The identity promoted for Brazil was that of an integrated country (reproducing a discourse previously used during the Estado Novo dictatorship between 1937 and 1945) under the command of a strong government. Again, discourse referring to the idealised past epitomised by the *bandeirante* (basically, pathfinder or trailblazer) was mobilised as a justification for the new projects (referred to as 'colonisation nuclei'). Ricardo (1970: xl), in a post-face addition to the third edition of his book in 1970 (the first edition dates from 1940), conceptualised these settlement projects as representing a 'larval State' (something like a germinal State), made up of miniature cities, at the origin of 'our self-government in the formation of the Brazilian State'. *Bandeirismo* was identified with the 'revolution' and with the defence and preservation of the fundamental values of the modern world: Christianity, international law, European capitalism and industrial revolution. National integration was the discourse that served as the focal point for this new identity, indissolubly linked to the ideology of national security under US policy towards Latin America. In this context, the figure of the *bandeirante* assumed a new importance, particularly in relation to what was called the Rondon Project,[1] for the exaltation of the Belém-Brasília motorway and the construction of the Trans-Amazonian Highway, which, according to Ricardo, was a result of the 'instinctive ferment of heroic times' (Ricardo 1970: 623). The new March to the West was announced. Once again, an organised Brazil moved westwards, realising its 'internal imperialism' inch by inch. This is an ongoing struggle to reinforce the first landmarks of the border conquests (Ricardo 1970: 622).

It is important to emphasise, first, that these discourses were reproduced differently in various texts and speeches over time, depending on specific political contexts. The military, supported by the Escola Superior de Guerra (ESG), used texts and speeches to argue in favour of implementing the National Integration Programme and, in particular, the Amazon operation (Joanoni Neto 2014: 107). Secondly, notwithstanding the inevitable peculiarities and displacements of ideas inherent in historical dynamics, the figure of the pioneer, a patriot who responds to the nation's appeal and migrates to the frontier, as a *bandeirante*, a civiliser, is present and constant both in official documents, such as company speeches and propaganda pieces, and in the memory of local residents. The objectives expressed in speeches were based on the idea of national integration, border security, progress, development and civilisation.

> With a view to the humanisation, integration and valorisation of the immense territory, still largely unproductive and deserted, the tri-peninsular scheme (...) is certain to indicate (...) three successive phases in the broad geopolitical manoeuvre to be performed (...). First phase: articulate the basis of our continental projection, connecting the Northeast and the South to the central part of the country, while ensuring the inviolability of the vast expanse of the countryside by the effective obstruction of possible routes of penetration. Second phase: stimulate the advance of the colonising

[1] Inter-ministerial action of the federal government that, in partnership with higher education institutions, aims to contribute to sustainable local development and the construction and promotion of citizenship. The Rondon Project thus prioritises developing actions that bring permanent benefits to communities, especially those related to the improvement of social welfare. It also seeks to consolidate the sense of social responsibility and collective—in favour of citizenship—national development among university students, contributing to their education and providing them with knowledge of the Brazilian reality.

> wave to the Northwest from the centre in order to fully integrate the Central West peninsula into the rest of the country. Third phase: civilise the Amazonian rainforest, relying on the occupation sites, starting from an advanced base built in the Central-West region (…). (Couto e Silva 1982: 13–2, written in 1960)

Proceeding with this project, Decree 61,330 of 11 September 1967 created the Working Group for the Integration of the Amazon, with significant military participation from the National Security Council and the General Staff of the Armed Forces. Shortly before this, Law 5173 had been approved, reformulating the Superintendence for the Economic Valorisation of the Amazon (SPVEA) and creating the Superintendence for the Development of the Amazon (SUDAM), in addition to Law 5174 of 27 October 1966 dealing 'with the concession of fiscal incentives in favour of the Amazon Region and other measures' (Brasil 1966). A short time later, Decree-Law 291 of 28 February 1967 was passed, establishing 'incentives for the development of the Western Amazon of the Border Strip covered by the Legal Amazon'. (Brasil 1967).

Under this political configuration, the entire Legal Amazon was home to only 3.7% of Brazil's population. According to official discourse, this situation meant that there was pressure on the government to act in the region. In this context, measures taken included the creation of the Ministry of the Interior by General President Costa e Silva through Decree-Law 200 of 25 February 1967. This ministry, under the leadership of General Afonso Augusto de Albuquerque Lima, was placed under the Superintendence for the Development of the Amazon and the Superintendence of Manaus Free Zone.

The Ministry of the Interior was responsible for regional development, population settlement, territorial occupation and internal migration, among other things. It is worth noting that Albuquerque Lima is cited in literature on the military governments as belonging to the most radical wing of the so-called hard line, a military group or faction that favoured keeping the military in power and taking a range of actions against opponents of the dictatorial regime. (Martins Filho 2004: 133). Supported by these measures, primary-sector investments in the Amazon grew quickly from 1966 onwards, with livestock businesses predominating. Five projects were approved in this year. In 1967, another 55 were approved, mostly for the primary sector, 10% for agriculture and 90% for livestock. The manufacturing industry continued to receive incentives, with 31 projects in 1967, corresponding to 27% of total incentives. The measures taken began to form the contours of what would become a dominant trend throughout the Amazon after 1970, that is, the investment of financial capital in cattle farming, particularly in the states of Pará and Mato Grosso.

The National Integration Programme

The National Integration Programme (NIP) was created by the Brazilian military government through Decree-Law 1106 of 16 July 1970, signed by President Medici. In order to avoid the risk of 'alien penetration' (Couto e Silva 1982: 43, text written in 1952), the government took out large loans from the Inter-American Development Bank (IDB) and the United States Agency for International Development for investment in infrastructure. Considered one of the main federal government programmes for the development of the Amazon, this programme 'had resources of US$ 430,000,000 (four hundred and thirty million dollars), to be constituted in the years 1971 to 1974, with the specific purpose of financing the plan for infrastructure works' (Resende 1971: 393). As result of this, the decree-law that created the NIP affirmed in its second article:

> The first stage of the National Integration Programme will be constituted by the immediate construction of the Transamazônica and Cuiabá-Santarém motorways.
>
> First Paragraph - Ten kilometres of land on both sides of the new federal motorways will be set aside for colonisation and agrarian reform, with the resources of the National Integration Programme, to carry out its occupation and adequate and productive economic exploitation. (Brasil 1970)

Around ten months later, another decree-law was published, Decree-Law 1164 of 1 April 1971, and the above text was substantially changed to expand the range of land designated for investment. For the most part, this was vacant government land:

> Article 1 - *The vacant lands located in the area of one hundred (100) kilometres wide, on each side of the axis of the following motorways, already built, under construction or projected*, are declared *indispensable for national security* and development, in the region of the Legal Amazon, defined in the second article of Law 5,173 of 27 October 1966. (Brasil 1971, emphasis added)

This new decree-law, in addition to including an adjustment to the NIP as originally published in 1970, demonstrated a strengthening of central power over regional political groups. Up to that point, land tenure had been an important currency for exchange and political bargaining among different local groups, in addition to, at various moments, financing State governments (Moreno 2007: 113). Among the motorways included in the text of the decree-law were three passing through the state of Mato Grosso: the north/south BR-163 motorway in the central region of the state, the north/south BR-158 motorway in the east and the north-west/south-east BR-364 motorway (which had been built earlier but was now enlarged). Also significant was the Trans-Amazonian Highway in the Pará state, which became a migratory corridor from the Brazilian Northeast to the Amazon, impacting the occupation of the western border region and the state of Mato Grosso (Fig. 13.1). It was in this space, now under the responsibility of the federal government, that colonisation and agricultural projects were implemented under conditions very favourable to the national business and rural settlement projects. The government formulated public policies that aimed at alliances with the business community in the expansion of a new market for land and economic exploitation based on monoculture plantations and cattle farming, adopting a discourse based on the idea that this would reduce pressure on the other regions of the country, always in the name of national security.

The 1973 report of the Secretariat of Planning and General Coordination of the state of Mato Grosso presented the goals of the NIP in its first phase, highlighting the construction of the Trans-Amazonian

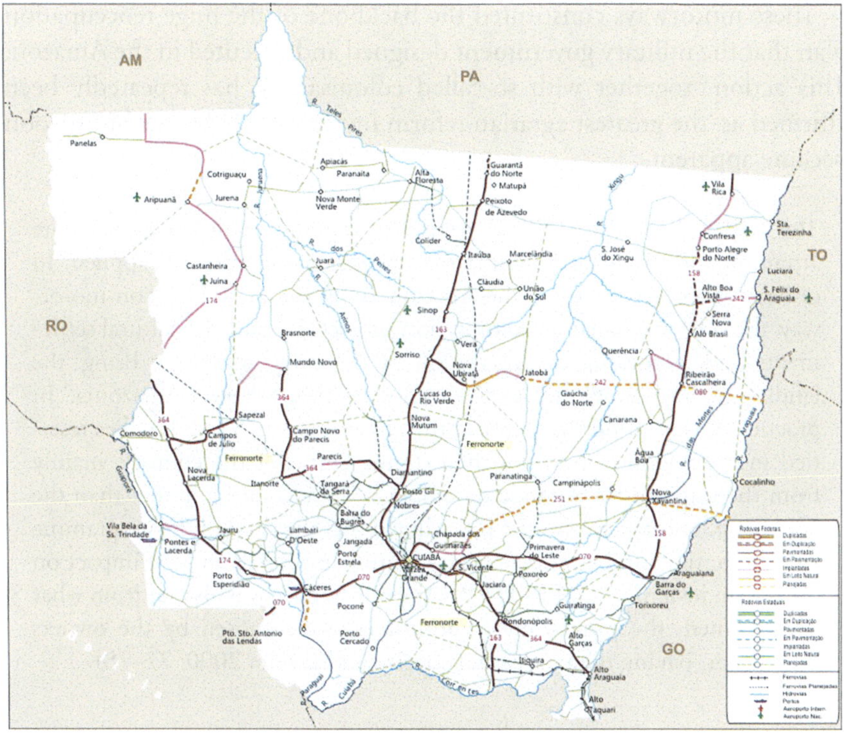

Fig. 13.1 Mato Grosso Road Map. (Source: http://www.infoescola.com/wp-content/uploads/2012/11/mapa-rodoviario-mato-grosso.jpg)

and Cuiabá-Santarém (BR-163) highways. Around this time, the first effects of the new model of land reoccupation were being felt, and road transport logistics were a major weak point in development and modernisation, hence the importance of integrating the state with the rest of the country:

> Based on this argument, the great road project was born, made up of the pioneer road—the Cuiabá-Santarém motorway—with a longitudinal axis of 1,750 km from the capital of Mato Grosso towards the north, reaching the Amazon rainforest and a vast native region of the Brazilian central plateau, and eventually reaching the most voluminous fluvial artery in the world, the Amazon River. (Mato Grosso 1973: 24)

These motorways constituted the backbone of the huge reoccupation plan that the military government designed and executed in the Amazon. This action, together with so-called colonisation, has repeatedly been affirmed as 'the greatest agrarian reform in the world', but its limits soon became apparent.

> The NIP and the PROTERRA were the programmes that received more attention and to which a significant amount of resources was applied. In order to populate a part of the Amazon along the Trans-Amazon motorway, the NIP was based on colonisation projects around agricultural settlements and, according to the version of the time, sought to bring 'the landless men of the Northeast to the lands without men in Amazonia.' In practice, it was found that most of the approximately 5,000 families resettled in the region came from the extreme South of the country, mainly from the states of Rio Grande do Sul and Santa Catarina, rather than the Northeast. Subsequent studies have shown that the costs of the programme were high, the number of beneficiary families minimal, and the impact on the region insignificant. PROTERRA's performance was also far from what was planned: the programme expropriated areas chosen by the owners themselves, paying them in cash. (Luceiro and Pereira 2000: 77–78)

Urbanisation as a Direct Result of Colonisation

The division of the state of Mato Grosso was only completed in 1977 through Complementary Law 31 of 1977, which created the state of Mato Grosso do Sul. Despite ongoing conflicts between local groups dating back to the nineteenth century, the division was only possible during the state of exception, which allowed the government to meet federal objectives while disregarding conflicts of interest between regional groups. The remaining part of the state of Mato Grosso had been a significant target for investments since 1967. The state had 38 municipalities in 1970, as established in the Legal Amazon. In the following four decades, that number went up to 141, of which 67 had fewer than 10,000 inhabitants and 45 had between 10,000 and 20,000. Only 29 had more than 20,000 inhabitants (Miranda 2016: 33).

To a large extent, the new cities were directly linked to several business investment projects, particularly colonisation projects, which were implemented in the state with strong encouragement from the federal government, particularly fiscal incentives. These colonisation projects, framed in government discourse as an alternative to the Brazilian agrarian question, were, ultimately, major urbanisation projects. Other government agencies that played a key role in this reoccupation of the Amazon were the Bank of the Amazon (BASA) and the Institute of Colonisation and Agrarian Reform (INCRA). To better understand the importance of both, it must be remembered that this project was necessarily endorsed by the large business community allied with the military government, which was a direct beneficiary of the financial resources made available. In the words of Planning Minister Reis Veloso in an official statement in 1973:

> ... the need to avoid a predatory occupation, with a consequent deforestation process, and to promote maintenance of the ecological balance, leads us to invite big companies to take on the task of developing this region. (Castro et al. 1994: 77)

According to Nepstad, Stickler and Almeida (2006: 42), 400,000 km^2 of forest were destroyed between 1970 and 1990, predominantly to make way for livestock activity, the industry most strongly stimulated by government projects aimed at large investors. This corresponds to 10% of the Amazonian rainforest. Entrepreneurs bought large areas of land from INCRA, submitting land use plans that would allow for the establishment of small landowners in lots ranging from 100 to 500 hectares or designate the land for farming activities (often beef cattle farming). The size of the areas passed on to these companies varied widely. Among the colonisation projects approved for the state of Mato Grosso, *Sociedade Imobiliária Noroeste do Paraná* (SINOP) received 400,000 hectares for a project with the same name; SINOP also developed other projects in the state. The company Integration, Development and Colonisation (INDECO) received 400,000 hectares for the Alta Floresta Project and also received thousands of hectares of land to develop other projects in the state. The *Cooperativa Tritícola do Iguaçu* (COTRIGUAÇU) received a million hectares and then, unable to continue with the project bearing its name, had to share this task with other companies. The creation and development of urban centres was a requirement of INCRA for the approval of all these projects and as propaganda was mostly aimed at small and medium rural producers in the South of Brazil, in order to stimulate the emergence of these centres.

Properties for both the so-called colonisation projects and agricultural purposes extended over large areas, which posed another problem. The sale of land and authorisation to settle over such large areas was based on the assumption that they were unoccupied, which did not correspond to local realities, as the federal government knew well. Reports of conflicts, deaths, threats and the expulsion of entire communities through the use of force became frequent (Guimarães Neto 2014). It is important to point out the historical analogy used in discourse that associates companies and entrepreneurs developing colonisation projects in the Amazon with the saga of the *bandeirantes*. Mainstream media called such developers 'modern' or 'twentieth-century' *bandeirantes*, referring to eighteenth-century incursions by the *sertanistas*, who were known as Indian murderers, destroyers of local communities.

The spatialisation of diverse territories, provoked by strong political and economic interests and produced in the name of the State and national security, intensified conflicts between different occupation projects and revealed the overlapping of different territorialities, in addition to forcing the re-territorialisation of some at the cost of de-territorialisation of others (Haesbaert 2004; Brasil, 1996). Again, the consolidation of what was seen as the integration of the Amazonian territory into Brazil took place from the outside, that is, disregarding the territorialities of indigenous people and the fragile equilibria between them and the *quilombola*, extractivist and squatter communities. What was presented as an economic and cultural need for survival, based on the supply of land and labour to the various groups who were encouraged to move to the Amazon, also called 'colonisation' of the Amazon, was manipulated so that small farmers and rural workers were blamed for resettlement failures.

Attempts to justify these policies took the form of a compelling discourse that affirmed them as a model of agrarian reform, associated with other agro-industrial valorisation policies in the South of Brazil (the so-called green revolution). Displaced people—especially small landowners, landlords and tenants from the South and landless rural workers from the Northeast—lived in violently precarious conditions, vulnerable to the growing concentration of land. Ianni (1979) analyses this process as an agrarian counter-reformation, and Martins (1994) highlights the disconnect between its actual characteristics, conservative and militarised, and discourse implying that this was a process of modernisation and democratisation of the land. Dispossessed of their basic needs for survival, migrants sold the few possessions they had and offered, in a new framework of labour formation, their labour in order to participate in the new economic projects that opened up in the Amazon. (Baumam, 2005). A whole field of possibilities was set up for the privatisation of land and business in the Amazon, now with the support of financial capital, and later landowners (Volkswagen Group, Banco Bamerindus, Banco de Crédito Nacional [BCN], among others), with large investments in livestock and mineral exploration (such as Vale do Rio Doce).

Since the 1980s, the State has seen its powers of management and control diminish throughout this territory. This condition was particularly aggravated by the privatisation processes that intensified after the 1990s, making the State one of the instruments of the aforementioned process of accumulation and, more recently, a hostage to discourse based on the income derived from the export of commodities.

> With growing public and private deficits, one of the main sources of foreign currency was exactly the export of soybean and a few other primary commodities (Hall et al. 2014). In a situation of challenging macroeconomic adjustment, rents from agribusiness greatly helped to maintain the Real Plan and fund growing government expenditures. The export market was also helped by the introduction of tax exemptions, such as the 1996 Kandir Law that removed provincial (state) taxes on the export of primary commodities, such as soybean, based on the need to mitigate the growing federal deficit created by the Real Plan (although this also penalised the state administration). Agribusiness exports became even more attractive when Brazil floated the national currency (Real) in 1999, which sent a shock across the economy that set Brazil's soybean boom in motion. This leverage effect of agribusiness rents was facilitated by favourable market prices in the first decade of the century, often described as the 'commodity

boom', due to rising demand (particularly from China). Between 2000 and 2005, the area devoted to soybean production doubled in Mato Grosso (from 3.12 million to 6.20 million hectares, cf. Cunha 2008) and it continued to expand throughout the decade. (Ioris 2016: 461)

It is necessary to add to this framework, in the field of market production, the monopoly of violence and definitions of what is legal or illegal in labour relations and land use. On this last point, the Brazilian Amazon's extensive monoculture plantations mean that it is among the most problematic agricultural frontiers in the world due to the emission of carbon dioxide and methane gas, meaning that Brazil is the largest emitter of greenhouse gases related to changes in land use (Coy and Klingler 2014: 2). In regard to the exploitation of the labour force, the model mentioned above plays a crucial role in the maintenance of degrading forms of worker insertion. 'Agribusiness replaced family farming. And slavery did not disappear' (Harvey 2013: 121). Companies have frequently been reported for using contemporary slave labour (the designation used by the International Labour Organization) since the end of the twentieth century, which, together with murders of rural workers, is evidence of a perverse scenario. Data from the Pastoral Land Commission (CPT) show that 1115 cases of homicide resulting from field conflicts were recorded between 1985 and 2014. Of these, only 12 were tried. According to the Working Group on Contemporary Slave Labour (GPTEC), 41,665 workers were rescued from conditions of slave labour between 1995, when the Special Task Force on Mobile Inspection (GEFM) was set up under the Ministry of Labour and Employment and the Executive Group for Suppression of Forced Labour (GERTRAF), and 2011.

As we have already pointed out, the data recorded indicate a model of occupation and economic development of the Amazon that benefits the economic power of large groups that have operated in the expansion of agro-industry and agriculture and have consumed natural resources unsustainably. The main strategy of these groups was initially to use irregular means to take over public lands and to perpetrate large-scale exploitation of labour, with a high incidence of slave labour, particularly on the southern and south-eastern borders of the Amazon, on a strip of land that includes the states of Rondônia and Mato Grosso to the west and the states of Pará and Maranhão to the east. This strip of land is called the Arc of Deforestation, and from the last decades of the twentieth century to the present day, it has seen the largest social tensions and violent conflicts over land, and invasions of indigenous lands and forest reserves, in the Amazon.

The reflections presented aim to provide new ways of looking at the Amazon, to show how the richness of Amazonian natural resources contrasts with economic precariousness and deficiencies in the areas of education (very low school performance) and health of the local populations (deprived of drinking water and sewerage). In addition, it is necessary to evaluate how the delicate biome can interact with Brazil's national development without causing further damage to the various ethnic groups which depend on the preservation of natural resources and economic sustainability strategies. A historical evaluation of the process of national integration and the creation of government agencies, in addition to an apparatus of laws and programmes with large budgets, but also dealing with the systematic project of forest reoccupation with a view to 'civilising' it, leads us to reflect on the 20% of forest cover lost in the last three decades. It also leads us to think about the thousands of people affected by actions taken by governments (municipal, State or federal) or the conflicts between large agricultural, mining and logging companies, with their predatory exploitation models and their policies regarding land ownership, which enclose paths and cut off access to former communal areas, along the banks of rivers that are a source of life for the original communities of the Amazon.

The conflicts taking place due to major interference by governmental institutions, bodies and agencies—in particular the repressive institutions of the State—are characterised by impunity and attacks on human rights, and have even led to the murder of people from various communities, such as in Eldorado de Carajás (1996), where 21 people were killed by military police in the state of Pará, or in Corumbiara (1995), where military police in the state of Rondônia killed ten people, including a nine-year-old child. On the Santa Lucia farm in Pará (2017), the military police also killed ten people, and we should highlight the Haximu massacre, where miners attacked a Yanomami settlement (1993), killing 16 people, including a one-year-old child, with machetes and guns. It is worth mentioning that this massacre is the only case where a crime has been judged as genocide in Brazil.

The modernisation project was begun in the mid-1960s by an illegitimate government, governed by laws and decree-laws passed in Congress under a state of exception, while the State was controlled by high-ranking military personnel and protected by a strong apparatus of repression. This structure implemented a set of public policies that produced a huge displacement of thousands of people, followed by the privatisation of public lands and the opening of federal roads which cut through the forest with no consideration for the existing territorialities. Examples include the case of the Panará, or Kreen-akarore, indigenous peoples who, due to the opening of BR-163 through their territory, had their population reduced from over 400 people to just 79 in less than ten years, or the Waimiri Atroari who, following the opening of BR-174 and the continuous conflicts caused by the invasion of their territory, saw their population fall from about 3,000 to only 350 in ten years. The construction of dams in the Amazon region is another side of the modernising project that has drastically interfered with the entire Amazon biome, with severe consequences for the local population. The construction of just one dam, Belo Monte on the Xingu River, directly affected hundreds of thousands of people, who lost their homes and land (Ribeiro 2018), and had incalculable environmental impacts.

Debates over possible development models, environmental conservation, socio-economic and political equality, sociocultural diversity, more sustainable bases of production and consumption and social and environmental justice are alive in the Amazon. It is important to understand that certain development models have now been directed, such as changes in traditional agriculture and sustainable agroforestry extraction proposals. Social exclusion is now seen as burdensome to municipalities—there are 771 municipalities within the nine states that constitute the Legal Amazon—because the low income of the populations causes disease, hunger and misery. The Amazon has been and continues to be violently affected by a Eurocentric, white, capitalist 'civilising' project. To look at the Human Development Index (HDI) of the 'soya cities'—note that the HDI in the state of Mato Grosso is higher than the Brazilian average—without considering the massive use of pesticides and the hundreds of thousands of hectares of forest turned into a green desert of private properties, is to overlook the real outcomes of this project which, along with so-called economic development, brought the disintegration of traditional livelihoods, turned indigenous people and rural squatters into the urban poor and attracted workers with the promise of work and money, only to reduce them to peasants. The wealth generated by agribusiness in Mato Grosso is concentrated in the hands of a few, as has occurred in other states where the same model was implemented.

The increasing violence and recent electoral trends have brought about deep and uncomfortable doubts regarding the direction of progress and democracy. In this sense, given their peculiarities and specific demands but also their sociocultural and socio-ecological wealth, the ongoing processes in the Amazon represent a very important chapter of contemporary history and geography. The populations of the region perceive and articulate such complexity in a very sharp and insightful manner, making use of a wonderful linguistic and narrative plurality. The international community is moving to put political and economic pressure on governments and business at various levels, trying to prevent the rapidly approaching tragedy. Ten thousand kilometres of forest were consumed by fire or predatory activity between August 2018 and August 2019 alone—an increase of 30% compared to the previous year. There were 18 murders of indigenous leaders, small farmers and environmentalists between January and August 2019, and none of these crimes have been officially investigated. Gold miners continue to advance into the forest, invading indigenous areas, using mercury to extract gold. Loggers, in turn, forcibly remove precious trees from environmental reserves and indigenous areas without any condemnation or action from the Brazilian government.

To conclude, we hope to have shown that the measures taken by the militarised State of the last decades of the twentieth century, when it organised public agencies and a whole legal apparatus to systematically 'civilise' the Amazon, produced much more than the so-called colonisation or agricultural projects that marked the 1970s and 1980s, especially in the southern part of the forest. What can be seen today is a direct result of such action, which, once implemented, has resulted in overlapping territories in constant conflicts of unequal force. The no-return march of 'progress', which advances and transforms without asking those affected if that is what they want for themselves and for the future of the new generations, reminds us of the words of philosopher Walter Benjamin in the *Theses on the Philosophy of History*: 'There is no document of civilisation which is not at the same time a document of barbarism' (Thesis 7). The discourse that presents development as 'good for the country' disregards traditional ways of life and transforms the original wealth of the Amazon into products for an economy that excludes as it grows.

Final Considerations

The analysis of the new spaces created by the force of capital, which counted on the structure of a State forged by the civil and military dictatorship (1964–1985), requires an open position by social scientists to understand the historical realities marked not by homogenisation, nor only by difference, but by new kinds of social life that presuppose interactions in which the other does not disappear. Spaces of displacement in diverse territorialities that extrapolate geographical definitions and dividing lines between political units affirm their importance as expressions of the space-time displacements of different groups in national society from occupied territories, notably indigenous peoples and traditional communities that live on extractivism or subsistence agriculture and which have diverse temporalities.

This great governmental project of reoccupation of the Amazon and protection of the geographical borders of Brazil, driven by the Cold War and carried out by the power of a dictatorship, is filled with paradoxes. (Martins, 1997). If the justifying discourse of this project brought nationalism and territorial defence to the foreground, the result was a growing openness to capital. If one of the pressing needs of the national agrarian elites was to contain rural conflicts throughout the country by displacing people from volatile areas to the Amazon with the promise of abundant land, its result was the urbanisation and expropriation of small farmers. Official discourse affirmed the realisation of the world's largest agrarian reform programme, and land was quickly concentrated, to the point where it is difficult to compare this process with others in the various regions of Brazil. There was much talk about modernisation and development, and indeed billions of dollars have been invested throughout the territory, but a quick glance at the data shows alarming levels of poverty and violence, very serious infrastructure problems and massive use of resources for the benefit of a few. If the policies of successive governments can be called public, their results are, at this time, highly privatised.

References

Arrighi, G. (1996). *O Longo Século XX*. São Paulo: Contraponto/EUNESP.

Baumam, Z. (2005). *Vidas Desperdiçadas*. Rio de Janeiro: Jorge Zahar.

Benjamin, W. (2009). *On the Concept of History*. Dallas: Create Space Independent Publishing Platform.

Bercovici, G. (2010). O Direito Constitucional Passa, o Direito Administrativo permanece: A Persistência da Estrutura Administrativa de 1967. In E. Teles & V. Safatle (Eds.), *O Que Resta da Ditadura: A Exceção Brasileira* (pp. 77–90). São Paulo: Boitempo.

Brandford, S., & Torres, M. (2017). *Soy Invasion Poses Imminent Threat to Amazon, Say Agricultural Experts*. https://news.mongabay.com/2017/02/soy-invasion-poses-imminent-threat-to-amazon-say-agricultural-experts. Accessed 9 Feb 2017.

Brasil. (1966). *Lei No 5,174, 27 Out 1966, Dispõe Sôbre a Concessão de Incentivos Fiscais em Favour da Região Amazônica e Dá Outras Providências*. http://www.planalto.gov.br/ccivil_03/LEIS/L5174.htm. Accessed 2 Dec 2019.

Brasil. (1967). *Decreto-Lei No 291, de 28 de fevereiro de 1967, Estabelece Incentivos para o Desenvolvimento da Amazônia Ocidental da Faixa de Fronteiras Abrangida pela Amazônia e Dá Outras Providências*. http://www.planalto.gov.br/ccivil_03/Decreto-Lei/1965-1988/Del0291.htm. Accessed 2 Dec 2019.

Brasil. (1970). *Metas e Bases para Ação de Governo. Ministério do Planejamento*. Brasília: Departamento de Imprensa Nacional.

Brasil. (1996). Portaria No 668, 1 Nov 1996. *Diário Oficial da República Federativa do Brasil* (p. 22, 599). Seção 1, Brasília.

Brasil, Presidência da República (1 abr. 1971). Decreto-lei N° 1.164. Declara indispensáveis à segurança e ao desenvolvimento nacionais terras devolutas situadas na faixa de cem quilômetros de largura em cada lado do eixo de rodovias na Amazônia Legal. Acesso em 04 jun 2012. Disponível em http://www.planalto.gov.br/ccivil_03/Decreto-Lei/Del1164.htm.

Castro, S., Barrozo, J. C., Covezzi, M., & Preti, O. (1994). *A Colonização Oficial em Mato Grosso: A Nata e a Borra da Sociedade*. Cuiabá: EdUFMT.

Couto e Silva, G. (1982). *Conjuntura Polícia Nacional: O Poder Executivo & Geopolítica do Brasil*. Rio de Janeiro: José Olympio.

Coy, M., & Klingler, M. (2014). Frentes Pioneiras em Transformação: O Eixo da BR-163 e os Desafios Socioambientais. *Territórios & Fronteiras, 7*(1), 1–26.

Figueiredo, L. (2015). *Lugar Nenhum: Militares e Civis na Ocultação dos Documentos da Ditadura*. São Paulo: Companhia das Letras.

Freitag, L., & Stanczyk Filho, M. (2010). *Região, Espaço e Poder*. São Paulo: Alameda.

Furtado, J. F. (2012). *Oráculos da Geografia Iluminista: Dom Luís da Cunha e Jean-Baptiste Bourguignon D'Anville na Construção da Cartografia do Brasil*. Belo Horizonte: EdUFMG.

Gomes, A. M. C. (2013). *Olhando para Dentro*. Rio de Janeiro: Objetiva.

Grosso, M. (1973). *Relatório da Secretaria de Planejamento e Coordenação Geral do Estado*. Cuiabá: Imprensa Oficial.

Guimarães Neto, R. B. (2002). *A Lenda do Ouro Verde: Política de Colonização no Brasil Contemporâneo*. Cuiabá: UNICEN.

Guimarães Neto, R. B. (2014). História, Trabalho e Memória Política: Trabalhadores Rurais, Conflito Social e Medo na Amazônia (1970–1980). *Revista Mundos do Trabalho, 6*(11), 129–146.

Haesbaert, R. (2004). *O Mito da Desterritorialização: Do "Fim dos Territórios" a Multi-territorialidade*. Rio de Janeiro: Bertrand Brasil.

Harvey, D. (2013). O novo imperialismo. Loyola: São Paulo.

Ianni, O. (1979). *Colonização e Contra-reforma Agrária*. Petrópolis: Vozes.

Ioris, A. A. R. (2016). Rent of Agribusiness in the Amazon: A Case Study from Mato Grosso. *Land Use Policy, 59*, 456–466.

Joanoni Neto, V. (2007). *Fronteiras da Crença: Ocupação do Norte de Mato Grosso após 1970*. Cuiabá: Tanta Tinta/EdUFMT.

Joanoni Neto, V. (2014). Em Busca de Outros Olhares: O Período Ditatorial Brasileiro e a Amazônia Legal. In X. C. Barbosa, M. H. M. Alcântara, & V. A. S. Aguiar (Eds.), *História, Memória e Direitos Humanos. 50 Anos da Ditadura Militar no Brasil* (pp. 102–115). Pontocom: Salvador.

Luceiro, E. R., & Pereira, A. J. (2000). Reforma Agrária no Brasil e a Aplicabilidade da Proposta do Movimento dos Trabalhadores Rurais sem Terra (MST): O 'Grupo Coletivo Força Jovem' como Forma de Assentamento–1997/1998. *Revista Economia e Desenvolvimento, 11*, 70–99.

Martins, J. S. (1994). *O Poder do Atraso*. São Paulo: Hucitec.

Martins, J. S. (1997). *Fronteira: A Degradação do Outro nos Confins do Humano*. São Paulo: Hucitec.

Martins Filho, J. R. (2004). A ditadura revisitada: unidade ou desunião? In: Daniel Aarão Reis; Marcelo Ridenti; Rodrigo Patto Sá Motta. (Org.). O golpe e a ditadura militar 40 anos depois (1964–2004). Bauru: Editora da Universidade do Sagrado Coração.

Miranda, L. (2016). *Atlas Geográfico de Mato Grosso*. Cuiabá: Entrelinhas.

Moreno, G. (2007). *Terra e Poder em Mato Grosso: Política e Mecanismos de Burla 1892–1992*. Cuiabá: EdUFMT/Entrelinhas.

Nepstad, D C; Stickler, Claudia M; Almeida, Oriana T .(2006). Globalization of the Amazon Soy and Beef Industries: Opportunities for Conservation. Conservation Biology, v. 20, p. 1595–1603.

Raminelli, R. J. (1996). *Imagens da Colonização: A Representação do Índio de Caminha A Vieira*. Rio de Janeiro: Jorge Zahar.

Raminelli, R. J. (2013). *A Era das Conquistas*. Rio de Janeiro: Editora FGV.

Resende, E. (1971). Estradas da Amazônia. In A. A. A. Lima et al. (Eds.), *Problemática da Amazônia*. Biblioteca do Exército: Rio de Janeiro.

Ribeiro, M. (2018). *Costs: Brasil/Inglaterra* (Documentary, 20 min.). https://vimeo.com/267872971. Accessed 18 Nov 2019.

Ricardo, C. (1970). *Marcha para Oeste* (4th ed.). São Paulo: José Olympio.

Souza, M. L. (2000). O Território: Sobre Espaço e Poder, Autonomia e Desenvolvimento. In I. E. Castro, P. C. C. Gomes, & R. L. Corrêa (Eds.), *Geografia: Conceitos e Temas* (pp. 77–116). Rio de Janeiro: Bertrand Brasil.

14

Unintended Consequences of 'Development' in the Amazon: Commercial Aquaculture and Malaria in Mâncio Lima, Brazil

Igor Cavallini Johansen and Marcelo Urbano Ferreira

Introduction

The Amazon Basin extends over nine countries in South America—Brazil, Bolivia, Peru, Ecuador, Colombia, Venezuela, Suriname, Guyana and French Guiana. Traditionally viewed as a pristine territory interspersed with remote, poorly interconnected, riverine villages, the Amazon Basin of Brazil is now home to a predominantly urban population, as a result of 'development'[1] policies implemented in the area since the early 1970s (Castro 2009; Trindade Júnior 2015; Vicentini 2004). Here we focus on how these policies have affected the environment and

[1] In this chapter we use the word 'development' in quotes to emphasise that this concept is a product of a specific historical context, most often understood as narrowly associated with economic growth.

I. C. Johansen (✉) • M. U. Ferreira
Department of Parasitology, Institute of Biomedical Sciences, University of São Paulo (USP), São Paulo, Brazil
e-mail: igorcavallini@usp.br; muferrei@usp.br

A. A. R. Ioris (ed.), *Environment and Development*,
https://doi.org/10.1007/978-3-030-55416-3_14

population health by creating favourable conditions for malaria transmission in densely populated urban spaces.

First, we summarise the main 'development' policies implemented in the Brazilian Amazon and their implications for the region's urbanisation. We then briefly present some impacts of these policies in terms of environmental and population dynamics, highlighting the case of malaria. Subsequently, we discuss how aquaculture, an economic activity promoted to bring 'development' to the region, unexpectedly resulted in increased malaria transmission and led to major urban malaria outbreaks in our study site in the Brazilian state of Acre. Next, we present other population and environment characteristics in the study site that favour sustained malaria transmission and suggest possible interventions to reduce their adverse effects. Finally, we discuss how to improve design interventions, considering and respecting local environmental and social characteristics, in the light of the lessons learned from the study case.

The Impact of 'Development' Policies in the Brazilian Amazon

Colonisation of the Brazilian Amazon began in the early sixteenth century, and urbanisation accelerated in the region during the boom in natural latex extraction for rubber production, which lasted from the mid-nineteenth century until the end of World War I, fostering the foundation of numerous new cities and towns and accelerating the growth of the extant urban hubs, especially Belém and Manaus (Vicentini 2004). Subsequently, the establishment of highly productive rubber tree plantations in Southeast Asia caused the collapse of latex extraction and rubber production in Brazil and severely affected the growing economy and urban dynamics in the Brazilian Amazon, which came to be characterised by economic and urban stagnation. This stagnation period lasted until 1960, with a short resurgence in rubber production in the Amazon during World War II, when Malaysia was occupied by Japanese troops and local rubber production became unavailable to the Allies (Neeleman and Neeleman 2015; Vicentini 2004). Since 1960, the new momentum has been characterised by an intense process of economic and social transformation, with a much faster urbanisation process (Castro 2009; Vicentini 2004).

Ribeiro (1993) argues that urbanisation in the Amazon between the 1960s and 1990s was closely linked to the capitalist appropriation of the frontier, which was intensified by official actions such as the Agromineral and Agricultural Poles Programme (*Programa de Polos Agropecuários e Agrominerais—Polamazônia*), the Great Carajás Programme (*Programa Grande Carajás* [PGC]) and, particularly, the implementation of the large projects encouraged by the Amazon Development Superintendence (*Superintendência para o Desenvolvimento da Amazônia* [SUDAM]). These strategies, together with other significant factors such as the creation of the Manaus Free Trade Zone in 1967[2] and the implementation of a large infrastructure of major axes based on the road system, ensured the viability of the proposed programmes.[3]

While early colonisation in the area relied on the rivers as the main routes for occupying the Amazon territory and creating the first cities, the occupation process in the late twentieth century was based on the opening of new roads (Hogan 2000). Consequently, recent urbanisation in the Amazon has been based on the new highways that cross the forest. The most significant are the Belém-Brasília Highway (opened in 1958), which links the capital of the country to the North, the Trans-Amazonian Highway (1972), from the Paraíba state in the Northeast to the Amazonas in the North; and the Cuiabá-Porto Velho Highway (1960), from the Mato Grosso state in the Centre-West to the Acre state in the North. Expansion into this new area was also supported by a discourse of

[2] The Manaus Free Trade Zone (MFTZ) is still in force. It establishes an area in Manaus city for free trade in imports and exports, with special tax incentives. Its purpose was to create an industrial, commercial and agricultural centre in the Amazon to boost development in the region (Brasil 1967). Aloise and Macke (2017) analyse the MFTZ as a study case of an eco-innovation policy in Brazil.

[3] State intervention in Amazon has been carried out using various instruments, as mentioned by Trindade Júnior (2015: 101), "The strong presence of the State composed the strategy of the regionalisation model based in the polarised regions, either by creating infrastructure; or by setting up planning and funding bodies; or additionally by tax / credit incentives / subsidies to big rural property and capital flows". In addition, it is important to note that "[t]he governmental role is varied with regard to both regional and social-class interests" (Godfrey and Browder 1996: 444).

obligation to offer land to those in need. In addition, this strategy responded to demands from economic actors to expand agricultural production from already consolidated areas (where land was expensive) to new axes of expansion. This monumental 'development' plan for the region would also expand markets, leading to urban and industrial growth (Monte-Mór 2004; Wood and Carvalho 1994). Table 14.1 shows that in 1970, the population of the Legal Amazon was 63% rural and only 37% urban. By 1991, this ratio was reversed, with a majority of the population living in cities (56%). The most recent data available show that by 2010, 72.5% lived in areas defined as urban and 27.5% in rural areas, meaning that over 17 million people in the Brazilian Amazon lived in cities.

Table 14.1 shows a decreasing trend in urban population growth in Brazil. However, although the proportion of people living in cities is smaller in the Amazon municipalities (72.5%) compared to the rest of the country (86.1%), proportional increases in the urban population are still occurring at a faster pace in the Amazon (2.4% per year versus 1.5% per year in other municipalities). See also Fig. 14.1. Monte-Mór (2004) developed the concept of 'extended urbanisation' for understanding the specific context of urban phenomena in the Amazon:

> Through extended urbanisation multiple urban centralities, from cities and towns to commercial and service centres, industrial plants, large ranches, local communities, rubber estates, and even(tually) indigenous areas combine to connect and (re)articulate local, regional, national and global forces and thus produce a variety of locales and populations more or less linked to urban-industrial capitalism. (p. 13)

According to this conceptualisation, even apparently less urbanised areas are eventually affected by a centrifugal movement that inevitably reaches them, disseminating aspects associated with the production structure, consumption and even lifestyle of urban areas. In the Brazilian Amazon, this expansion of the urban started with the aforementioned major 'development' projects, particularly with the advance of the frontier for agricultural production, livestock and mineral exploration (Monte-Mór 2004).

Table 14.1 Population (in thousands) and by urban/rural residence (%) in Legal Amazon municipalities, municipalities outside the Legal Amazon and total municipalities in Brazil: 1970–2010

		1970		1980		1991		2000		2010	
		Pop.	%	Pop.	%	Pop.	%	Pop.	%	Pop.	%
Legal	Urban	2,623	36.9	5,010	45.5	9,005	55.8	13,875	68.9	17,660	72.5
Amazon	Rural	4,490	63.1	5,992	54.5	7,126	44.2	6,269	31.1	6,710	27.5
(*n* = 771)	Total	7,114	100	11,002	100	16,131	100	20,145	100	24,371	100
Other municip.	Urban	49,473	57.5	75,427	69.8	101,985	78.0	124,049	82.9	143,274	86.1
(*n* = 4794)	Rural	36,546	42.5	32,581	30.2	28,708	22.0	25,677	17.1	23,110	13.9
	Total	86,020	100	108,008	100	130,694	100	149,727	100	166,384	100
Total Brazil	Urban	52,097	55.9	80,437	67.6	110,990	75.6	137,925	81.2	160,934	84.4
(*n* = 5565)	Rural	41,037	44.1	38,573	32.4	35,834	24.4	31,947	18.8	29,821	15.6
	Total	93,134	100	119,011	100	146,825	100	169,872	100	190,755	100

Source: IBGE (1970, 1980, 1991, 2000, 2010)

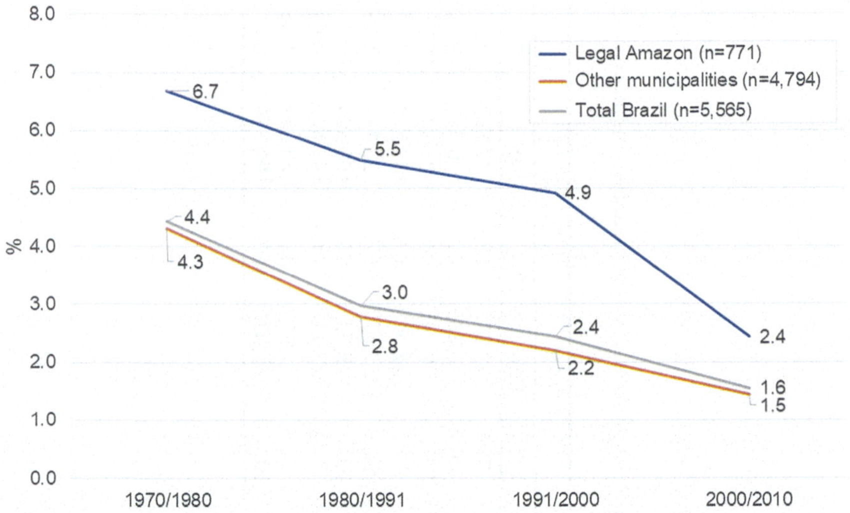

Fig. 14.1 Urban population annual geometric growth rate (%), Legal Amazon, other municipalities and total Brazil: 1970–2010. (Source: IBGE 1970, 1980, 1991, 2000, 2010)

A factor that favours this extended urbanisation process is the existence of multi-sited households, meaning those which have more than one residence, in both rural and urban areas (Padoch et al. 2008). Whether the household stays in one place or another depends on the season (high or low precipitation), occasion (city holidays attract people to the urban centres) and period of the month, as residents in distant areas also go periodically to banks in the urban centres to withdraw social benefits such as *Bolsa Família*, *Benefício de Prestação Continuada* (BPC), pensions and *Seguro-Defeso*, which is paid to fishermen during the fish reproduction season, when fishing is prohibited. This mobile population also uses urban centres to buy agricultural and industrialised products to take back to their rural localities.

Although extended urbanisation is a concept that can be applied to explain the general pattern of urbanisation in the Brazilian Amazon, it is important to highlight that this process is highly diverse at the local scale. This makes it difficult to find an 'urban pattern' in the region, which is shaped by a multiplicity of social, environmental, cultural and historical conditions that are present in the cities in the area. To analyse this process more effectively, Godfrey and Browder (1996) developed the concept of disarticulated urbanisation:

> Our perspective begins with a recognition that the settlement systems of Amazonia are disarticulated from any single master principle of spatial organisation. A wide diversity of settlement types confounds any single typology. The spatial organisation of regional settlement is largely asymmetrical and provides scant evidence of the orderly, nested hierarchies predicted by central-place theory. This diversity is compounded over time by an overlapping of historic settlement systems: The distinctive logic of contemporary settlement location, enabled by road-based transportation through the uplands, is superimposed on the relict urban systems of the rubber boom, consisting basically of a string of riverine settlements. (p. 443)

This concept helps us to understand the specificity of the urban in the Amazon, where evidence of disarticulated urbanisation shows that the Brazilian Amazon should be viewed not as a homogeneous entity but as a variegated region characterised by internally divergent processes resulting from different historical moments and diverse government interests and incentives. Here we find the link between urbanisation and 'development' projects in the region. During the military regime (1964–1985), 'development' projects in the Amazon were mainly led by the federal government, and dialogue with the public and civil society organisations was limited. These projects often resulted in a series of negative consequences for local society and the environment. Indeed, a large body of studies has already detailed the environmental impacts, particularly deforestation, caused by the opening of major highways in the Amazon (Alves 2002; Brondizio and Moran 2012; Fearnside 1987; Godar et al. 2012).

Other studies have highlighted the social impacts of these major 'development' projects. In terms of public health, one of the main threats was, and still is, malaria. Castro et al. (2006) investigated a prototypic example of frontier malaria evolution in the Machadinho settlement project in the state of Rondônia, which comprised three stages: epidemic, transition and endemic. The first stage is characterised by high malaria rates associated with massive environmental changes, such as deforestation, which favour human clustering close to aquatic habitats of the mosquito vectors (mainly *Anopheles darlingi* [*A. darlingi*]) and limit the efficacy of traditional control measures. The transition period is marked by declines in malaria rates that are partially explained by improved health infrastructure and reduced mobility of settlers. Malaria transmission becomes less intense and more stable in the final stages, with the advancement of forest clearing and the establishment of agriculture, livestock or urban areas. Although malaria risk in rural Amazonia in the early phases of frontier settlements in the mid-1980s and the 1990s has been thoroughly described, there is substantially less information regarding more consolidated settlements, after extensive social and ecological transformations had taken place. In the mid-2000s, longitudinal studies in the Granada farming settlement in Acre State, originally opened in 1982, showed how the continuous influx of new settlers to the forest fringes of the consolidated settlements created a heterogeneous pattern of malaria transmission with both epidemic and endemic features (Da Silva-Nunes et al. 2008).

These 'development' projects implemented in the Brazilian Amazon, with their impacts on the environment and population health, were based on a principle of economic growth as the ultimate goal. In the 1980s, as this 'development' model was extensively criticised, the concept of sustainability emerged (WCED 1987) and began to be used to qualify development. Even after its legitimisation following the 1992 United Nations Conference on the Environment and Development in Rio de Janeiro, the notion of sustainability has come under suspicion, both from supporters of ultra-liberalism and from its opponents, who feared that competition in terms of environmental issues could have an impact on their social priorities. In summary, the idea has been rejected by various political organisations on both the Right and the Left. More recently, however, the concept of sustainability "[...] has come to serve Greeks and Trojans when they want to express vague ambitions for continuity, durability or perpetuity. All referring to the future" (Veiga 2010: 12). Despite the popularisation of the term, there have been no significant changes in understanding the determinants of progress, prosperity or development, all of which are still usually seen as direct results of economic performance as measured by the gross domestic product (Veiga 2010).

In this chapter we deal with the notion of environmental and social determinants of health (CSDH 2008; Prüss-Ustün et al. 2016). We reaffirm that the two are closely interrelated. While social determinants are consequences of the circumstances in which people live, work and grow, they are largely constrained by the distribution of environmental resources across society, and also affected by improper management of these resources. In this way, 'development' projects in the Amazon changed the environment and, by inserting people into this modified environment, subjected them to health consequences that still persist. Nowadays, 99% of Brazilian malaria cases are concentrated in the Amazon region (Ferreira and Castro 2016). This disease is therefore the clearest example of the interconnection between 'development' policies, environmental change and unintended health consequences for the population. In this specific section, we have discussed how government policies promoted the expansion of the economic frontier in the Brazilian Amazon. The consequence was the increasing urbanisation these 'development' projects produced, which, on the one hand, linked rural and urban dynamics locally (extended urbanisation) and, on the other, produced an urban tissue highly disarticulated between one city and another (disarticulated urbanisation). These projects led to environmental and social impacts in the region. Next, we will present a study case that is an example and a consequence of these processes.

Characterisation of the Study Area

Mâncio Lima municipality covers an area of 5453 km^2 in the Northwest of the Acre state in the Brazilian Amazon (Fig. 14.2). Its urban area is located next to the Japiim river (07°36'28.6" S, 72°54'23.0" W), and this is where more than half of its 18,977 inhabitants reside (IBGE 2019). Like many cities in the Amazon, the occupation of Mâncio Lima began in the late nineteenth century during the period of rubber exploration. However, the community only officially became a city, politically independent from the neighbouring municipality of Cruzeiro do Sul, in 1976 (Prefeitura de Mâncio Lima 2019). Mâncio Lima has a tropical climate, with an average annual temperature of 26 °C and average annual

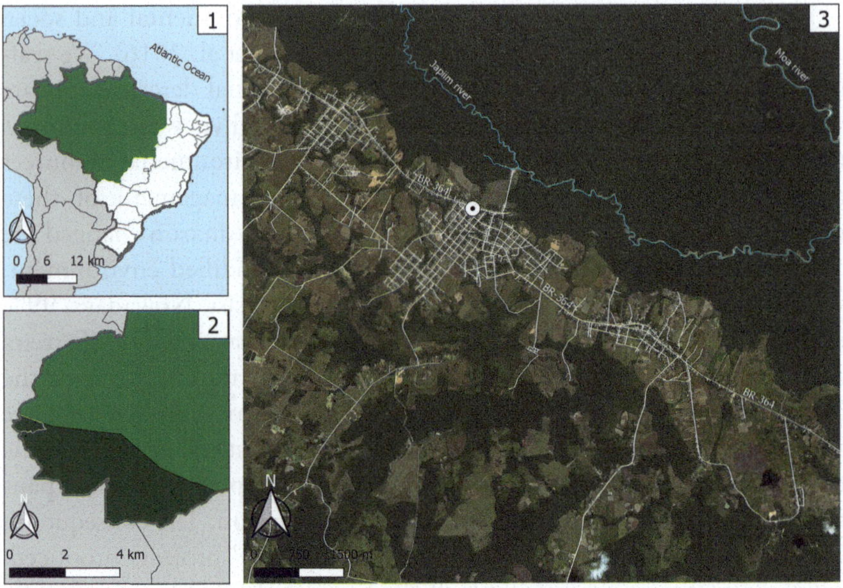

Fig. 14.2 Location of study area: Mâncio Lima, state of Acre, Brazil. 1: Brazilian Amazon; 2: Acre state and Mâncio Lima municipality highlighted; 3: Mâncio Lima urban area. White circle in Fig. 2.3 shows the location of city hall, in the centre of the city. Figure created with QGIS software version 3.8, an open source Geographic Information System (GIS) licensed under the GNU General Public License (https://bit.ly/2BSPB2F). Publicly available shape files provided from the Brazilian Institute of Geography and Statistics (IBGE) website (https://bit.ly/34gMq0S). Google satellite imagery reproduced observing the conditions for free use (http://bit.ly/2PuYuHt)

precipitation of 2,167 mm, with rains concentrated between the months of November and April (INMET 2019). In terms of environmental characteristics, the city is distributed along an expanded axis from the South-east to the North-west, following the path of the BR-364 highway. The northern portion of the municipality is mainly covered by forest, with some fragments in the south, as can be seen in Fig. 14.2.

In terms of economic activities, Mâncio Lima did not follow other Brazilian municipalities, whose urbanisation was closely associated with an intensive process of industrialisation, as was the case in the South-east of the country, in particular during the second half of the twentieth century (Cano 2011). Mâncio Lima, far from the 'developed' urban centres, remained dependent on agriculture and livestock production, with non-intensive use of technologies and low added-value production. In the city, formal jobs are scarce, mainly restricted to activities related to public administration. Part of the urban population engages in occupations within the city based on seasonal private services, such as building and repairing houses, pruning trees, cleaning gardens, caring for children and elderly people, etc. Another group of the urban population is involved in rural activities, such as commercial agriculture or production for subsistence, working for themselves or for others, which greatly encourages population mobility between urban and rural areas.

The nearby forest is seasonally flooded with fresh water. In addition, the city is crisscrossed by small streams. These settings provide ideal conditions for the construction of fish ponds, either by damming rivers or by digging ponds beside them (Fig. 14.3). This urban environment pervaded by rural characteristics (since fish farming is primarily a rural activity) greatly favours the reproduction of *A. darlingi*, the main malaria vector in the Brazilian Amazon (Dos Reis et al. 2015a, b; Hiwat and Bretas 2011).

Constant population mobility facilitates the flow of malaria parasites from rural areas, where the disease was previously more common, to the urban area, where the vector density grows as a result of the availability of breeding sites, especially fish ponds (Dos Reis et al. 2015a, b). As a result, the annual parasite incidence in Mâncio Lima was estimated at 422 cases per 1000 inhabitants in 2018, the highest level for a municipality in Brazil (Brasil 2019). Besides the economic constraints faced by Mâncio Lima and considering the way its environmental characteristics favour aquaculture, a specific regional development policy was designed for the region, promoting fish ponds and aquaculture. The specificities of this policy are presented in the next section.

Commercial Aquaculture in Mâncio Lima

In the beginning of the 2000s, the Acre state government created the Program for the Development of Fish Farming in the Juruá Valley (the Juruá is the river that gives its name to the region where Mâncio Lima is

Fig. 14.3 Different Water Bodies within the Mâncio Lima Urban Area. A: Typical natural wetland with *moriche* palm trees; B: Commercial aquaculture—a fish pond empty after sale of fish; C: A group of fish ponds filled with water from a river (top left) that has been dammed to artificially create the ponds (top-right and bottom-left). (Pictures taken during fieldwork [September 2019])

located). The programme covered the following municipalities: Cruzeiro do Sul, Feijó, Jordão, Marechal Thaumaturgo, Porto Walter, Rodrigues Alves, Tarauacá and Mâncio Lima (Alves 2017). Fish farming began in Mâncio Lima in 2001, when 21 residents formed a group to farm fish in dams. In April 2003, with the support of the Acre Department of Agriculture and Livestock, the first course in Fish Farming Management took place in the Juruá Valley. In the same year, an association was formed with the support of Acre's Micro and Small Business Support Service (SEBRAE-AC). Technicians from the State Department of Agriculture and Livestock began to make regular visits to the municipality to support the development of this economic activity (Alves 2017; Duarte 2007).

However, as already noted in previous studies, fish production eventually faced a set of obstacles in the region, as production was largely restricted to consumption in the community and its surroundings, mainly due to the difficulty of transporting fish for commercialisation in more distant urban centres. In addition, there were difficulties in obtaining financial resources to support the activity, fish feed acquisition and technical assistance (Sarah et al. 2003). In Mâncio Lima, several ponds were opened without planning, both with and without the support of the local government (Alves 2017). Even today, new ponds are being built. Although the activity remains, given that government support for it is volatile, many ponds ended up being abandoned or lacking the necessary care, thus becoming potential breeding grounds for malaria vectors.[4]

At the time most fishponds were opened, in the mid-2000s, very little was known about their potential as larval habitats for malaria vectors. All Amazonian anophelines that are tolerant of turbid water, such as *A. albitarsis s.l.*, *Anopheles benarrochi* (*A. benarrochi*), *Anopheles oswaldoi* (*A. oswaldoi*), *Anopheles nuneztovari* (*A. nuneztovari*) and *Anopheles rangeli* (*A. rangeli*), could be found in fishponds of South America. Surprisingly, however, *A. darlingi,* a species that usually prefers clear water, had also occasionally been found in fishponds in Brazil and Peru, although not in Colombia (Maheu-Giroux et al. 2010). Subsequent studies confirmed

[4]The most recent Acre State Water Resources Plan, which also presents information about fish ponds as an economic activity, does not mention either malaria or *Anopheles* management (SEMA 2012).

the association between commercial aquaculture, mosquito vector breeding and urban or peri-urban malaria transmission in Brazil and Peru (Ahumada et al. 2013; Arruda et al. 2018; Dos Reis et al. 2015a, b; Maheu-Giroux et al. 2010).

Since the strong association between ponds and vectors was discovered, the opening of new ponds in Mâncio Lima has been subject to greater supervision by city staff. The ponds are frequently inspected by Health Department personnel, and owners are informed about proper management to avoid the reproduction of malaria vectors. However, the profusion of ponds, the limited number of health professionals and the size of the urban area make the work of monitoring and eliminating mosquito-breeding sites in Mâncio Lima fish ponds a challenging task. It is important to emphasise that not all fishponds are sources of *A. darlingi*. This vector relies on fish-pond borders that are shaded and covered by vegetation. Cleaning these borders reduces the likelihood of these ponds becoming potential breeding sites for anophelines (Dos Reis et al. 2015a). Furthermore, fish ponds are not the only element keeping the disease endemic in the municipality. There are a number of other factors, both environmental and social, that favour this malaria-prone scenario.

Among the social elements favouring malaria in Mâncio Lima, population mobility is key. As mentioned earlier, a significant section of the population has residences in both urban and rural areas. These are known as multi-sited households (Padoch et al. 2008). This facilitates the spread of malaria parasites in both directions. Previous studies on this topic have demonstrated that local transmission and imported cases from rural or forest areas are bases for the maintenance of malaria in Mâncio Lima city. Regarding importation, it is known that male residents engaged in subsistence farming in nearby settlements serve as a source of new parasite strains that are continuously being introduced in the town (Corder et al. 2019; Lana et al. 2017). Consequently, having mosquito-breeding sites in the urban area increases the likelihood that a new parasite, once it has reached the city, will rapidly spread among people who have not travelled. In summary, fish ponds, based on a natural environment with plenty of flooded areas and potentiated by a policy of regional 'development', together with population dynamics inserted in a social and economic context that binds together rural and urban dynamics—of which the most significant element is population mobility, favouring the constant exchange of malaria parasites—are the most fundamental nodes of the chain that fosters malaria transmission within the Mâncio Lima urban area.

Possible Interventions

Many strategies are already in use in Mâncio Lima to control malaria occurrence, such as: environmental surveillance of vector-breeding sites; distribution of bed nets (since *A. darlingi* bites most often occur during the night); and indoor residual spraying, to reduce vector density inside households. Following the 1988 Brazilian Constitution, health system decentralisation increasingly transferred the responsibility for controlling malaria to municipalities (Ferreira and Castro 2016). In this context, federal and state administrations were compelled to provide the necessary resources, through a variety of health programmes, to strengthen local malaria control strategies (Griffing et al. 2015). The analysis presented in this chapter, as well as fieldwork experiences in the area, point to some specific interventions that could be applied through local policies to contribute to malaria control. Considering that a substantial part of Mâncio Lima's malaria is reported in the urban area and by urban residents, there is a unique opportunity to focus efforts and scarce resources in this area, compared to distant and more-difficult-to-reach rural and riverine communities.

Consequently, we suggest: (a) larval source management, which can be achieved by permanent or recurrent habitat modification, biological control with natural predators and chemical or biological larviciding in fish ponds and natural flooded areas; (b) identification of the most mobile populations, who can be targeted with more intensive control interventions (distribution of insecticide-treated bed nets, periodic malaria screening and treatment); and (c) house screening and other strategies to make houses more 'mosquito-proof'. Strategies based on (natural and built) environmental management should be combined with an efficient diagnosis and treatment system, preventing new occurrences from multiplying and, therefore, immediately interrupting the transmission cycle.

Urbanisation and Malaria Links in the Context of Development and Post-development Studies

The age of 'development' began in January 1949, when the United States president Harry S. Truman, in his inauguration speech, declared the Southern Hemisphere to be 'underdeveloped'. With this discourse, Truman put the United States and other industrialised nations at the top of the evolutionary scale. The idea of 'development' provided a comforting vision of a global order where the United States would naturally rank first, the era of American hegemony. In the post–World War II context, the US would be the country to lead others along the bright path to 'development' (Esteva 2010). Since the 1980s, 'development' has come to be more intensely questioned, as have many of its impacts. Under these circumstances, 'development' came to be seen by some as a 'new religion of the West' (Rist 1990), an imposition of science as power (Nandy 1988), a theory that does not work (Kothari 1988), which brings environmental destruction (Martine 1993), which means cultural Westernisation and homogenisation (Constantino 1985), and which is an instrument of hegemony benefitting global capitalist and economic elites (Moore 1995). This critical revision is what Ziai (2007) defends, by stating that it is necessary to denaturalise the 'development' discourse and encourage the discussion of alternatives. Sachs (2010: xv) agrees:

> Today, the lighthouse shows cracks and is starting to crumble. The idea of development stands like a ruin in the intellectual landscape. Delusion and disappointment, failures and crimes, have been the steady companions of development and they tell a common story: it did not work. Moreover, the historical conditions which catapulted the idea into prominence have vanished: development has become outdated. But, above all, the hopes and desires which made the idea fly are now exhausted: development has grown obsolete.

However, there are powerful divergences from this post-development wave. Pieterse (2000) points out that the post-development movement has no positive programme, that it is based on critiques of development but there is no constructive proposal to replace the latter. Additionally, from the author's perspective, post-development assumes a single and narrow meaning of development, rather than penetrating its polysemic realities. Finally, the critique states that during the period when development was the concept driving most countries, especially during the second half of the twentieth century, many aspects of society improved. The reduction in child mortality rates worldwide is particularly emphasised.

Despite conceptual and political divergences, it must be recognised that one perspective should not erase another. 'Development' policies in the Amazon theoretically offered more access to land for people living in areas where it was scarce or not suited for agricultural production. It is also widely accepted today that technology is present in most parts of the 'urban forest', connecting people and offering access to information and, to some extent, public services. However, it is nevertheless a fact that the environmental impacts of this 'development' process have been immense.[5] Furthermore, the results in terms of the built environment still demand much improvement, as the Amazon cities are among the most deficient in Brazil in terms of access to sanitation, including potable water, despite, counter-intuitively, being located in the richest freshwater reservoir on the planet. The deficit in environmental services in the Amazonian cities makes vultures a common feature of the landscape, since the weather conditions (hot and humid) mean that improperly managed garbage quickly rots. Finally, during the frontier expansion in the 1970s, the disease burden directly affected populations that migrated to the area. Malaria was one of the most important diseases affecting the lives and deaths of these populations.

Encouraged by 'development' policies in the region, malaria is still an unsolved problem, despite many efforts to control it. After a period of decrease, there was an upsurge in the disease in Brazil in 2017 (WHO 2018). Considering the specificities of population dynamics, environmental characteristics and urban settings presented in this chapter, with regard to policies for the Amazon, we defend the need to move away from

[5] The environmental effects of this occupation continue to be seen today. Forest fires increased substantially in 2019, as has been widely reported by international media (NY Times 2019). Investigations linked the fires to large-scale agricultural producers and their takeover of public land that had previously been covered by forest (El País 2019; Guardian 2019).

simplistic strategies that are weakened by a tendency to overly generalise complex realities. With regard to organising action in the region to promote social and environmental welfare in the light of development/post-development studies, it is important to reconsider the thoughts of Becker (2001). Analysing the conjuncture of the Amazonian frontier in the twenty-first century, she points out that approaches should feature:

> [...] consideration of multiple interests and actions implemented at several scales, making them compatible to a logic that preserves the planet's health, respects national and regional interests, and promotes the development of local populations. [...] [This can be done] by aggregating value to forest products and by incorporating the effective demands and projects of involved populations, eliminating innocuous and paternalist actions. (p. 322)

Consequently, a successful long-term policy for improving social and environmental conditions in the region needs to take several elements into account. First, social participation should be the basis for any new intervention. Second, a well-planed strategy can only succeed with different administrative departments working together (health, education, infrastructure, social protection, etc.). It would also be desirable, although not mandatory, for distinct levels of the administration to work in partnership—the municipal, state and federal governments. Third, the environmental specificities of the region must be examined in depth and carefully considered. Fourth, any intervention should consider lessons learned from the unintended consequences of previous 'development' interventions in the Amazon. Fifth, the health status of the population must be prioritised, with a particular focus on reducing the burden of the most challenging diseases present in the locality.

Concluding Remarks

Malaria remains an unsolved public health threat, especially in tropical countries in Africa, Asia and the Americas. In Brazil, the specific process of economic frontier expansion, based on an ideal of 'development' and conducted mainly by federal and state policies, caused malaria to increase and become an endemic disease in the Amazon, where the overwhelming majority of cases in Brazil occur today. In terms of the social impacts of this disease, symptoms usually result in at least two to three days of work absenteeism, reducing economic production. Considering that outside the biggest cities the Amazon population is concentrated in areas where formal jobs are scarce, the consequence of malaria symptoms for seasonally employed people is even more devastating. Since most workers depend on daily payment for occasional labour, those who are not available to work do not receive any compensation for sick days.

Furthermore, a considerable proportion of the population relies on fishing and agricultural production outside the city, generating constant population flows between urban and rural areas. Malaria can impede people from performing their regular rural activities, putting their subsistence and that of their families at risk. Additionally, although only in a small fraction of cases, severe malaria can cause death. In terms of the political repercussion of the problem, it is worth highlighting the already complex structure to combat malaria in the Amazon, based on joint actions implemented by federal, state and local administrations. The persistence of this disease monopolises significant professional, economic and political resources. Even when the situation is brought under control, consistent investment will have to continue in order to guarantee that success is maintained. Regarding political dynamics, we also should emphasise that failures in malaria control are occasionally due to discontinuities in malaria programmes when a new local administration assumes office.

This chapter adds to the current literature by exemplifying, with a case study in the twenty-first century, how 'development' policies in the Brazilian Amazon, from the 1970s to the present, are interlinked with urbanisation in the area, presenting environmental and social implications. For this purpose, we took malaria occurrence (and persistence) as an example. The case of fish ponds in Mâncio Lima urban area is significant, as it shows the interlinked relationship between population and environmental dynamics, mediated by public policies.

References

Ahumada, M. L., Pareja, P. X., Buitrago, L. S., et al. (2013). Comportamiento de Picadura de *Anopheles darlingi* Root 1926 (Diptera: Culicidae) y su Asociación con la Transmisión de Malaria en Villavicencio (Colombia). *Biomédica, 33*(2), 241–250. https://doi.org/10.7705/biomedica.v33i2.1492.

Aloise, P. G., & Macke, J. (2017). Eco-innovations in Developing Countries: The case of Manaus Free Trade Zone (Brazil). *Journal of Cleaner Production, 168*, 30–38. https://doi.org/10.1016/j.jclepro.2017.08.212.

Alves, D. S. (2002). Space-time Dynamics of Deforestation in Brazilian Amazonia. *International Journal of Remote Sensing, 23*(14), 2903–2908. https://doi.org/10.1080/01431160110096791.

Alves, M. R. (2017). *Análise Socioeconômica do Território da Piscicultura e da Malária em Mâncio Lima, Acre.* PhD Thesis, Escola Nacional de Saúde Pública Sergio Arouca, Fundação Oswaldo Cruz.

Arruda, R. A., Araujo, F. M., Schlosser, A. R., et al. (2018). Determinantes Ambientais e Não-ambientais da Transmissão do Plasmódio na Paisagem Urbana Amazônica e suas Consequências Clínicas: Estudo de Base Populacional em Mâncio Lima, Acre. *Comunicação em Ciências da Saúde, 28*(1), 12–22.

Becker, B. K. (2001). Amazonian Frontiers at the Beginning of the 21st Century. In D. J. Hogan & M. T. Tolmasquim (Eds.), *Human Dimensions of Global Environmental Change* (pp. 299–322). Rio de Janeiro: Academia Brasileira de Ciências.

Brasil. (1967). Decreto-lei No. 288, de 28 de Fevereiro de 1967, Altera as Disposições da Lei Número 3.173 de 6 de junho de 1957 e Regula a Zona Franca de Manaus. https://bit.ly/2PtTZgt. Accessed 15 Oct 2019.

Brasil. (2019). *Casos Autóctones de Malária por Município de Infecção 2003 a 2019.* Ministério da Saúde. https://tabsoft.co/2WvWIaO. Accessed 15 Oct 2019.

Brondizio, E. S., & Moran, E. F. (2012). Level-dependent Deforestation Trajectories in the Brazilian Amazon from 1970 to 2001. *Population and Environment, 34*(1), 69–85. https://doi.org/10.1007/s11111-011-0159-8.

Cano, W. (2011). Urbanização: Crise e Revisão de Planejamento. In *Ensaios sobre a Crise Urbana do Brasil* (pp. 117–145). Campinas: Editora da Unicamp.

Castro, E. M. R. (Ed.). (2009). *Cidades na Floresta.* São Paulo: Annablume.

Castro, M. C., Monte-Mór, R. L., Sawyer, D. L., et al. (2006). Malaria Risk on the Amazon Frontier. *Proceedings of the National Academy of Sciences, 103*(7), 2452–2457. https://doi.org/10.1073/pnas.0510576103.

Constantino, R. (1985). *Synthetic Culture and Development*. Quezon City: Foundation for Nationalist.

Corder, R. M., Paula, G. A., Pincelli, A., et al. (2019). Statistical Modeling of Surveillance Data to Identify Correlates of Urban Malaria Risk: A Population-based Study in the Amazon Basin. *PloS ONE, 14*(8), e0220980. https://doi.org/10.1371/journal.pone.0220980.

CSDH (Commission on Social Determinants of Health). (2008). *Closing the Gap in a Generation: Health Equity Through Action on the Social Determinants of Health*. Final Report. Geneva: World Health Organisation.

Da Silva-Nunes, M., Codeço, C. T., Malafronte, R. S., et al. (2008). Malaria on the Amazonian Frontier: Transmission Dynamics, Risk Factors, Spatial Distribution, and Prospects for Control. *American Journal of Tropical Medicine and Hygiene, 79*(4), 624–635. https://doi.org/10.4269/ajtmh.2008.79.624.

Dos Reis, I. C., Codeço, C. T., Degener, C. M., et al. (2015a). Contribution of Fish Farming Ponds to the Production of Immature Anopheles spp. in a Malaria-endemic Amazonian Town. *Malaria Journal, 14*(1), 452. https://doi.org/10.1186/s12936-015-0947-1.

Dos Reis, I. C., Honório, N. A., De Barros, F. S. M., et al. (2015b). Epidemic and Endemic Malaria Transmission Related to Fish Farming Ponds in the Amazon Frontier. *PLoS ONE, 10*(9), e0137521. https://doi.org/10.1371/journal.pone.0137521.

Duarte, R. B. A. (2007). *Histórias de Sucesso: Agronegócios – Aquicultura e Pesca*. Brasília: Sebrae.

El País. (2019, August 27). *O Negócio Milionário dos Incêndios na Amazônia*. https://bit.ly/31WAH5T. Accessed 21 Oct 2019.

Esteva, G. (2010). Development. In W. Sachs (Ed.), *The Development Dictionary: A Guide to Knowledge as Power* (pp. 1–23). New York: Palgrave Macmillan.

Fearnside, P. M. (1987). Causes of Deforestation in the Brazilian Amazon. In R. F. Dickinson (Ed.), *The Geophysiology of Amazonia: Vegetation and Climate Interactions* (pp. 37–61). New York: John Wiley & Sons.

Ferreira, M. U., & Castro, M. C. (2016). Challenges for Malaria Elimination in Brazil. *Malaria Journal, 15*(1), 284. https://doi.org/10.1186/s12936-016-1335-1.

Godar, J., Tizado, E. J., & Pokorny, B. (2012). Who is Responsible for Deforestation in the Amazon? A Spatially Explicit Analysis Along the Transamazon Highway in Brazil. *Forest Ecology and Management, 267*, 58–73. https://doi.org/10.1016/j.foreco.2011.11.046.

Godfrey, B. J., & Browder, J. O. (1996). Disarticulated Urbanisation in the Brazilian Amazon. *Geographical Review, 86*(3), 441–445. https://doi.org/10.2307/215507.

Griffing, S. M., Tauil, P. L., Udhayakumar, V., et al. (2015). A Historical Perspective on Malaria Control in Brazil. *Memórias do Instituto Oswaldo Cruz, 110*(6), 701–718. https://doi.org/10.1590/0074-02760150041.

Guardian. (2019, August 26). *Brazil Officials Failed to Act After Warning of 'Fire Day' in Amazon, Prosecutors Say.* https://bit.ly/2NDfGIz. Accessed 26 Oct 2019.

Hiwat, H., & Bretas, G. (2011). Ecology of *Anopheles darlingi* Root with Respect to Vector Importance: A Review. *Parasites & Vectors, 4*(1), 177. https://doi.org/10.1186/1756-3305-4-177.

Hogan, D. J. (2000). A Relação entre População e Ambiente: Desafios para a Demografia. In H. Torres & H. Costa (Eds.), *População e Meio Ambiente: Debates e Desafios* (pp. 21–52). São Paulo: Editora SENAC.

IBGE (Instituto Brasileiro de Geografia e Estatística). (1970–2010). *Population Census.* http://www.ibge.gov.br. Accessed 7 Oct 2019.

IBGE (Instituto Brasileiro de Geografia e Estatística). (2019). *Estimativas da População Residente no Brasil e Unidades da Federação com Data de Referência em 1° de julho de 2019.* https://bit.ly/2PuW6R1. Accessed 7 Oct 2019.

INMET (Instituto Nacional de Meteorologia). (2019). *Normais Climatológicas do Brasil.* https://bit.ly/34CT5Ti. Accessed 21 Nov 2019.

Kothari, R. (1988). *Rethinking Development: In Search of Humane Alternatives.* Delhi: Ajanta.

Lana, R. M., Riback, T. I. S., Lima, T. F. M., et al. (2017). Socioeconomic and Demographic Characterisation of an Endemic Malaria Region in Brazil by Multiple Correspondence Analysis. *Malaria Journal, 16*(1), 397. https://doi.org/10.1186/s12936-017-2045-z.

Maheu-Giroux, M., Casapía, M., Soto-Calle, V. E., et al. (2010). Risk of Malaria Transmission from Fishponds in the Peruvian Amazon. *Acta Tropica, 115*(1–2), 112–118. https://doi.org/10.1016/j.actatropica.2010.02.011.

Martine, G. (1993). População, Meio Ambiente e Desenvolvimento: O Cenário Global e Nacional. In *População, Meio Ambiente e Desenvolvimento* (pp. 21–41). Campinas: Editora da Unicamp.

Monte-Mór, R. L. M. (2004). *Modernities in the Jungle: Extended Urbanisation in the Brazilian Amazonia.* PhD Thesis, University of California Los Angeles.

Moore, D. B. (1995). Development Discourse as hegemony: Towards an Ideological History – 1945–1995. In *Debating Development Discourse* (pp. 1–53). London: Palgrave Macmillan.

Nandy, A. (Ed.). (1988). *Science, Hegemony and Violence: A Requiem for Modernity*. New Delhi: Oxford University Press/United Nations University.
Neeleman, G., & Neeleman, R. (2015). *Soldados da Borracha: O Exército Esquecido que Salvou a II Guerra Mundial*. Porto Alegre: EDIPUCRS.
NY Times. (2019, October 10). *Why Amazon Fires Keep Raging 10 Years After a Deal to End Them*. https://nyti.ms/333Vypa. Accessed 11 Oct 2019.
Padoch, C., Brondizio, E., Costa, S., et al. (2008). Urban Forest and Rural Cities: Multi-sited Households, Consumption Patterns, and Forest Resources in Amazonia. *Ecology and Society, 13*(2), 1–16.
Pieterse, J. N. (2000). After Post-development. *Third World Quarterly, 21*(2), 175–191. https://doi.org/10.1080/01436590050004300.
Prefeitura de Mâncio Lima. (2019). *Sobre Mâncio Lima*. https://www.mancio-lima.ac.gov.br/sobre. Accessed 15 Nov 2019.
Prüss-Ustün, A., Wolf, J., Corvalán, C., et al. (2016). *Preventing Disease Through Healthy Environments: A Global Assessment of the Burden of Disease from Environmental Risks*. Geneva: World Health Organisation.
Ribeiro, M. A. C. (1993). Amazônia Legal: Rede e Fluxos na Nova Divisão Territorial do Trabalho. In *Annals III Simpósio de Geografia Urbana* (pp. 177–178). Rio de Janeiro.
Rist, G. (1990). Development as the New Religion of the West. *Quid Pro Quo, 1*(2), 5–8.
Sachs, W. (2010). *The Development Dictionary: A Guide to knowledge as Power* (2nd ed.). New York: Palgrave Macmillan.
Sarah, M. G. M., Santos, M. I. S., Souza, L. P., et al. (2003). Aspectos da Atividade de Piscicultura Praticada por Produtores Rurais no Município de Cruzeiro do Sul – Acre. *Enciclopédia Biosfera*, Centro Científico Conhecer – Goiânia, *9*(16).
SEMA (Secretaria de Estado de Meio Ambiente). (2012). *Plano Estadual de Recursos Hídricos do Acre*. SEMA, Rio Branco. https://bit.ly/2BYe1I7. Accessed 12 Oct 2019.
Trindade Júnior, S. C. C. (2015). Thinking about Territorial Modernisation and Diffuse Urbanisation in the Amazon. *Mercator, 14*(4), 93–106. https://doi.org/10.4215/RM2015.1404.0007.
Veiga, J. E. (2010). *Sustentabilidade: A Legitimação de um Novo Valor*. São Paulo: Editora SENAC.
Vicentini, Y. (2004). *Cidade e História na Amazônia*. Curitiba: Editora UFPR.
WCED (World Commission on Environment and Development). (1987). *Our Common Future*. Oxford: Oxford University Press.

WHO (World Health Organisation). (2018). *World Malaria Report 2018*. WHO Press. https://bit.ly/331BnbE. Accessed 15 Oct 2019.

Wood, C. H., & Carvalho, J. A. M. (1994). Colonização e Expansão da Fronteira Agrícola da Amazônia. In *A Demografia da Desigualdade no Brasil* (pp. 247–262). Rio de Janeiro: IPEA.

Ziai, A. (2007). *Exploring Post-development: Theory and Practice, Problems and Perspectives*. London: Routledge.

15

Political Economy of Amazon Development and Hydropower Construction

Antonio Augusto Rossotto Ioris

Introduction

We may start with a graphic example of heavy water engineering that, although distant from the Amazon, has nonetheless a number of important and revealing parallels. The scene is the Russian summer of August 1933. A group of 120 leading writers went on a boat tour along the recently inaugurated White Sea Canal, one of the evidences of the supposed success of the first Five-Year Plan of the Soviet regime. The canal was built at breakneck speed between 1931 and 1933 using the manual labour of 120,000 workers, most of them prisoners sentenced for minor political faults. As in the case of many other projects, such as the Moscow-Volga canal and Aral Sea irrigation, water infrastructure in the Soviet Union costed lives, destroyed reputations and drained scarce national resources. But the abundant signs of cruelty and state propaganda associated with the White Sea Canal did not affect the mood during the festive

A. A. R. Ioris (✉)
School of Geography and Planning, Cardiff University, Cardiff, UK
e-mail: IorisA@cardiff.ac.uk

A. A. R. Ioris (ed.), *Environment and Development*,
https://doi.org/10.1007/978-3-030-55416-3_15

trip, and the guests were in high spirit, enjoying the best food and drinks, oblivious to the national famine across most of the country (Figes 2008). What happened in the Union of Soviet Socialist Republics (USSR) regime anticipated or mirrored many other situations in which the control of water became closely associated with economic development goals, nationalist ideologies and the affirmation of modern values. Comparable examples include the Marathon Dam in Greece, Nehru's dams in India, Roosevelt's projects along the Tennessee Valley, interconnected hydropower schemes in the Scottish Highlands, dams in the Upper Douro in Portugal and the Aswan Dam in Egypt, among many others. The construction of dams and other related infrastructure is time and again used to propel and celebrate country modernisation. But it also reveals the ontological impurity of modernity, always partial, fraught with gaps and contradictions.

This chapter is intended to broaden the discussion on the economic shortcomings of development and socio-ecological impacts of dam construction in the Brazilian section of the Amazon (around 80% of the area covered by the Amazonian biome is located within Brazilian borders). More than 140 hydropower schemes are already functional or under construction in the Amazon, with another 428 under consideration, which are going to result in cumulative and monumental impacts (Latrubesse et al. 2017). There is an evident mismatch between benefits being accrued by construction companies and well-off socio-economic groups outside the Amazon, whilst most of the negative consequences are borne by peasants and communities located where dams are built. Scientific evidence is still patchy but consistently demonstrates the serious and often misunderstood ecological and hydrological impacts associated with dam construction in the region ever since, as well as the disruption of riparian communities, changes of local habits and interpersonal relations and destruction of historic and archaeological sites (relevant examples of development-related impacts in the region are included in the Environmental Justice Atlas, available at: https://ejatlas.org).

Our goal here is not to revisit economic statistics or the details of engineering projects but to assess the main direction and the internal contradictions of prevailing development trends. Lived Amazonian spaces have been dramatically transformed by sustained migration, new production technologies, urban expansion and new relations of production. Growing towns, farms, industries and households all need electricity, which means that the construction of hydropower schemes is one of the top infrastructure priorities of national, regional and local policymakers and politicians. It shall become clear below that the disjuncture between the formal enunciation of water projects in the name of national development and the actual intricacies of their implementation has produced a highly unequal distribution of opportunities and gains, whilst new meanings and values have been embodied in concrete life experiences related to country modernisation. The evolution of water management has been an integral element of the imposition of hegemonic ideologies and the mobilisation of labour and resources for the purpose of economic growth. As argued by Reis and Mollinga (2015), what is in place here is a crucial dialectic of discursivity and materiality. Such unevenness of development and the perverse socio-ecological processes of change need to be further examined, which is done here through the lens of cultural political economy (CPE), which aims to contribute to better understanding the dialectic between what is unique and local and the more widespread and global features of water management in modernity.

Political Economy and Socio-Culture

The departure point of this discussion is the claim that CPE represents a privileged entry point to the problems of water allocation, use and conservation in the world today. Political economy, against a background of intense market globalisation, mounting interdependencies and striking socio-spatial inequalities, is inescapably cultural, and those processes cannot be reduced to narrow economic thinking and simplistic formulas advanced by many right- or left-wing authors alike. On the contrary, the cultural basis of cross-scale interactions should be at the centre of any analysis of global or local politico-economic trends. Aiming to overcome reductionist formulations, CPE emerged in the 1990s in the context of the so-called cultural turn of social sciences against historicist or trans-historical analyses of structuralist authors. It was an attempt to take more seriously the complex relations between meanings and practices, as much as values and traditions, that underpin a way of life and the associated economic relations (Jessop and Oosterlynck 2008). CPE overlaps and synergies with similar approaches, including cultural studies, critical geography and social anthropology, which share the same interest in unfolding the complex political, economic and sociocultural conditions at stake as a result of global integration and socio-ecological change. Best and Paterson (2010) rightly point out that the cultural political economy of contemporary Latin America, with its flows of migration, money and information, is just not amenable to simplistic models. Still, CPE has been rarely employed in the study of the politics of natural-resource management and environmental change, though it can help to connect economic activity with discourses, symbolisms and political contestation.

A main strength of CPE authors has been their ability to make sense of the impacts and tensions of uneven development without reducing everything to the economic sphere. It means recognising that there are gaps and interstices in human relationships, notably in the practices of daily life, family traditions and aspirations that flourish and proliferate beyond the economy. Work on CPE is concerned with the variety and non-linearity of the lived economic experiences, what Mann (2012) calls the 'sidewalk path' dimension that complements the 'main flow' going down the street. Closely associated with the cultural turn, CPE has helped to restore the weight of identities, discourses, work cultures and the social and cultural embedding of economic action (Sayer 2001). Two influential authors in this debate, Jessop and Sum (2006), combined Gramscian and Foucauldian ideas to explore the relationships between economic imaginaries and inter-subjective meanings, arguing that CPE occupies a sort of middle ground between reified economics (the naturalisation of economic categories) and sociological approaches that address sociocultural activities and micro-level interactions.

On the other hand, however, the work on CPE has also reflected the exaggerations common to post-structuralist approaches, in particular the tendency to privilege the immaterial and symbolic dimensions of reality. Most post-structuralist authors typically emphasise the primacy of flat ontologies and specific circumstances but at the expense of generalisations and commonalities that also need to be recognised. It makes sense to reject rigid, all-encompassing, grand narratives, but the (dynamic) structures of local, national or global reality also need to be considered in the analysis of socio-ecological relationships and historical agency. As observed by Chang (2011: 120), because 'a factor is structural...does not mean that the outcome of its influence is predetermined'. Even if CPE scholars have rightly condemned the narrow thinking behind conventional political economy, their contribution have too easily disregarded the decisive social influence of economic forces, market transactions and institutional settings. After all, there is not only difference but also necessity in the world (Sayer 1992), bearing in mind also that necessity is 'the becoming-necessary of the encounter of contingencies' (Althusser 2006: 261). There is another, even more serious distortion derived from the obsession of post-structuralists, including many scholars working on CPE, with the inadequacies of grand theories. It is the risk of reinforcing mainstream economic approaches through the 'back door', that is, through an inadvertent association with neoclassical economic thinking. Culturalists and marginalists both attack the alleged narrowness of many political economy studies, which are found wanting and struggle to deal with world intricacies.

Cultural studies and marginalist economics have thus concurred in the attempt to challenge basic politico-economic categories, particularly those of the Marxist tradition rightly or wrongly associated with infrastructure and economic growth in the USSR mentioned above (i.e. the labour theory of value, class-based disputes, labour power as commodity, surplus value as exploitation, etc.). Because marginalists concentrate on the determination of prices through the consideration of the marginal utility of goods and services, in contrast to the focus of classical political economists on value, order and economic growth, one can often find connections with culture-sensitive and location-focused explanations of economic processes. However, here lies a trap: despite its methodological emphasis on individual preferences and people's specific circumstances, marginalist economics is in fact erected upon a clear, positivistic paradigm that proclaims the superior rationality of market freedoms to impose a business rationality on all dimensions of human experience (this is intensely present in contemporary policies adopted for the development of the Amazon region and dam construction). Standing in evident contrast with the gains obtained with the cultural turn, marginalism is equally or even more rigid, ideological and schematic than the politico-economic argument that it tries to subvert, especially because of the reduction of the complexity and non-linearity of the world to the mono-dimensionality, alienation and utilitarianism of market transactions (commonly mystified through the uncritical use of statistics and mathematical models). In other words, the prominence of subjectivism and individualism did not prevent marginalist thinking from de-contextualising economic processes from their sociocultural context.

Nevertheless, it is not enough to denounce the deceptive positivism of marginalist approaches, since it is crucially important to retain CPE firmly within the politico-economic critique of disembedded markets and of capitalist relations of production and reproduction. If CPE is a great beneficiary of the cultural turn, it is possible to argue that the emphasis on subjectivities, differences and uniqueness may have gone too far. This must serve as an alert to the overindulging of the cultural turn and the danger of eroding the political and economic dimensions in favour of a messy and shallow 'cultural bazar'. That is because the illuminating focus on the social and cultural embedding of economic activities cannot merely ignore the contribution of political economists to unlock the complexities of capitalist socio-economy (Sayer 2001). It means that the full potential of CPE can be realised through a permanent and creative association between theory and practice, between critical philosophy and philosophising for change. Consequently, there is no need to ditch the critical basis of political economy to be able to encompass sociocultural issues and subjectivities. On the contrary, the explanation of the mounting problems of capitalism and search for economic alternatives must necessarily embrace the more-than-economic elements of reality (beyond the reductionist thinking of previous scholars). It can be argued that a genuine transformation will require multiple forms of mobilisation, global alliances and ideological and practical prefigurations, which undeniably opens up space for the contribution of CPE. One particular chapter of that controversy is the fate of the Amazon and its modernisation, subjection and exploitation in the name of water development. It is evidently impossible to deal with the full picture of water development in the Amazon here, and the next pages briefly consider policies and strategies related to dam construction in the region. The intention is basically to make use of CPE to offer an epistemological critique of the politico-economic context of water in the Amazon, which has been firmly based on the simplification of socio-natural systems according to the powerful discourses of resource exploitation, territorial occupation and economic growth.

Development and Dam Construction in the Amazon

The contradictory side of dam construction in the Brazilian tracts of the Amazon, even according to mainstream economic reasoning, has been denounced, among others, by Carvalho (2006), Fearnside (2016), Hanna et al. (2016), McCormick (2011) and Scholz (2005). These authors have criticised the fact that many new, large projects are being planned, and several have already been built, but all have suffered from the lack of transparency, proper risk assessment and convincing cost-benefit analyses. They have also pointed out that the range of issues related to water engineering in the Amazon are not merely economic, political or cultural but spiral exactly at the interface between those three realms. Dams, as part of the infrastructure of development, are not only 'temples of the modern world' (paraphrasing Prime Minister Nehru) but are actually one of the most emblematic manifestations of certain politico-economic values at the expense of culture, ecology and resourcefulness of those affected by the advance of modernity. Dams work as one of the most powerful forces of the commodification and oversimplification of nature. As indicated by Marx (1976), when the products of labour are brought into relation with each other as equivalent values, concreteness and specificities are lost, and the material expressions of labour are converted into 'hieroglyphics'. A discussion of the generalities and commonalities of dam construction that underpins regional development in the Amazon reveals a great deal about the contradictions of economic trends and the tragedy of the joint exploitation of society and the rest of nature covered by the hieroglyphics of capitalist modernity.

It is important to critically revisit the political and socio-economic basis of the evolving national development experience. Brazil and the few other Latin American countries that had some nascent industrial infrastructure already in place in the 1940s benefitted from the opportunities provided by the war economy and continued to pursue import substitution policies. Those strategies were adopted in a context of fragile, formal democratic rule, increasing urban migration and persistent sociopolitical hierarchies. The Amazonian section of Brazil was still at that point a vast economic frontier with a sparse population mostly involved in subsistence production and extractivism. Gradually, the national government started to introduce development policies, including new banks, research centres and regulatory agencies, that were primarily associated with rubber production. An emblematic milestone was the famous speech delivered by President Vargas (the champion of Brazilian industrialisation) in Manaus, in 1940, calling for the occupation and modernisation of the region (promising that it was time for the Amazon to become a new chapter in the 'history of civilisation'). One important consequence was the Plan of the Economic Valorisation of the Amazon in 1953 and the creation of the Manaus Free Trade Zone (*Zona Franca*) in 1957, initially as a trade port and after 1967 as an important industrial centre (aimed to lessen the concentration of industries in the South-east and South of Brazil).

The first phase of large-scale infrastructure construction in the Amazon can be described as the 'time of the bulldozer', between the 1960s and 1980s, when Brazil was ruled by a military dictatorship that worked to preserve conservative interests through a plan of centralised and autocratic modernisation. National development policies benefitted from cheap international credit, initially provided by bilateral and multilateral banks and since the 1970s increasingly by private banks holding Eurodollars associated with the recycling of Organization of the Petroleum Exporting Countries (OPEC) money after the dramatic oil price increases (O'Brien 1991). Brazil became particularly addicted to low-interest loans, which served to produce temporary economic rents, build expensive (often questionable) engineering works and accumulate capital in the hands of the well-off groups (soon leading to serious inflation and the collapse of the state's ability to repay or serve loan interests, with the cessation of 'friendly' credit). The incorporation of the Amazon region into national development was essentially an exogenous and uncompromising process, formulated in Brasília, exalted by economic groups in the South-east of the country and encouraged by multilateral financial organisations. Once again, foreign politico-economic influences and extraterritorial demands would blend with local cultures and contribute to forge economic, social and spatial hierarchies.

A main politico-economic vector associated with modernity and eagerly advanced by the military dictatorship in the Amazon was road construction, but not any road—it had to be schemes of Amazonian proportions, such as the Trans-Amazonian and BR-163 motorways. New roads assisted the opening of farms, which was encouraged by fiscal benefits and by a legislation that considered the removal of vegetation as 'improvement' of private property (Ioris 2016). Bulldozers were not only used to build roads and clear land for pastures and crops, but also to divert rivers and erect dams for hydroelectricity generation. Various large-scale projects were planned without much regard for local socio-ecological and sociocultural features. The objective was to occupy and rapidly grow the economy before other nations (USA and China, in particular) could seize the Amazon. Most of the existing hydropower infrastructure in Brazil today was actually introduced by the military dictatorship, which then benefitted from the availability of international development money (especially petrodollars) and the repression of social and political opposition (Ioris 2007). As a result of coordinated politico-ideological and economic strategies, around 74% (it used to be more than 90% a few decades ago) of electricity consumed in Brazil comes from hydropower schemes (ANEEL 2008), which makes the country one of the leaders of renewable energy generation in the world. The Amazon was not ignored in the programme of dam construction; however, despite the incredible amount of surface water (around a fifth of the planet's freshwater), the region is quite ill-suited for dams (due to the relatively flat landscape), and swathes of land and forest had to be flooded in exchange for relatively little electricity.

Among the projects introduced by the military government were Coaracy Nunes, in Amapá, and Curuá-Una, in Pará, but the worst failure was Balbina, a plant that furnishes Manaus with very low efficiency of creation in relation to the flooded area. Balbina Dam, inaugurated in

1989 near Manaus, which flooded 2360 km^2 for a very low power generation capacity of only 250 MW and long periods of low operation due to low flows. Balbina is often described as the 'worst hydroelectric scheme in the world', as it releases more methane (a powerful greenhouse gas that is accumulated at the bottom of the reservoir) than very inefficient coal power plants; in addition, Manaus is now connected to the national grid, which makes Balbina largely redundant. Samuel Dam also had questionable efficiency, and the site of the dam is so flat that engineers had to build 30 miles of dikes to help create a lake of 520 km^2; Samuel Dam has a capacity of 217 MW, which was recognised even before the construction as insufficient to serve the growing cities of Porto Velho and Ji-Parana. The national electricity utility Eletrobrás conceded that Balbina and Samuel were mistakes, and it was affirmed by one of its directors that 'air surveys and topography measurements were not accurately made because of the tree cover' (Simons 1989).

Tucuruí, inaugurated in 1984, was the largest Amazon scheme built during this first phase and, despite its size, most environmental and social impacts were simply disregarded. The construction of the dam destroyed 2430 km^2 of forest and displaced more than 33,000 people, besides negative consequences for the indigenous population in three indigenous areas, namely Parakanã, Pucuruí and Montanha (Comissão Pró-Índio de São Paulo 1991). The filling of the artificial lake caused the extinction of various biological species and to the proliferation of waterborne diseases. As a result, Tucuruí attracted severe national and international criticism, but most socio-ecological problems were never addressed, and the next phase of dam construction was characterised by equality devastating impacts. The traditional acceptance of environmental destruction changed very little after the introduction of more comprehensive water and environmental regulation in the 1990s, as demonstrated by the example of Belo Monte.

The military regime ended in 1985 in a context of political discontent, economic instability and great uncertainty. The regime was increasingly undermined by macroeconomic turbulence created by spiralling public debts, high inflation and exhaustion of state funds, which had been the main economic driver of the so-called 'Brazilian economic miracle' (between 1968 and 1973) fuelled by international loans. The Eletrobrás

Plan 2010—published in 1987 by the Energy Ministry and a cause of great controversy due to early leaks—listed 297 sites suitable for the installation of new plants with more than 30 MW in the country; 68 of those were located in Amazonia but had to wait until the economic turbulence was overcome. After a lengthy transition, and benefitting from the legitimacy earned since the presidential election of 1989 (the first after 29 years), the historical circumstances were ripe for pro-market reforms and the reorganisation of the state by the new governments. The economic stability and inflation control offered by the Real Plan of 1994 provided the basis for legal and institutional adjustments. The Cardoso administration (1995–2002) promoted successful monetary and fiscal adjustments, which fuelled an ambitious reform of the state apparatus along the lines of liberalisation and growing integration into global markets.

The Real Plan was complemented by institutional restructuring that included the removal of trade barriers (and at times the promotion of imports to avoid inflation), rigid monetary regulation, adjustments in the labour market and reorientation of the state apparatus. Energy policies and the electric sector were directly affected by the powerful neoliberalising agenda; for instance, the reduced tariffs charged by public utilities, a mechanism largely used during the 1980s to contain inflation, were removed in favour of private suppliers. Water and environmental regulation were also transformed due to the introduction of a new legislation directly informed by the doctrine of ecological modernisation, such as the 1997 Water Law, centred on river basin committees, water licences and correspondent charges (Ioris 2010). The reorientation of government agencies and the introduction of the new water legislation paved the road for the expansion of private gains based on natural resources publicly or collectively owned (Ioris 2009). New energy and water regulatory agencies, among others, were created to oversee the growing involvement of the business sector in infrastructure construction and operation. A private energy market was created in 1995 (reorganised in 2003) in which consumers and providers could freely trade energy with each other.

The second phase of dam engineering was mainly focused on politico-economic adjustments, which coincided with the neoliberal reform of the Brazilian State in the 1990s. The contradictions of the state-led model of development—implemented in Brazil along most of the twentieth century and intensified by the military governments making use of foreign savings through loans from multilateral banks—had resulted in growing economic inefficiencies and reliance on the continuous injection of capital by the state. Along the lines of the institutional changes introduced by the Real Plan, both generation and commercialisation of energy then became available to private national entrepreneurs, increasingly associated with international investors or energy companies. Because of the changing role of the state apparatus, which was increasingly focused on pro-private sector incentives instead of direct construction and operation (which nonetheless continued to happen, despite the neoliberal discourse), together with national and international criticism, very few hydropower schemes were built during the second period. Coordination and decision-making became significantly diffused across many agencies without the presence of a centralised, well-equipped agency as during the military dictatorship. It represented a real tension between ambitious efficiency and operational goals and the reality of institutional fragmentation and diminished investments of the national government (Goldenberg and Prado 2003). The proclaimed virtues of the open energy market were never translated into investments and coordinated efforts. It culminated in the national energy crisis over nine months (June 2001 to March 2002), when energy offer was significantly reduced, caused by months with low rainfall, which required energy rationing at a very high political cost (leading to the loss of the 2002 presidential election by Cardoso's political group). According to a technical report of the National Energy Policy Council (2001), the failure to invest in new dams was responsible for two-thirds of the energy-rationing solution (Kelman 2001).

The neo-liberalising adjustments adopted in the 1990s—with all the idiosyncrasies and contingencies of any neo-liberalisation process (Ioris 2013)—led to the third, still-unfolding phase of dam engineering in the Brazilian Amazon. It has been a period of neo- developmentalist strategies and formalist environmental and socio-economic regulation. Between 2003 and 2016, under Presidents Lula and Dilma, national policies and government initiatives were characterised by a peculiar combination of populist measures (such as conditional cash transfers, as in the case of the Family Stipend [*Bolsa Família*] programme) and reassured guarantees to financial and business sectors. During this third phase, the country benefitted from the commodity boom and increased the reliance on foreign goods, at the cost of de-industrialisation and over-appreciation of the national currency. In that context, the state renewed the programme of infrastructure construction, with increasingly larger dams in the Amazon, through associations between public and private construction companies, closely supported by local and national politicians and the hegemonic economic sectors. National dam construction policies had to be adjusted because the best sites, situated closer to the main industrial and urban areas in the South and South-east of Brazil, were all taken in the process, which means that there is now a pressure to move to new areas, particularly in the Amazon, and even plans to reduce hydroelectric production in other river basins in favour of irrigation or other uses. Despite mounting impacts and growing controversies, the hydroelectric exploitation of Amazonia has never been as much on the agenda as it is now.

Since 2003, the Amazon hydropower sector has provided an emblematic demonstration of the choreography of continuities and adjustments (Broggio et al. 2017). In the Madeira Basin, after a lengthy political dispute, the Jirau and Santo Antônio plants were licensed in July 2007, allowing the overflow of up to 529 km^2. The two schemes have affected several indigenous groups, such as the Karitiana, Karipuna, Urueu-Wau-Wau, Katawixi and also other isolated, unnamed tribes (some living only six miles from the Jirau dam), who were forced to migrate to new lands, where they clashed with '*garimpeiros*' (wildcat miners) logging companies and were exposed to diseases against which they had no immunity (Le Clézio and Razon 2010). Another highly contentious project was Belo Monte, which has attracted great publicity in the international media (e.g. 29 articles posted by BBC News alone between 2010 and 2017). Belo Monte, still under construction, demonstrates that the neo-liberalising platform was being cleverly adjusted to fulfil neo-developmentalist goals. It also epitomised the revival of the construction of large hydroelectric power plants in the Amazon and the encroachment of Brazilian energy demands upon neighbouring nations (because some schemes have flooded Bolivian land, and Brazilian construction companies are deeply involved in projects in Peru, Venezuela and other countries). The project was partly inaugurated in 2016, with 11,233 MW of installed capacity and a cost of more than US$ 13 billion. After its completion, Belo Monte became the fourth largest power station on the planet, just behind Itaipu (inaugurated in 1984 on the border between Brazil and Paraguay) and Xiluodu and Three Gorges in China.

The Belo Monte project, with another name, Cararaô, was originally conceived by the military in 1975 but could not go ahead due to the collapse of the regime and its growing financial and political problems. The residents of Altamira (near the site of the dam) and the indigenous people keep up a resistance that is organised and quite influenced by the traumatic experience with Tucuruí. There was a large gathering in 1989, with more than 1000 participants, including more than 600 indigenous people, which managed to attract international attention to the dispute that led to the cancellation of a World Bank loan under negotiation (Carvalho 2006). The proposal was revisited by the neoliberal administration of President Cardoso but only really implemented by the government of Presidents Lula and Dilma. A consortium of state-owned companies (as discussed in a previous chapter, this is called *Norte Energia* [more details at http://norteenergiasa.com.br]) won the contract to build Belo Monte and manage it for 35 years. It is actually bewildering that the Lula government included many energy and environmental authorities who had for a long time fiercely opposed this and other similar projects in the past (e.g. Pinguelli Rosa 1989). The government had to respond to the criticism raised by environmental groups and local communities (including important Indian groups) who managed to form an international coalition against Belo Monte (that included Hollywood stars and famous artists). Capacity and transmission lines were reduced, and the new design removed the sizeable reservoir, in order to minimise the negative impacts. Moreover, the controversial features of Belo Monte continue to stir protest and serious resistance.

The project was the subject of a lengthy battle in court, which led to repeated interruptions of the construction and operation (the last interruption ordered by a judge happened in 2017, but local groups also regularly occupy the site of the dam to protest against the unresolved problems associated with the project), and the final construction phase was only concluded in 2019 (inaugurated with great pomp by the neo-fascist president Bolsonaro, who was elected in the previous year). The most controversial issue is related to actual viability of the project without other supporting dams upstream. Due to the long dry season and the resulting long period of low flows in the Xingu River Basin (a common feature of eastern Amazon rivers), Belo Monte has a low operation and economic performance if operating as a single dam. It means that a cascade of dams is necessary, but this would magnify the impacts on indigenous land, natural parks and farms. The national government has guaranteed, on many occasions, that Belo Monte is the only large-scale dam of the Xingu River, but the wider problem for the Brazilian society is the low legitimacy and eroded trust in public authorities. This is a global phenomenon that seriously affects the quality of the formal democracy of many countries, Brazil included. Similar reassurances were given in relation to the construction of other schemes in the Amazon, such as the 43 large dams to be constructed along the (even more vulnerable and controversial) Tapajós River by 2022 (Fearnside 2015). Those dams directly benefit from the friendly interpretation of the legislation and lax enforcement of water and environmental regulation, which means that several indigenous reserves and conservation areas will be flooded and degraded.

Abers et al. (2017) argue that the interests of the construction and business sectors for new infrastructure projects in the Amazon contrast with the persistent disregard for public participation and social demands, which reveals the contradictory and asymmetric priorities of the state. The assessment of applications for the environmental and water licences, in 2011, was notoriously undemocratic (public consultation was minimal and rigidly controlled by the developers) and their approval strongly pushed forward by the federal government under the justification that economic growth required additional sources of energy (Sevá Filho 2005). It is estimated that Belo Monte displaced between 20,000 and 50,000 people (including many families left behind after the collapse of rubber production in the early 1900s), while only one-third of the compensation promised by the authorities was ever paid (Anderson 2017).

Among many activists and local communities, there is a distinctive feeling of betrayal and deception associated with Belo Monte (Diversi 2014), especially because of the strange new alliances formed between left-wing politicians and conservative, economic and political groups (Melo 2016). The election of Lula, an important union leader, had raised great expectations, but once in office, his government (2003–2010) was allied with the most archaic and corrupt politicians and construction companies in the country. It was not coincidence that the approval of Belo Monte involved systematic diversionary operations employed to divide the opposition and, at the same time, to legitimise the controversial project (Atkins 2017). During the Dilma government (2010–2016), which succeeded Lula's, the secret service was even asked to spy on indigenous leaders, anthropologists and champions of social movements who opposed the construction of Belo Monte (Valente and Aragon 2017).

Also disturbing were the dishonest practices employed for the construction of Belo Monte and many other oil and nuclear power plants, Olympic Games facilities, football stadiums and similar other projects championed by Lula and his loyal successor Dilma (who was removed from office by the congress in 2016). Corruption and manipulation on an industrial scale were only properly revealed through the work of the *Lava-Jato* (Carwash) Operation of the federal police and the public prosecutor. Numerous politicians, operators and directors of engineering companies decided to collaborate with the judicial authorities, in order to secure a reduction in criminal penalties (plea bargain agreements), and exposed a monumental scheme involving bribing, illegal financial of electoral campaigns, flawed costing and engineering details, irregular granting of environmental and water licences and cartelisation of the construction sector (Calixto 2017). With the increasing arrests of top politicians, senior civil servants and owners of construction companies, affecting a significant proportion of the congress, including the speakers of the Senate and of the House, and several ministries (Casado et al. 2017), it became clear that water and energy infrastructure in the Amazon are ingrained in the multiple operations conducted by the national state in the name of development. A similar approach and involving the same construction companies (more notoriously, Odebrecht, South America's largest construction group) was responsible for the construction of hydropower projects in other countries, as in the case of Lauca Dam in Angola (Leahy 2016).

In economic terms, Belo Monte stimulated a negotiation with Chinese investors to install an aluminium factory in the region of Barcarena (in the state of Pará) and the Canadian company Belo Sun to extract gold in the same area of the hydropower plant (the environmental licence granted in March 2017, but critics—such as Amazon Watch—say that the risks were ignored by the authorities; see Poirier 2017), among other similar projects. But there are further ramifications of the construction of new dams beyond the mere appropriation of public funds. The most relevant is the nexus between energy, mining, agribusiness and navigation. The justification of the new hydropower schemes seems to be increasingly both for the purpose of energy generation for cities, industries and mining operations and for the creation of new navigation routes. The new dams allow large-scale barges to pass rapids and cross waterfalls, which significantly reduces transport costs to ports along the Amazon River, accessible to transcontinental transportation ships, and ultimately increase the profitability and the areas of soybean cultivation (Fearnside 2015). At face value, it seems that there is a fantastic potential for the expansion of river navigation, considering that it only accounts for 4% of domestic Brazilian trade, but it is almost certain that navigation plans will follow speculation and financialisation pressures, leading to land grabbing, chaotic migration and deforestation (Becker 2012). The growing politico-economic relevance of agribusiness in Brazil is directly connected with the challenges of water management and Amazonian development. The sector is commonly considered a great Brazilian achievement, due to technological improvements and production growth, but the sector is also responsible for ecosystem degradation, contamination of water reserves, socio-spatial inequalities and macroeconomic vulnerability (Ioris 2015). Apparent techno-economic success has encouraged further land and water grabbing, as the agribusiness sector has been unable to resolve long-lasting problems such as economic development that benefits small groups of large-scale landowners, commercial companies and transnational corporations at the expense of mounting socio-natural impacts.

Cultural Politico-Economic Insights: Distribution, Recognition and Resignification

As discussed so far, what happened in the last four decades transformed the Amazon region into the main energy and agricultural frontier in Brazil, particularly because of the abundance of land and water and, notably, the limited ability of local groups to resist the strong advance of engineering infrastructure and plantation farms. One persistent feature of the process of socio-spatial change has been the overestimation of benefits and underestimation of impacts. Construction projects have created only a limited number of temporary jobs but left a serious legacy of social disruption, violence, prostitution and many other evidences of human degradation (as in the recent case of the periphery of Altamira, massively transformed by the construction of Belo Monte). The benefits of those projects are grabbed in other parts of the country and abroad (in the form of energy and crop exports), while in the location of the dam many suffer from waterborne diseases, unemployment, lowered income and sheer abandonment (regarding basic public services and human rights). The size of infrastructure might have changed, but there is a disturbing qualitative continuity between the colonial past, the developmentalist period and the neo-liberalised present, which is also translated into serious imbalances between different Brazilian regions. The design and construction of hydropower schemes in the more recent development phase, centred on production of primary commodities (minerals and agribusiness goods) for the global market, basically replicates the disruptive and hierarchical practices earlier adopted in the name of colonisation, nation-building and, in the post-war decades, national development.

The politico-economic trends of regional development are complex, uneven, gendered and racialised, which have important differences and also commonalities with the experience in other parts of the world. On the one hand, the Amazon is quite unique because it contains the largest tropical rainforest in the world, but it still has many indigenous groups with distinctive cultural and linguistic traditions (increasingly being eroded), a unique history of colonisation and a marginal role in the various national economies. On the other hand, water development and dam-building have been important elements of elite consolidation and hegemonic development plans, as described by Mohamud and Verhoeven (2016) for the Nile Basin. There has been a comparable nexus between dams, political authority and ideological influences. The strong appeal of development influences interpersonal relations and permeates cultural representations and lived experience on the ground (increasingly commodified and mediated by money). As in the case of Chile, conflicts over proposed dams reveal the different meanings and values of nature, even as contemporary water management approaches are permeated by market-based approaches that exacerbate uneven power relations (Palomino-Schalscha et al. 2016). In China and Vietnam, just as in the Amazon, the local and disorganised populations endure the greatest impacts of water projects, whereas the positive results of hydropower schemes are primarily appropriated by energy companies and urban and industrialised parts of the country (Rousseau et al. 2017). The Chinese mega project of the Three Gorges Dam has also dispossessed local communities, contributed towards the formation of capital—in the form of contemporary primitive accumulation—and been benefitted by the help of politicians and bureaucrats, who have also reinforced their positions in the structures of power (Webber 2012).

As in many other parts of the world, water management in the Amazon has been based on rent-seeking strategies at the expense of socio-ecological impacts and devastation of indigenous culture and local knowledge. What is more, local cultural elements are not really considered for the development of policies and decision-making that affect large segments of the local population. The aggressive economic transformation of the Amazon has also resulted in novel lifestyles and consumer preferences that increasingly mirror the cultural patterns elsewhere in the main urban centres of Brazil (with the associated levels of consumerism, processed food, the spread of Evangelical churches, new music and dance styles born in the urban peripheries, etc.). The appeal of development and the imperative of economic growth at any cost have been embraced by both dictatorial right-wing governments and populist-nationalist presidents, who all share the same ideological agenda. Moreover, as important as recognising the political and economic dimensions of dam construction and its commonalities with other similar water development experiences is to comprehend how culture is seen as a realm which mediates all experience and can be mobilised for political action. Disputes and controversies around water are never simple or unidirectional but reveal the ability of local actors—indigenous and non-indigenous groups in the case of the Amazon—to learn from past incidents and creatively react through a range of channels and with many different allies, just as communities affected by new dams have been reacting in the Upper Douro in Portugal (Ioris 2012). The practical consequence is that the analytical potential of CPE—considering subjectivities, historical agencies and cultural specificities of political and economic processes—is particularly relevant to reject both the biases of rigid structuralist approaches and the individualist excesses of post-structuralist authors.

Finally, several useful CPE lessons can be learned from reflecting on the Amazon experience. The emphasis on the symbolic and rhetoric components cannot diminish the materiality and physical aspects of culture. According to Peet and Watts (2004), in the final decades of the last century, there was a postmodern obsession with reflexivity, which vainly emphasised words more than deeds, discourse rather than development, hype over reality. There is even room here for a new epistemological 'turn' that properly emphasises materialist sensibilities and the transformative work of meaning-making cultural processes in the concreteness of the world (Kirsch 2012). Related to that, CPE studies are certainly associated with the quest for justice, but this cannot be equated with the mainstream economist treatment of social and economic recognition as a mere epiphenomenon of distribution; in effect, the two—recognition and distribution—need to be theorised together (Fraser and Honneth 2003). It means that CPE approaches are useful to redress the balance between the intricate dimensions of the lived and politicised reality. Recognition, including multiple struggles for identity and difference, is a central concept in a context of accelerated trans-cultural contacts, as in the Amazon, and should be articulated together with the claims of subaltern groups affected by regional development. Sayer (2001) rightly advocates a CPE based on a fuzzy distinction between the system (formal structures beyond the subjective experience of actors) and lifeworld (related to experience, informality and the culture in which actors socialise) that cuts across the culture-economy and recognition-distribution binaries.

Yet, it is not sufficient to remove the artificially created binary between distribution and recognition; an accomplished CPE procedure requires extending the analytical framework to the politics of resignification. This will be necessary because discourses and symbols constantly change, and the historicity of complex dynamics needs to be carefully considered. Resignification—in the sense of a collective re-imagination of the public and private trends—can work as a mediator between the radical modification of economic processes and a serious and a politically meaningful appreciation of the cultural diversity of the Amazon. Moving beyond the micro-politics of post-structuralism—under the assumption that the specific and local experiences of minorities and marginalised groups hold the moral reservoir that macro-politics and large-scale changes apparently lost—the concept of resignification represents, first of all, a commitment to reinterpret and help to overcome a perverse reality that is systematically reinforced and cuts across nested scales of interaction (from forest-dependent communities in remote catchments disrupted by hydropower schemes to national political coalitions and globalised markets). This is needed to subvert the hegemonic interpretative order predicated upon the exclusion of other symbolic possibilities, and such counter-hegemonic practices of resignification help to challenge the status quo in the effort to reject 'economisms' and bring together cultural questions, ecological challenges and grassroots politics.

Conclusions

The three phases of dam construction in the Brazilian section of the Amazon since the 1960s have been marked by conflicting interests and socio-ecological impacts associated with the advance of national development plans. In order to make sense of the complexity and unevenness of modern water management, a discussion of CPE was employed to explore the local specificities and wider political and economic 'conjunctures' that have facilitated the conversion of water into an economic resource. As in the case of canal construction for the aforementioned plans of the Stalinist state, dams in the Amazon had the double aim to foster agro-industrial production and, also, to serve as an emblematic symbol of the modernisation of the Brazilian economy. Particularly, the Belo Monte hydropower scheme, built along the Xingu River, reveals the persistence of authoritarianism, technocracy, unfairness and corruption behind narratives and projects that are supposed to boost national and democratic economic development.

The main conclusion to be drawn is that sensibility to culture, agency and social identity must be connected with the wider macroeconomic policies and national politics. In other words, the focus on culture and local circumstances cannot be invoked at the expense of the equally important social, economic and political phenomena at the regional and national scales. There have been three critical moments of hydropower projects in the Amazon, evolving from a period of more direct state intervention to the more recent partnership with private construction companies, but several important features have always remained, such as the disregard for the existing socio-ecological features of the river basins involved and the attempt to convert shared waters into private sources of economic gain. Economic growth and regional development have largely depended on water, but first water has to be recreated as a hydropower resource divested of its other important features. The large-scale and speedy incorporation of water into economic development strategies demonstrates the persistent and systematic advance of modernity in the Amazon through the powerful resignification of social relations associated with water. Water has lost most of its sociocultural, ecological and geo-morphological functions, mainly to fulfil exogenous demands for energy, navigation and agricultural expansion. It has been a phenomenon of epic proportions, involving both material and structural transformations of ecosystems, communities, lifestyles and socioeconomic arrangements. At the centre of those processes of change lies the appropriation of territorial resources, such as water, and the conversion of the complexity and heterogeneity of the Amazon into the abstracted realm of market transactions and the circulation of capital. Overall, it has been a widespread tragedy of the non-commons.

References

Abers, R. N., Oliveira, M. S., & Pereira, A. K. (2017). Inclusive Development and the Asymmetric State: Big Projects and Local Communities in the Brazilian Amazon. *Journal of Development Studies, 53*(6), 857–872.

Althusser, L. (2006). *Philosophy of the Encounter: Later Writings 1978–1987* (G. M. Goshgarian, Trans.). London/New York: Verso.

Anderson, M. (2017). *Displaced by Brazil's Giant Belo Monte Hydroelectric Dam, 'River People' Reoccupy Reservoir.* https://news.mongabay.com/2017/03/displaced-by-brazils-giant-belo-monte-hydroelectric-dam-river-people-reoccupy-reservoir. Accessed 30 Dec 2018.

ANEEL. (2008). *Atlas de Energia Elétrica do Brasil* (3rd ed.). Brasília: Agência Nacional de Energia Elétrica.

Atkins, E. (2017). Dammed and Diversionary: The Multi-dimensional Framing of Brazil's Belo Monte Dam. *Singapore Journal of Tropical Geography, 38*(3), 276–292.

Becker, B. K. (2012). Reflexões sobre Hidrelétricas na Amazônia: Água, Energia e Desenvolvimento. *Boletim do Museu Paraense Emílio Goeldi (Ciências Humanas), 7*(3), 783–790.

Best, J., & Paterson, M. (Eds.). (2010). *Cultural Political Economy.* New York: Routledge.

Broggio, C., Droulers, M., & Pallamar, J.-P. (2017). Les Barrages Hydroélectriques Amazoniens, un Enjeu de Géopolitique Interne au Brésil. *Hérodote, 165*(2), 185–203.

Calixto, B. (2017). *O Que as Delações da Odebrecht Dizem sobre Corrupção nas Hidrelétricas da Amazônia.* http://epoca.globo.com/ciencia-e-meio-ambiente/blog-do-planeta/noticia/2017/04/o-que-delacoes-da-odebrecht-dizem-sobre-corrupcao-nas-hidreletricas-da-amazonia.html. Accessed 13 Apr 2019.

Carvalho, G. O. (2006). Environmental Resistance and the Politics of Energy Development in the Brazilian Amazon. *Journal of Environment and Development, 15*(3), 245–268.

Casado, L., Mattoso, C., & Megale, B. (2017). *Ao TSE, Delator Cita Lobão e PMDB em Esquema de Propina de Belo Monte.* http://www1.folha.uol.com.br/poder/2017/03/1869389-ao-tse-delator-cita-lobao-e-pmdb-em-esquema-de-propina-de-belo-monte.shtml. Accessed 21 Feb 2019.

Chang, H.-J. (2011). *23 Things They Don't Tell You about Capitalism.* London: Penguin.

Comissão Pró-Índio de São Paulo. (1991). *Segundo Tribunal Internacional das Águas. Caso: As Hidrelétricas na Amazônia Brasileira*. Comissão Pró-Índio: São Paulo.

Diversi, M. (2014). Damming the Amazon: The Postcolonial March to the Wicked West. *Cultural Studies—Critical Methodologies, 14*(3), 242–246.

Fearnside, P. M. (2015). Amazon Dams and Waterways: Brazil's Tapajós Basin Plans. *Ambio, 44*, 426–439.

Fearnside, P. M. (2016). Environmental and Social Impacts of Hydroelectric Dams in Brazilian Amazonia: Implications for the Aluminium Industry. *World Development, 77*, 48–65.

Figes, O. (2008). *The Whisperers: Private Life in Stalin's Russia*. London: Penguin.

Fraser, N., & Honneth, A. (2003). *Redistribution or Recognition? A Political-Philosophical Exchange* (J. Golb, J. Ingram, & C. Wilke, Trans.). London/New York: Verso.

Goldenberg, J., & Prado, L. T. S. (2003). Reforma e Crise do Setor Elétrico no Período FHC. *Tempo Social – USP, 15*(2), 219–235.

Hanna, P., Vanclay, F., Langdon, E. J., & Arts, J. (2016). The Importance of Cultural Aspects in Impact Assessment and Project Development: Reflections from a Case Study of a Hydroelectric Dam in Brazil. *Impact Assessment and Project Appraisal, 34*(4), 306–318.

Ioris, A. A. R. (2007). The Troubled Waters of Brazil: Nature Commodification and Social Exclusion. *Capitalism Nature Socialism, 18*(1), 28–50.

Ioris, A. A. R. (2009). Water Reforms in Brazil: Opportunities and Constraints. *Journal of Environmental Planning and Management, 52*(6), 813–832.

Ioris, A. A. R. (2010). The Political Nexus between Water and Economics in Brazil: A Critique of Recent Policy Reforms. *Review of Radical Political Economics, 42*(2), 231–250.

Ioris, A. A. R. (2012). The Political Geography of Environmental Regulation: Implementing the Water Framework Directive in the Douro River Basin, Portugal. *Scottish Geographical Journal, 128*(1), 1–23.

Ioris, A. A. R. (2013). The Adaptive Nature of the Neoliberal State and the State-led Neoliberalisation of Nature: Unpacking the Political Economy of Water in Lima, Peru. *New Political Economy, 18*(6), 912–938.

Ioris, A. A. R. (2015). The Production of Poverty and the Poverty of Production in the Amazon: Reflections from those at the Sharp End of Development. *Capitalism Nature Socialism, 26*(4), 176–192.

Ioris, A. A. R. (2016). Rent of Agribusiness in the Amazon: A Case Study from Mato Grosso. *Land Use Policy, 59*, 456–466.

Jessop, B., & Oosterlynck, S. (2008). Cultural Political Economy: On Making the Cultural Turn without Falling into Soft Economic Sociology. *Geoforum, 39*(3), 1155–1169.
Jessop, B., & Sum, N.-L. (2006). *Beyond the Regulation Approach: Putting Capitalist Economies in their Place*. Cheltenham: Edward Elgar.
Kelman, J. (Ed.). (2001). *Relatório da Comissão de Análise do Sistema Hidrotérmico de Energia Elétrica*. Brasília: Conselho Nacional de Política Energética.
Kirsch, S. (2012). Cultural Geography I: Materialist Turns. *Progress in Human Geography, 37*(3), 433–441.
Latrubesse, E. M., et al. (2017). Damming the Rivers of the Amazon Basin. *Nature, 546*, 363–369.
Le Clézio, J. M. G., & Razon, J-P. (2010). *Un Projet de GDF-Suez Met en Danger les Dernières Tribus Isolées d'Amazonie*. http://www.lemonde.fr/idees/article/2010/04/07/un-projet-de-gdf-suez-met-en-danger-les-dernieres-tribus-isolees-d-amazonie-par-jean-marie-g-le-clezio-et-jean-patrick-razon_1329918_3232.html. Accessed 04 Jan 2018.
Leahy, J. (2016). *A Brazilian Bribery Machine*. https://www.ft.com/content/8edf5b2c-c868-11e6-9043-7e34c07b46ef. Accessed 28 Dec 2018.
Mann, G. (2012). Release the Hounds! The Marvelous Case of Political Economy. In T. J. Barnes, J. Peck, & E. Sheppard (Eds.), *The Wiley-Blackwell Companion to Economic Geography* (pp. 61–73). Oxford: Blackwell.
Marx, K. (1976 [1867]). *Capital* (Vol. 1) (B. Fowkes, Trans.). London: Penguin.
McCormick, S. (2011). Damming the Amazon: Local Movements and Transnational Struggles over Water. *Society & Natural Resources, 24*(1), 34–48.
Melo, M. A. (2016). Crisis and Integrity in Brazil. *Journal of Democracy, 27*(2), 50–65.
Mohamud, M., & Verhoeven, H. (2016). Re-engineering the State, Awakening the Nation: Dams, Islamist Modernity and Nationalist Politics in Sudan. *Water Alternatives, 9*(2), 182–202.
O'Brien, P. (1991). Debt and Sustainable Development in Latin America. In D. Goodman & M. Redclift (Eds.), *Environment and Development in Latin America* (pp. 24–47). Manchester/New York: Manchester University Press.
Palomino-Schalscha, M., Leaman-Constanzo, C., & Bond, S. (2016). Contested Water, Contested Development: Unpacking the Hydro-social Cycle of the Ñuble River, Chile. *Third World Quarterly, 37*(5), 883–901.
Peet, R., & Watts, M. (2004). *Liberation Ecologies: Environment, Development, Social Movements* (2nd ed.). Abingdon: Routledge.

Pinguelli Rosa, L. (1989). Hidrelétricas e Meio Ambiente na Amazônia: Análise Crítica do Plano 2010. *Revista Brasileira de Energia, 1*(1), 7–24.

Poirier, C. (2017). *Toxic Mega-mine Looms over Belo Monte's Affected Communities.* http://amazonwatch.org/news/2017/0404-toxic-mega-mine-looms-over-belo-montes-affected-communities. Accessed 04 Apr 2019.

Reis, N., & Mollinga, P. P. (2015). Public Policy and the Idea of the Vietnamese State: The Cultural Political Economy of Domestic Water Supply. *Asian Studies Review, 39*(4), 628–648.

Rousseau, J.-F., Orange, D., Habich-Sobiegalla, S., & Van Thiet, N. (2017). Socialist Hydropower Governances Compared: Dams and Resettlement as Experienced by Dai and Thai Societies from the Sino-Vietnamese Borderlands. *Regional Environmental Change, 17*, 2409–2419.

Sayer, A. (1992). *Method in Social Science: A Realist Approach* (2nd ed.). London/New York: Routledge.

Sayer, A. (2001). For a Critical Cultural Political Economy. *Antipode, 33*(4), 687–708.

Scholz, I. (2005). Environmental Policy Cooperation Among Organised Civil Society, National Public Actors and International Actors in the Brazilian Amazon. *European Journal of Development Research, 17*(4), 681–705.

Sevá Filho, A. O. (Ed.). (2005). *Tenotã-Mõ: Alertas sobre as Consequências dos Projetos Hidrelétricos no Rio Xingu.* São Paulo: International Rivers Network.

Simons, M. (1989). *The Amazon Forest: Brazil Wants its Dams, but at What Cost?* http://www.nytimes.com/1989/03/12/weekinreview/the-world-the-amazon-forest-brazil-wants-its-dams-but-at-what-cost.html. Accessed 25 Mar 2019.

Valente, R., & Aragon, R. (2017). *Abin Espionou Indígenas e ONGs no Governo Dilma.* http://www1.folha.uol.com.br/poder/2017/05/1882257-abin-espionou-indigenas-e-ongs-no-governo-dilma.shtml. Accessed 09 May 2019.

Webber, M. (2012). The Political Economy of the Three Gorges Project. *Geographical Research, 50*(2), 154–165.

16

Water Governance and the Hydrosocial Territory of the Teles Pires River Basin in the Brazilian Amazon

Daniela Maimoni Figueiredo
and Antonio Augusto Rossotto Ioris

Introduction

Water allocation, use and conservation remain extremely relevant for the debate on the interface between environment and development. The surge of interest in water governance over the past three decades and its conversion into a central environmental management issue are related to a series of global initiatives, particularly the 1977 UN Conference on Water in Mar del Plata, the 1992 International Conference on Water and the Environment in Dublin and the 1992 UN Conference on Environment and Development in Rio de Janeiro. These resulted in important documents such as the Dublin Principles and Agenda 21, which include guidelines on the reform of water management. These

D. M. Figueiredo (✉)
Federal University of Mato Grosso (UFMT), Cuiabá, Brazil

A. A. R. Ioris
School of Geography and Planning, Cardiff University, Cardiff, UK
e-mail: IorisA@cardiff.ac.uk

A. A. R. Ioris (ed.), *Environment and Development*,
https://doi.org/10.1007/978-3-030-55416-3_16

initiatives were aimed at promoting institutional changes, particularly regarding the affirmation of the river basin as the space for integrated water management, the implementation of public participation in decision-making as a key mechanism to foster decentralisation and the pursuit of a holistic approach. Concepts like integration, the river basin, holism and public participation, among others, have increasingly influenced water policies and management strategies around the world, albeit with uneven results, and, more importantly, many continuities with previous interventions that were considered inadequate but continue to undermine improvements in the water sector.

New concepts and novel management approaches have been adopted by many countries and, in the case of Brazil, resulted in the approval of the Water Resources National Policy Act (Law No 9433/1997), which directly followed the international principles of decentralised management at the hydrographic basin level, participation of a wide range of actors through the River Basin Committees and the valuation of water as a public asset with economic value (Veiga and Magrini 2013). River Basin Committees (RBC) are part of the Water Resource System, which is also composed of the National Water Resources Council, State Water Resources Councils, State Water Resources Management Institutions and Water Agencies (Veiga and Magrini 2013; see Fig. 16.1). According to the aforementioned Law 9433, problems related to water must be solved through a process of political, economic and social negotiation, involving different government and social actors at different levels, following the guidelines of the Brazilian Constitution from 1988 and the French water management model (Campos and Fracalanza 2010).

Since the passing of Law 9433, tangible progress has been made in the management of Brazil's water resources (Veiga and Magrini 2013). The new Brazilian management model has been primarily implemented in the more developed (industrialised) regions of the country, which normally face greater problems related to water quality and/or quantity (the Southeast, South and Northeast, according to Veiga and Magrini 2013). These regional asymmetries have been largely justified by the main water stakeholders on the grounds of economic development ideology, lack of financial resources and public perception of a supposed good water availability. This idea prevailed for decades throughout Brazil, where water

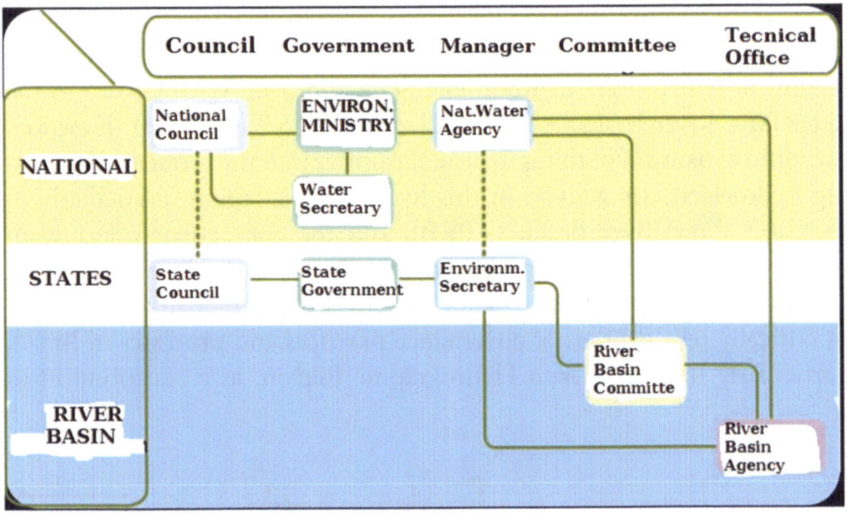

Fig. 16.1 Brazilian National Water Resource System Framework. Note: in January 2019, this system was transferred from the Environment Ministry to the Regional Development Ministry, but in the majority of States, water management is still controlled by Environment Secretaries. (Modified from ana.gov.br)

was mistakenly treated as an abundant and inexhaustible natural resource, an erroneous view that for many years concealed water's true sanitary, cultural, social, economic and geopolitical value, which has since become clear and will become even more pronounced during the twenty-first century (Bordalo 2017). Likewise, there has been a lack of effective implementation in terms of democratic and more inclusive socio-spatial goals (Mesquita 2018; Siegmund-Schultze et al. 2015; Martins 2015).

Failures and contradictions in the legislation itself have become evident, such as top-down regulation and narrow, biased, resulting in a situation of great political and regional asymmetries that favour some sectors and locations at the expense of the wider community of water users and civil society at large. This is particularly evident in the Amazon Hydrographic Region, where water management is lagging behind other regions of the country. The Amazon River, the world's largest by water volume, flows through this region, which has a low demographic density and one of the highest rainfall indices on the planet (Freitas and Freitas

2018). One of the Law 9433 delay indicators in this region, among others, is the number RBC installed: of the 224 in Brazil, only 9 are in this region, of which 3 are in Rondônia State and 3 in Mato Grosso in the Teles Pires River Basin (ANA 2018; see Figs. 16.2 and 16.3). However, the rapid expansion of the agricultural frontier into the forest and increasing hydroelectricity activity in this hydrographic region, particularly in the Teles Pires River Basin, (TPRB), and the consequent pressure on water resources, means that more and better preventive policies are needed.

The asymmetric implementation of Law 9433 demonstrates the need to critically revaluate water governance practices and processes in Brazil, particularly in the Amazon Hydrographic Region, as a contribution to

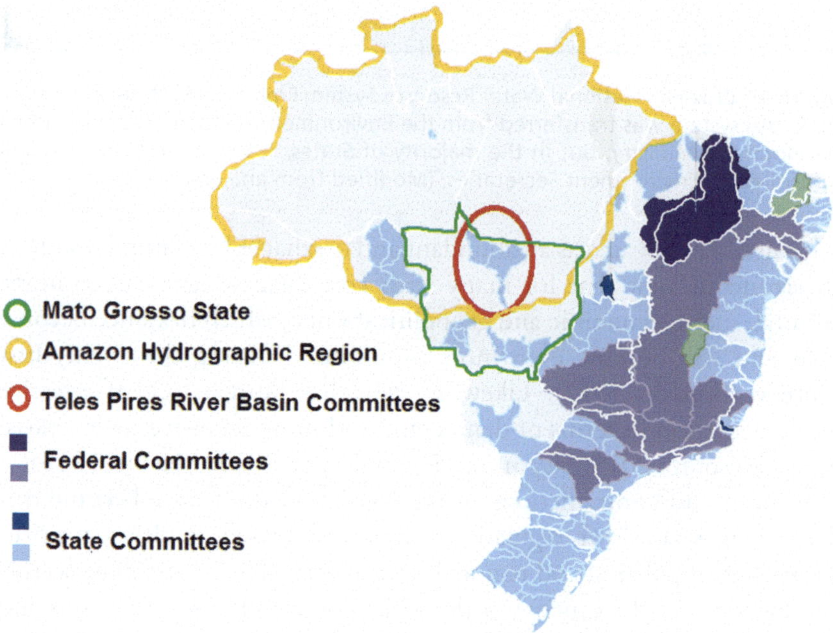

Fig. 16.2 River Basin Committee locations in Brazil, with emphasis on the Teles Pires River Basin committees in the Amazon Hydrographic Region. (Modified from ANA 2018) Note: Federal Committees are installed in rivers whose drainage area occupies two or more Brazilian States (Federal Rivers); State Committees are installed in rivers whose drainage area is located only in one Brazilian State (State Rivers)

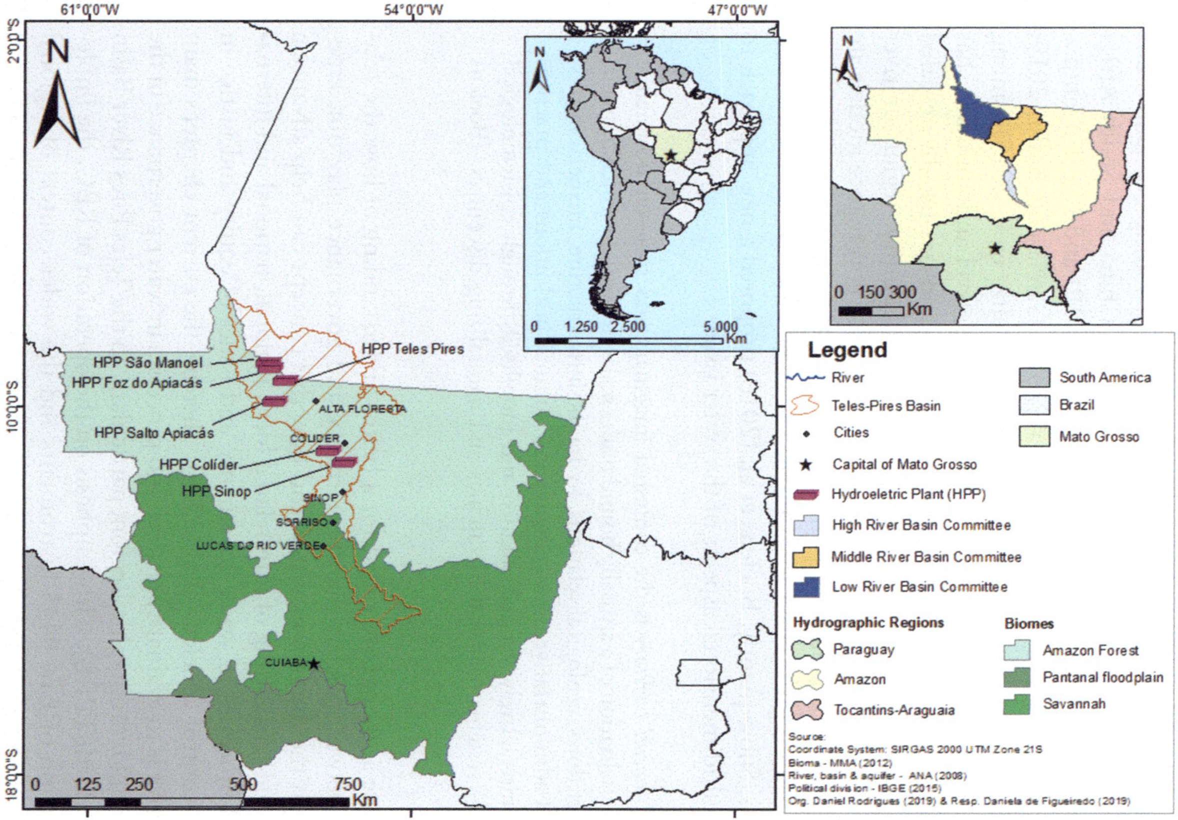

Fig. 16.3 Mato Grosso State, with its biomes and hydrographic regions, highlighting the Teles Pires River Basin and its main urban areas, hydropower plants, and river basin committees

inquiry and analysis relating to the water resources systems. Water governance has been evaluated by scientists in many countries, for example, in the United Kingdom (Rollason et al. 2018), including Scotland (Ioris 2008), the Netherlands, Germany, Australia, China and South Africa (Pahl-Wostl 2019), Peru (García and Bodin 2019) and Brazil (Campos and Fracalanza 2010; Empinotti et al. 2018; García and Bodin 2019; Martins 2015; Mesquita 2018, Siegmund-Schultze et al. 2015), among others. This research, varying in terms of scale, level and approach, has attempted to understand governance systems and has usually revealed certain deficiencies or contradictions in each country, region or river basin context. These studies have also shown that governance models are related to social, natural, political and historical factors in a territory, demonstrating the need for a broad vision of water issues in territorial studies. Subsequently, Boelens et al. (2016) proposed a new approach to the multidimensional aspects of the water issue:

> *Hydrosocial territory* is the contested imaginary and socio-environmental materialisation of a spatially bound multi-scalar network in which humans, water flows, ecological relations, hydraulic infrastructure, financial means, legal-administrative arrangements and cultural institutions and practices are interactively defined, aligned and mobilised through epistemological belief systems, political hierarchies and naturalizing discourses. (Boelens et al. 2016: 2)

These authors have called for a better understanding of how the relations between society, nature, territory and governance play out, especially in the domain of water, and how this concept can help advance a better understanding of interrelated local, regional, national and international processes of water governance and issues of equity and justice in water control (Boelens et al. 2016). Thereby, the current chapter's main objective is to evaluate practices and processes in water governance in the Amazon Hydrographic Region, particularly in the Teles Pires River Basin (henceforth TPRB), with emphasis on the operation of RBC, the implementation of Law 9433 instruments and the wider context of regional development and new economic frontiers (in particular, agribusiness and energy generation operations). The analysis is based on empirical research carried out in the TPRB and the text also offers a descriptive and analytical approach to understanding the *hydrosocial territory* on the Teles Pires River Basin and significant factors directing its construction.

Water Governance

The evaluation of water governance in a country, State or river basin may be carried out through various approaches, considering broad and multidimensional features and the many tendencies in literature and debate that have influenced concepts, approaches and governance practices (Ribeiro and Johnsson 2018). Most contributions to the governance debate and related concepts share a common concern for the relationship between government intervention and societal autonomy, but different strands of the literature highlight different facets of this continuum (Treib et al. 2007). As such, 'governance' is defined as a function, that is, the function of collectively solving societal problems, as opposed to government (local, national and, to a limited extent, international), which defines a structure (Finger et al. 2006).

Therefore, there is no ideal and universal governance model, but whichever approach is adopted, some common premises are present, such as the pursuit of fair governance though transparency, integrity and accountability and other ingredients, such as collaboration and cooperation (Ribeiro and Johnsson 2018). Treib et al. (2007) propose a classification of water governance according to three dimensions: (1) water polity (the institutional framework), (2) water politics (power relations between political actors) and (3) water policy (the mechanisms and instruments used to achieve certain outcomes). Likewise, Theodoro (2017) details these dimensions as follows:

> *Polity* deals with how political power is organised in society, how a political system is ordered and designed by the legal system and by institutional structure. *Politics* deals with the political process itself, with political actions by social and political stakeholders. *Policy* deals with political programs, with technical problems and the content of political decisions; it is how public policies are proposed and implemented.

These three dimensions are the point of departure to develop for governance modes (Treib et al. 2007). Governance modes refer to the various forms through which governance can be realised, and understanding what determines the performance of governance systems and how they can be designed—if this is possible at all—is a major challenge for science and policy (Pahl-Wostl 2019). In the process of water governance, it is essential to recognise how the flows of power and water are interconnected, transforming society and the environment together (Empinotti 2016).

Case Study's Methodological Approach

Regarding water as an eminently political resource (López et al. 2019), the three dimensions proposed by Treib et al. (2007) were adopted in the present study to inform the evaluation of the water management experience in the TPRB: *Polity* refers to the water resources system and the evaluation of its structures, particularly power relations and integration, and its institutional role in this basin; *Politics* refers to the committee's function and role in the TPRB and its political relations with others stakeholders; and *Policy* refers to the implementation of water resources instruments, as defined by Law 9433, including water use permits. These three political dimensions were analysed together, since the interrelationship between internal processes is non-linear, that is, there is a relationship between political structures (polity) that influences political negotiation (politics), which in turn influences the outcomes achieved (policy) (Theodoro 2017).

Departing from this conceptualisation of policy and politics, the specific water policies were evaluated taking into account the implementation of Law 9433 instruments and identification of the mechanisms adopted to achieve existing outcomes. This national legislation defines five management instruments and works towards the integration of these instruments, aiming towards more systemic management of water resources (Veiga and Magrini 2011). As defined by the law, these instruments are: the Water Resource Plans, at national, state and river basin levels; a system for classification of water bodies according to their preponderant use; water permits (licences); water usage charges; and the Water Resources Information System. All five instruments established by Law 9433 were evaluated in this study, except for water usage charges, which were not implemented in TPRB or in Mato Grosso State, and in Brazil, as a whole only around 20% were implemented on river basins with committees (ANA 2018), where River Basin Agencies (see Fig. 16.1) were created to manage this charges and funds.

The operation of the RBC and their relationship with other stakeholders and institutions in the system were assessed through the evaluation of management outcomes. Semi-structured interviews were carried out with 17 key members of the three committees installed in the TPRB (that amounts to about 23% of the total committee membership), comprising water users, civil society organisations (NGOs [non-governmental organisations]), public agencies at the federal, state and municipal levels and academics/researchers from public universities. The interviews covered the following ten themes: interviewees' knowledge about the basin, water users and conflicts; water conflicts addressed by the committees; committee relations with public water management institutions; committee relations with local governments (municipalities); failures in water resource management; information and democratic functioning of the committee; representativeness and effective member participation in committee meetings; and River Basin Water Resource Plan instruments. It is important to mention that the three committees installed in the TPRB represent around 80% of its drainage area (which in total is around 141,483 km^2), especially: (1) Upper Teles Pires, where the drainage area is made up of Teles Pires River right side tributaries, (2) Medium Teles Pires and (3) Low Teles Pires, where the drainage area is made up of Teles Pires River left side tributaries (Fig. 16.1). Together, the committees represent around 550,000 inhabitants in the TPRB, and they are located in Mato Grosso State, which is part of the Centre-West Region of Brazil and the Amazon Hydrographic Region.

In addition to the interviews with relevant members of the RBC, two meetings, respectively, of the Upper and Low committees, were observed and recorded, evaluating the issues presented, the flow of discussion, interactions between committee members and which members participated most. The minutes of all previous committee meetings (between

2012 and 2017) were also appraised through content analysis to contribute to understanding water use trends and emerging conflicts, the functioning of the three committees in this basin, the operation of governmental institutions and the attitude and demands of other stakeholders. Furthermore, documents from the water resources management regulators (Mato Grosso Environment Secretariat-MTES and the National Water Agency-NWA) were analysed using NVivo coding, including Water Resources Plan. A survey of water permits was produced using data from MTES (official website) and NWA (formally obtained) between 2007 and 2017. The surveys mentioned above, together with empirical knowledge obtained through the various research methods employed for this research, allowed us to understand not only the three political dimensions of water governance, but also the social and natural factors related to the water governance system in the TPRB, and further insights were provided by focusing on the historical perspective, as discussed below.

Regarding the *hydrosocial territory* scale, the meso-scale of the TPRB in Mato Grosso State was adopted, considering the hydrographic division adopted by government, where committees are installed and management instruments have been implemented, that is, where water governance has been evaluated. The interconnection between all the above elements of water policy and politics was evaluated using the lens of the *hydrosocial territory* concept (Boelens et al. 2016) and the guidelines adopted by López et al. (2019) to identify the *hydrosocial territory* configuration in this basin, seeking to answer four questions: (1) who produces the territory?—which individual or collective stakeholders produce the territory? (2) How is the territory produced?—which concrete actions produce the territory? (3) What type of territory does this produce?—what specifically characterises the organisation of the hydrosocial territory and its associated ownership, distribution and access to natural resources? (4) Why produce the territory?—what is the purpose of producing the territory? Before we can present and discuss the empirical results, it is required to contextualise the history and geography of the TPRB.

Brief Comments on the Geographical and Historical Perspective

Historically, the first official programme promoted by the Federal Government to colonise and occupy the Amazon Region, particularly Mato Grosso State, was called the 'March to the West' (*Marcha para o Oeste* in Portuguese), beginning in the 1930s, which gradually increased land disputes and discrepancies between land use value and market value (Silva and Sato 2012). This gradual occupation continued during the 1950s and 1960s together with increased road building in the Amazon Region. However, it was only in the 1970s that a great phase of spatial change began in Mato Grosso State, quite evident in the TPRB, with family migration from the South and Southeast regions of Brazil and decisive government support. This reinforced the idea that land was 'empty space' that must be occupied, but in fact this land was never empty, it was only referred to as such because local populations of indigenous people and small farmers, as well as local ecosystems, were not taken into account (Silva and Sato 2012).

This occupation meant that vast tracts of rainforest and savannah vegetation were converted into places dominated by farming, and that more than 20 new municipalities (counties) in the TPRB have been created in the last 40 years. Place changes were directly affected by centralised policymaking centred on the rapid and simultaneous transformation of urban and rural areas, particularly in private colonisation schemes (Ioris 2017). This phase, which Ioris (2017) calls *Displacement*, was not a purely social process, but involved both the transference of land to the hands of new farmers from other Brazilian regions—at the expense of the livelihoods of squatters and indigenous groups—and the conversion of savannah and forest into pastures and cropland. Between 1990 and 2000, this author identified a new phase in the TPRB, called *Replacement*, characterised by land concentration and the proletarianisation of farms, along with the beginning of farm technification, but still with precarious infrastructure and low levels of government support.

The next phase, up to the present, has been characterised by agribusiness insertion in regional and global markets, with strategic government support and the influence of national and international corporations, together with inequalities and the increasing socio-environmental impacts. This has been called *Misplacement* (Ioris 2017). The concept of agribusiness was originally introduced in the 1950s with the introduction of Fordist agriculture in the USA, and is increasingly associated with globalised industry and economic neoliberalisation, characterised by contradictions, failures and limitations at local, national and global scales, despite its results in terms of additional areas under cultivation, intensification of production, and complex market integration (Ioris 2016a). This occupation model and its results in terms of socio-environmental conflicts have also been observed in other Brazilian regions. Anchored in the assertive ideology of agribusiness, this has led to the insertion of the country, particularly Mato Grosso State, as a global protagonist in the geopolitical scenario (Silva and Sato 2012) in terms of agriculture commodities production and export. Overall, the occupation and transformation of TPRB occupation are relatively recent phenomena in terms of their politico-economic and social-ecological track records emerging and in terms of their intersection between powerful global markets and regional development (Ioris 2018). Of the ten products most exported by Brazil, seven are of agricultural origin, with soybeans in first place (Bombardi 2017), as around 10% of global soybean production comes from Mato Grosso State (Ioris 2016a), particularly the TPRB.

Assessment and Discussion

Water Policy Implementation and Conflicting Water Demands

The Water Resources Plan, one of the water management instruments defined by Law 9433, can be elaborated at three levels: national, state and river basin. As there is no River Basin Plan for the TPRB, the State and National Water Resources Plans are applicable in the river's drainage area. The State Plan was developed by the Mato Grosso Environment Secretary (SEMA 2009) following methodologies and directives from the Water

National Agency (WNA). This plan has 19 programmes, of which the majority have been partially executed. The most advanced are the water quality monitoring programme, which feeds into the Water Information System (another instrument), regulation and implementation of water use permits and promotion of the RBC creation programmes (Marchetto et al. 2018). All of these are based in the Mato Grosso State area, including TPRB.

Committee members' perceptions of the Water Resource Plan (the concept and its importance) were recorded for 5 of the 17 interviewees, all of them government technicians and teachers from universities. On the other hand, the River Basin was mentioned in many meetings, as reported in meeting minutes. In the current version of the National Plan, to be executed from 2016 to 2020, there are 16 proposed priorities (MMA 2017), of which 6 apply to the TPRB. However, guarantees to the plan's execution are fragile, as in previous versions and other plans in Brazil (at state and basin levels), which presented failures in execution demonstrated by low achievement rates related to a lack of planning and strategies, monitoring difficulties and a lack of financial resources. Regarding the classification of water according to its quality and uses, under the legal superficial water framework instrument, no water body was formally classified, that is, all of them belong to Class 2 (Brazil 2005). However, in practice, water quality standard violations have occurred, particularly related to agricultural pollution, for example, sediment, nutrient and agrochemical run-off (Carneiro et al. 2015; Prado 2018; SEMA 2017).

Water use permits are the management instrument most implemented in Brazil. They have been applied in 24 of the 27 federal provinces (states), including Mato Grosso, and 88% of the State Governmen are involved in their operationalisation (Ribeiro and Hora 2019). They were implemented in Mato Grosso State in 2007; thus, there is no register of water use prior to this, particularly in the TPRB. It is known empirically that the main water uses before 2007 were low scale irrigation, animal supply, and domestic water use (rural and urban supply), particularly after 1970 when the intense occupation of the TPRB began, but this is not officially recorded. Between 2007 and 2017, around 670 water permits were granted in the TPRB with 540 granted by the State Government for water use from State rivers (Teles Pires tributaries: 66%) and groundwater (34%). In the Teles Pires River around 130 water permits were granted by the Federal Government, because part of its drainage area lies in Para State, although most is in Mato Grosso State. As each water permit grants between one and three different water uses to the same user, total water use is greater than the total of water permits granted (Table 16.1).

The main water use in the TPRB is irrigation (36.0%), granted to areas greater than 30 hectares, followed by human supply (24.5%), which refers mainly to the rural supply and is different from urban public supply (4.0%, together with waste water dilution). This is carried out by private (50%) or public (50%) sanitation companies. Animal supply (11.7%) relates mainly to aviculture and pig farming for the slaughterhouses of transnational agribusiness companies located in the basin, such as BRFoods in Lucas do Rio Verde, and JBS in Colíder municipalities, both food industries. Other unspecified water use permits (9.0%) are granted for hydropower plants, hydraulic construction, pesticide spraying and dilution and mining uses (Table 16.1). Despite low numbers of

Table 16.1 Main water uses permitted in the Teles Pires River Basin for each type of waterbody, between 2007 and 2017, published as *permit acts* on the official State Government Site for the Teles Pires tributaries and groundwater or ceded by Federal Government on the Teles Pires River

			Water body		
Water uses	Total number of water uses	Water uses (%)	Affluent	Groundwater	Teles Pires River
Irrigation	237	36.0	214	7	16
Agroindustrial supply and waste water	55	8.4	15	31	9
Pisciculture supply and waste water dilution	31	4.7	24	0	7
Animal supply	77	11.7	1	72	4
Energy generation	11	1.7	9	0	2
Public supply and waste water dilution	26	4.0	6	13	1
Human consumption	164	24.5	2	147	15
Others	59	9.0			

water use permits for hydropower generation (1.7%), this requires a high water volume to feed reservoirs (although this volume is not clear from the water use permits). There are now five large hydropower plants installed in the TPRB (four located in the Teles Pires River), three of them controlled by transnational companies and two by Brazilian public companies, generating around 3650 MW and flooding an area of about 800 km^2 area for dam construction. In addition, there are seven small hydropower plants installed on smaller tributaries (<30 MW; see Fig. 16.3).

Water Use and the Construction of a Hydrosocial Territory

The two main water users in the TPRB in terms of number of permits granted, water withdrawal and/or water volume are the agribusiness and energy sectors. Agribusiness water uses include irrigation, aquaculture, agroindustry and animal supply. Taking into account the interviews with committee members, eight interviewees, all of them from High Teles Pires, mentioned that there are water conflicts related to agribusiness, particularly related to irrigation, which account for 60% of water permits granted for this portion of the river basin. They mentioned that in small rivers, some requests for water use permits are being denied because of the quantity of water in use for irrigation or for feeding hydropower reservoirs. All irrigation water permits are granted during the dry season (May to October), when river volumes and hydropower reservoirs are naturally low due to rain scarcity. In the high portion of the Teles Pires River, for example, water volume decreases by around 56% in dry season in relation to the rainy season (Alcantara 2009). Conflicts related to agribusiness and to the rural and urban models of river basin occupation were reported in the minutes of the first meeting of this committee:

> The committee [high portion] was created because of the increasing demand for water resources for irrigation projects, the decreasing water quality in some rivers, mainly near urban areas due to industrial waste water and domestic dilution, in addition to the agrochemicals which are

> broadly used on a large scale by agribusiness in this region. The picture in main and tributary rivers in the basin is of conflict about the quality and supply of water due to use for irrigation projects, restricted industrial projects, and the launch of waste domestic water into some rivers without support for its dilution, for example. On the other hand, as this committee is located in the main agribusiness development area of the country, its creation occurs at an extremely favourable time, as is evident from the support expressed by many public and civil society entities including those representing water users. In the last few decades, this region has grown at a higher rate than the national average which has contributed to problems such as deforestation in legal reserve areas, and in headwater and riparian forest regions that are permanent conservation areas [according to the law]. (SEMA 2017: 108)

However, in the minutes of this meeting, there is no report on water use for energy generation, despite the many hydropower plants installed, under construction or in the planning stages in the TPRB (including in the high river basin portion) when this first meeting occurred, and even the historical negative legacy of hydroelectric dams in the Amazon Region (Gallardo et al. 2017) and its socio-environmental sensitivities, which mean it remains vulnerable to large hydropower projects (Moretto et al. 2012). The emphasis on hydroelectric power generation (renewable), which makes up 63.7% of Brazil's energy matrix (ANA 2018), is nowadays justified using the rationale of ecological modernisation, that is, to promote opportunities for production practices and consumption adjustments, with some residual environmental sensitivity, but without challenging political-economic and governance bases or the primary responsibilities for environmental and hydrological degradation resulting from the changes in natural river flow and the flooding of large areas of forest for reservoir construction.

Governance foundations from hydropower plants start with the environmental permit process, which, in the case of TPRB, has presented many failures related to its run over process, which made in-depth social participation and discussion about socio-environmental impacts difficult (Fórum Teles Pires 2017), as well as to the delimitation of its area of influence, to the integrated and systemic approach (Rostirolla 2018), and to assessment of the synergic effects of hydropower on the river basin (Gallardo et al. 2017). In addition, hydropower environment studies in the TPRB were restricted to the local effects of dams, with a limited systemic approach to the whole drainage area, and were driven in a fragmented way by environment agencies at national and state levels. These aspects mean that the decision-making process for hydroelectric development has been perverted in a variety of ways, with the result that the environmental and human impacts of dams have had very little effect on the actual decisions to implement the projects, as a result of the influence of construction firms and foreign financiers and equipment suppliers who have contributed to giving minimal consideration to the environmental and social impacts of the project (Fearnside 2001).

In the low TPRB area, many conflicts between the energy sector and indigenous communities or small farmers have been reported, in addition to the environment impacts, particularly to ichthyofauna, which are vital for many fishing families, including indigenous people whose main source of protein is fish (Fórum Teles Pires 2017). In the interviews, only one committee member recognised conflicts occurring between such people and the energy sector; thus, members of this sector did not recognise any conflict. In the meeting minutes, no conflicts related to hydropower were registered. Just one mention was made by an energy sector technician when he gave a lecture about the implementation of the environmental programme in one TPRB' hydropower plant. The environmental licencing process and the contradictions of social participation, as observed in the TPRB, limit the actions to be taken and the responsibility taken by the hydropower sector, as was also observed in other hydropower plants in the Amazon Hydrographic Region (Sousa Junior and Reid 2010). Public discussion is needed of the laws that currently guarantee completion of any dam or other large infrastructure project irrespective of environmental and social impacts and violations of licensing requirements (Fearnside 2015). Future research on Amazonian hydropower might focus on deepening understanding of how to define direct and indirect impacts and who and what areas are impacted by different hydropower projects. This should be addressed by research frameworks and licensing instruments, which could alleviate the mismatch between scientific findings, policy instruments and the reality on the ground (Athayde et al. 2019).

Thus, access to water resources depends on institutional arrangements that ultimately reflect the social structures present in each region and sociocultural context (Empinotti 2016). In the TPRB, agribusiness and energy sectors, mediated by government, appropriate and make use of water through infrastructure technology that continuously changes the territory and landscape and intervenes in the water and biophysical cycles, and in social relations. Thus, the agribusiness and energy sectors, with government support, are fundamental to the emergent construction of *hydrosocial territory* in the TPRB (Boelens et al. 2016), including in urban areas, where building and constant change are mainly related to agribusiness and where water use is appropriated by sanitation companies. In other words, river basin management, water flows, water use systems and hydrological cycles have recently come to be mediated by governance structures and human interventions that entwine biophysical, technological, social and political aspects (Boelens et al. 2016) in the TPRB. Indeed, water technology, society and nature are intrinsically interrelated and mutually determining elements that together organise as specific socio-natural networks (Boelens et al. 2016). Regarding TPRB socio-natural context, from the present research, it is proposed that the *Misplacement* phase should include emergent *hydrosocial territory* construction, where agribusiness, the energy sector and government are the main construction agents, answering the first question asked by López et al. (2019): who produces the *hydrosocial territory*? Following these authors' guidelines, it is also possible to answer other questions: how is the territory produced? It is produced by current and past concrete action by these agents and from legal, institutional and financial government frameworks and governance.

What type of *hydrosocial territory* does this produce? The territory produced is characterised by water use concentration and centralised water governance. The current water governance model is part of this territory construction, characterised by fragile water management and planning (as mentioned above) and by centralised decision-making (as detailed below). Guiding ideas from Amazon planning and development, in the TPRB in particular, reduced nature to utilitarian and financial dimensions, where food and river come to be measured by kilowatts and quantities of commodities (Batista and Miranda 2019). Water as a common and public good (Law 9433) turns into a commodity, understood through a 'virtual water' perspective, that is, hydro-demand needs to produce a market unit (Ioris 2010). Thus, the proposed reason for creating the *hydrosocial territory* (Why produce the territory?) is to use water to insert and strengthen the TPRB, Mato Grosso State and Brazil in regional and global markets in agriculture and energy commodities, and therefore guarantee and concentrate economic and political power.

Water Polity and Water Politics: River Basin Committees

Historically, River Basin Committees in (RBC) Brazil have been set up for various reasons, including financial reasons, water conflicts, negotiations by those involved, legal determination or critical water events (Mesquita 2018). In the case of the three committees in the TPRB, creation started in 2012 and was promoted and articulated by the Environment Secretary of Mato Grosso State. According to Law 9433, committees must have a certain proportion of members from the public sector (State Secretaries and Counties Secretaries), from water users (those holding permits for water uses) and from civil society organisations or NGOs (with proven track records on water resources and the environment). During the observed meetings and based on analysis of meeting proceedings, NGO participation was found to be low. In addition, there are few counties (municipalities—local government) who continually participate, but usually those who participate most are counties with larger populations and those where meetings usually take place. These are also the same counties that carry out water management and headwater recovery actions. Municipalities are charged with performing a broad range of water management activities and have numerous administrative responsibilities; however, they are not always internally structured to manage complex social-ecological issues spanning sectors and administrative boundaries (García et al. 2019). Regarding each county administrative structure, it was mentioned in a meeting that county committee members are changed periodically, meaning that participation is discontinuous, which leads to a lack of motivation among county members to remain on committees.

According to the Law 9433, the Low Teles Pires committee must have a delegate from the indigenous community, since there is an indigenous territory in its area. However, as mentioned in a meeting, this delegate has not attended the meetings. Despite a recent move towards greater decentralisation in Brazil, indigenous communities are conspicuously absent from formal systems of water quality governance, and indigenous initiatives tend to be constrained by government actions and inaction (Berry et al. 2018). This gap can stem from a many factors, including high logistical costs, interaction failures with the RBC and low comprehension of the technical terms discussed in the meeting (Mesquita 2018). Appling the *Water Polity* and *Water Politics* dimensions, it was possible to identify the central State Government's role in TPRB water governance, as well as the roles of the Water Resource State Council and the committee's advisor. This centralised approach to polity and politics follows the conventional *top-down* model, an established practice which is prevalent in many countries (Ioris 2008; Pahl-Wostl 2019), as well in other Brazilian river basins (Abers 2010; Mesquita 2018). It was recognised in the TPRB by a majority of RBC members who responded, as reported by one of them:

> We have another isolated nucleus which is Cuiabá [Mato Grosso State capital, where the State Environment Secretary is based] which controls water management today. Any water use demands don't come here, don't come to the committee, these demands go to the State Environment Secretary [...]. So, the committee's role is partially suppressed, and partially not stimulated.

On the other hand, a government delegate considered the following situation in relation to the TPRB committee's action:

> We realise that they [RBC members] have not yet integrated the committee into their activities. These committees are still young and their members have not seen their power.

In this same sense, one of the RBC member respondents regarded government centralisation as natural. He argued that RBC are not prepared, and that they need to be strengthened for the future, when there will be water conflicts. However, he considered the RBC's existence to be unnecessary:

> It is not necessary for committees to exist (...). When a committee is created, is because the responsible entities, the institutions, are not competent.

A delegate from a small water user mentioned:

> As we are still not conscious about the committee's role, what it represents, about its power and representativeness, this process for us is still about knowledge and exchange.

The advisory role of the three TPRB's committees was also evident in the meeting minutes and during the meetings, where internal bureaucratic issue (statute, elections), and general technical discussions were predominant. This role has been observed in many Brazilian committees which have an advisory tendency; they meet infrequently and do relatively little behind the political-administrative decision-making scenes, so their boundaries are sparse and discontinuous, even in regions with older and more active committees (Theodoro 2017). Water permit criteria were discussed in only one TPRB committee meeting, on the request of the Public Ministry, no-system member institution, but no decision was taken. The majority of the respondents considered that technical knowledge is an important factor for participation in committees. This knowledge is concentrated mainly in the government and universities delegates, as well as some county and water user delegates. Access to technical knowledge is important to the inclusive RBC member issue; this means that those who have access to such knowledge and manage to frame their interests in technical discursive terms can have a say (García and Bodin 2019). Analysing this issue, Martins (2015) takes into account that, in general, the relationship between experts and non-experts in the Brazilian committees fails to express different points of view, which is healthy for governance, resulting in unequal political practices. Of the 17 committee member respondents, 11 said that they believe this forum is democratic, despite recognising that knowledge is an important factor in increasing participation, and also their perception about the centralisation of government roles.

One challenge related to asymmetrical water governance on the TPRB involves ensuring the equitable and democratic inclusion of all stakeholders (Table 16.2), because the governing capacity of committees is increasing all the time, developing and refining the negotiation and the administrative means (Mesquita 2018) of stakeholders at the many governance levels. Stakeholders and levels are important to the water resources management participatory system, which requires an institutional and integrated system, where all participants need to have the same information and capacity for influence on the decision-making process (Theodoro

Table 16.2 Identification and roles of main water governance stakeholders in the Teles Pires River Basin

Institution	Institution level	Performance and function
State Managing Institution (Environmental Secretary)	State	Centralised decision-making; water permits approve on States rivers (Teles Pires tributaries, State Rivers, see Fig. 16.2); usually conducts committee meetings
Water Resources State Council	State	Define and approve water use permit criteria
River Basin Counties	County	Asymmetric participation in the committees (particularly small counties).
Federal Managing Institution (National Water Agency)	Federal	Low participation and fragile committees and State integration. Water use permit approve on Teles Pires River (Federal River, see Fig. 16.2)
Tree River Basin Committees (High, Medium and Low)	River basin	Advisory role. Low integration between the three committees and State and Federal Managing Institutions.
Water users	State and Federal	Active committee participation
Non-Governmental Organization	State and County	Low committee participation

2017). In the TPRB, this integration has many barriers, particularly in relation to the connection between the three committees, as mentioned by two respondents, and between State and National levels. This integration, as in many other countries where contention between these levels is a concern, Brazil and Mato Grosso State in particular, has been facing difficulties in effective coordination of the integrated approach involving all stakeholders (Veiga and Magrini 2013), which was recognised by four interviewees, including a government technician.

Despite failures and contradictions in water governance, TPRB committee members were found to be motivated by the following factors: the possibility of changing support structure policies and incentives (e.g. water charges) in future; the current need to create opportunities for negotiation between multiple stakeholders around water problems and issues; Brazilian water technicians' belief in the river basin water management model; and the growing legitimacy of social and political participative management (Abers and Jorge 2005). In relation to water use charges and river basin integration as incentive factors for committee continuity, a TPRB committee member mentioned that:

> Our committee is preparing for when water charges will be implemented, as then we will have financial resources for our actions. Then, many people will want to participate and will understand the committee's importance [...]. When we have more integration, I want our committee to be a reference, we are working for this.

On the other hand, the majority of committee members demonstrated implicitly or explicitly their perceptions about water abundance on the TPRB, a factor that does not demand committee decisions or actions. The perception of water abundance is described as a source of conflict among members of the Mato Grosso water sector, as it allegedly contributes to a state of carelessness and lack of awareness of the importance of good water governance and investment in water infrastructure, given the simplistic view that 'water cannot be a problem' in an area with abundant natural water resources (Schulz and Ioris 2017). Therefore, this perception does not take into account the seasonal rain regime during the year, and eventual water crises during the dry season, as occurred in 2018 in

Planalto da Serra county (high river basin portion; personal information), and in 2010 in Alta Floresta county (in the low river basin portion; Globo 2010). Both crises affected public water supply and were related to environmental degradation (headwater and riverine deforestation) as well as the dry season. Considering climate change estimates for the region (Mohor et al. 2015), and intensive deforestation in the TPRB, which has decreased infiltration and run-off rates in the basin (Alcantara 2009), as well as accelerated increases in water use, crises are likely to be recurrent, reinforcing the importance of preventive water management actions (Veiga and Magrini 2013).

Conclusion

The current water governance model in the TPRB, as shown by results from the three political dimensions analysed (Treib et al. 2007), is centralised at State Government (*Polity*) level. Relations with river basin level (the three committees) are conducted in a top-down way, while relations with the National level are horizontal, both with fragile practices and process (*Politics*). This centralisation comes partly from traditional government structures, resulting in a situation where committees have only advisory role and where there is low articulation between different committees. Committees are under-representative in terms of civil society-NGOs (Politics). Water management instruments are partially implemented, with emphasis on water use permits and on the information system (*Policy*). This governance model, despite its recognised advances, gaps and contradictions, has been presented as rational, technical and objective, together with the 'sustainability' discourse adopted both by public institutions and by major water users (the agribusiness and energy sectors). Despite the rhetoric of environmental sustainability, official initiatives continue to subject socio-natural water systems to economic exploitation (Ioris 2016b) and to water commoditisation, as a result of the agriculture and energy commoditisation characteristic of the neoliberal model. Despite promoting a genuine change in public policies, as in Brazilian Water Resources Policy since 1997 (under Law 9433), these laws have largely preserved the hegemonic interests of landowners, companies and real estate investors to the detriment of ecological recovery and the majority of the population (Ioris 2016b).

Preserving this hegemony at the expense of any other socio-economic alternative remains directly based on the original mechanisms of territorial conquest and political control, which have been in practice since the middle of the last century (Ioris 2018). There are two main ways that the power of this hegemonic model continues to expand: one is at the spatial level, as verified by land and water use and appropriation in the TPRB, and the other is at the subjective level, that is, in the context of individual reasoning (Ivars 2019). The aim is to limit collective resistance and close off the possibility of considering other notions of development and to install a comprehensive-historical threshold in regard to alternatives (Svampa 2015). Thereby, despite some advances in water governance in the TPRB, these are continually frustrated by neoliberal socio-economic priorities, which are supposedly for the common good, and which strengthen among committee members in particular, and also society in general, the idea that water ownership and centralised management are inexorable, that is, that there is no other possible water governance model. This is indicative of the flexible character of Law 9433, which on the one hand has sustainability, multiple water use and social participation as principles, but on the other hand, implementation can be adapted to the unfair TPRB socio-environmental model, which emphasises the principle of water's economic value. In addition, the perception of water abundance (Schulz et al. 2017) is an obstacle to understanding the political governance dimensions and the importance of long-term planning, including the issue of climate change and its relation to deforestation. This perception is one more factor driving water use concentration and the partial implementation of governance instruments.

Those aspects mentioned above, together with the water governance model in the TPRB, are the key factors that shape the construction of the *hydrosocial territory* and define its purpose, type and identity, as illustrated and summarised by Fig. 16.4. It is important to highlight that these factors are the outcomes of the Mato Grosso *hydrosocial territory*, which in turn is influenced by the Brazilian territory. Even though some factors are not the same in each Brazilian context, the traditionally centralised government structure, flexibility of Law 9433 and water and/or land

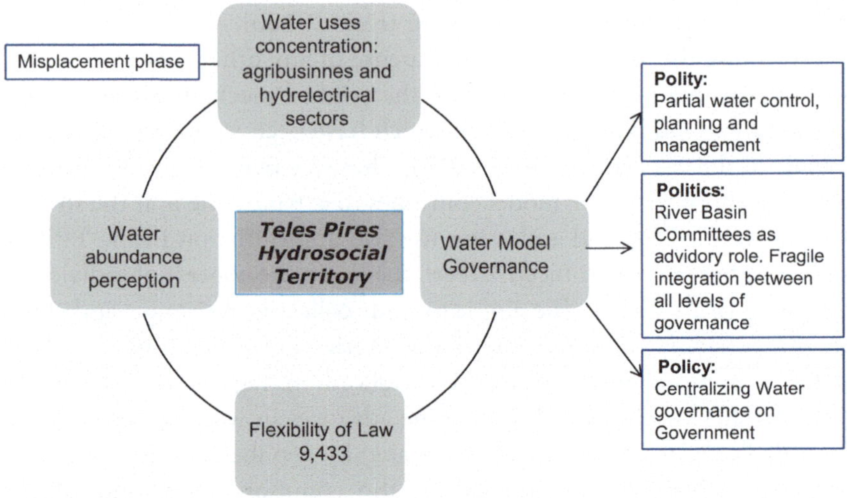

Fig. 16.4 Key factors that create and shape the emergent *Hydrosocial Territory* on the Teles Pires River Basin (Boelens et al. 2016), including the three political dimensions of the water governance model (Polity, Politics and Policy; Treib et al. 2007) and water concentration, as a reflexion of historical land concentration, currently in the *Misplacement* phase. (Ioris 2017)

concentration by stronger economic sectors mostly apply in most of Mato Grosso and Brazil as a whole. *Hydrosocial territories* at a specific scale exist and are deeply enmeshed in other territories that exist and operate at broader, overlapping, counterpoised and/or hierarchically embedded administrative, cultural, jurisdictional, hydrological and organisational scales (Boelens et al. 2016).

Finally, it is important to mention that current environmental policy tendencies in Brazil will strengthen government centralisation in water governance, weaken social movements and social participation forums and increase private water appropriation and concentration by stronger economic sectors. This is particularly evident in the transfer of all water management systems (see Fig. 16.1) from the Environment Ministry to the Regional Development Ministry (Decrees 9666 and 9672 of 02 January 2019), which is also responsible for the irrigation and sanitation sectors, so that the National Water Agency is now responsible for water permits for this two water users. As already cautioned by Figueiredo (2019), this change tends to reinforce water's economic value, reduce ecosystem approach to river basins and compromise previous advances and the future improvements needed to guarantee fair governance and water access for multiple users.

Acknowledgements The authors thank the members of the Teles Pires River Basin Committees for the interviews and CAPES-PNPD for the postdoctoral fellowship of the first author, who is a permanent professor and associate researcher of the Master´s Degree in Water Resources at Federal University of Mato Grosso.

References

Abers, R. (2010). Pensando Politicamente a Gestão da Água. In R. Abers (Ed.), *Água e Política: Atores, Instituições e Poder nos Organismos Colegiados de Bacia Hidrográfica no Brasil* (pp. 13–36). São Paulo: Annablume.

Abers, R., & Jorge, K. D. (2005). Descentralização da Gestão da Água: Por que os Comitês de Bacia Estão Sendo Criados? *Ambiente & Sociedade, 8*(2).

Alcantara, L. H. (2009). *Análise Espaço-temporal do Escoamento na Bacia do Alto e médio Rio Teles Pires*. MSc Dissertation, Federal University of Mato Grosso.

ANA. (2018). *Conjuntura Recursos Hídricos Brasil 2018*. Brasília: Agência Nacional de Águas.

Athayde, S., Mathews, M., Bohlman, S., Brasil, W., Doria, C. R. C., Dutka-Gianelli, J., Fearnside, P. M., Loiselle, B., Marques, E. E., Melis, T. S., Milikan, B., Moretto, E. M., Oliver Simith, A., Rossete, A., Vacca, R., & Kaplan, D. (2019). Mapping Research on Hydropower and Sustainability in the Brazilian Amazon: Advances, Gaps in Knowledge and Future Directions. *Current Opinion in Environmental Sustainability, 37*, 50–69.

Batista, I. M. S., & Miranda, L. M. (2019). The 'Hydrobusinesses' in the Rivers of the Amazon. *Revista Brasileira de História, 39*(81). https://doi.org/10.1590/1806-93472019v39n81-06.

Berry, K. A., Jackson, S., Saito, L., & Forline, L. (2018). Reconceptualising Water Quality Governance to Incorporate Knowledge and Values: Case Studies from Australian and Brazilian Indigenous Communities. *Water Alternatives, 11*(1), 40–60.

Boelens, R., Hoogesteger, J., Swyngedouw, E., Vos, J., & Wester, P. (2016). Hydrosocial Territories: A Political Ecology Perspective. *Water International, 41*(1), 1–14.

Bombardi, L. M. (2017). *Geografia do Uso de Agrotóxicos no Brasil e Conexões com a União Europeia*. São Paulo: FFLCH-USP.

Bordalo, C. A. L. (2017). Novos Olhares da Crise Hídrica Mundial. In E. S. Chavez, C. A. Di Mauro, & E. C. Moretti (Eds.), *Água, Recurso Hídricos: Bem Social Transformado em Mercadoria* (pp. 16–33). ANAP: Tupã.

Brazil. (2005). *Resolução Conama No. 357 de 17 de Março de 2005*. Brasília: MMA.

Campos, V. N. O., & Fracalanza, A. P. (2010). Governança das Águas no Brasil: Conflitos pela Apropriação da Água e a Busca da Integração como Consenso. *Ambiente & Sociedade, 13*(2), 365–382.

Carneiro, F. F., Rigotto, R. M., Augusto, L. G. S., Friedrich, K., & Búrigo, A. C. (2015). *Dossiê Abrasco um Alerta sobre os Impactos dos Agrotóxicos na Saúde*. Rio de Janeiro and São Paulo: EPSJV and Expressão Popular.

Empinotti, V. L. (2016). Politics of Scale and Water Governance in the Upper Xingu River Basin, Mato Grosso. In A. A. R. Ioris (Ed.), *Agriculture, Environment and Development: International Perspectives on Water, Land and Politics* (pp. 197–220). Basingstoke: Palgrave Macmillan.

Empinotti, V. L., Gontijo, W. C., & Oliveira, V. L. (2018). Federalism, Water, and (de)Centralisation in Brazil: The Case of the São Francisco River Water Diversion. *Regional Environmental Change, 18*(6), 1655–1666.

Fearnside, P. M. (2001). Environmental Impacts of Brazil's Tucuruí Dam: Unlearned Lessons for Hydroelectric Development in Amazonia. *Environmental Management, 27*(3), 377–396.

Fearnside, P. M. (2015). Amazon Dams and Waterways: Brazil's Tapajós Basin Plans. *Ambio, 44*, 426–439.

Figueiredo, D. M. (2019). Uncertainties in Water Resource Management in Brazil after Recent Institutional Restructuring. Paper presented at the *Third Symposium Sustainable Development in Latin America*, London, 4–7 May 2019.

Finger, M., Tamiotti, L., & Allouche, J. (2006). Introduction: Conceptual Elements. In M. Finger, L. Tamiotti, & J. Allouche (Eds.), *The Multi-governance of Water: Four Case Studies* (pp. 1–41). New York: State University of New York Press.

Fórum Teles Pires. (2017). *Barragens e Povos Indígenas no Rio Teles Pires*. Brasília/Cuiabá/Alta Floresta: Fórum Teles Pires.

Freitas, M., & Freitas, M. C. S. (2018). Regional Development for Sustainability in Amazonia: Controversies and Challenges. *Geography, Environment, Sustainability, 4*(11).

Gallardo, A. L. C. F., Silva, J. C., Gaudereto, G. L., & Sozinho, D. W. F. A. (2017). Avaliação de Impactos Cumulativos no Planejamento Ambiental de Hidrelétricas na Bacia do rio Teles Pires (Região Amazônica). *Desenvolvimento e Meio Ambiente, 43*, 22–47.

García, M. M., & Bodin, Ö. (2019). Participatory Water Basin Councils in Peru and Brazil: Expert Discourses as Means and Barriers to Inclusion. *Global Environmental Change, 55*, 39–148.

García, M. M., Hileman, J., Bodi, Ö., Nilsson, A., & Jacobi, P. R. (2019). The Unique Role of Municipalities in Integrated Watershed Governance Arrangements: A New Research Frontier. *Ecology and Society, 24*(1).

Globo. (2010). *Córregos Secam e Cidade da Amazônia Matogrossense Fica sem Água*. https://oglobo.globo.com/brasil/corregos-secam-cidadde-da-amazonia-mato-grossense-fica-sem-agua-2948257#ixzz5SDZQFzWy. Accessed 30 July 2019.

Ioris, A. A. R. (2008). Water Institutional Reforms in Scotland: Contested Objectives and Hidden Disputes. *Water Alternatives, 1*, 253–270.

Ioris, A. A. R. (2010). Da Foz às Nascentes: Análise Histórica e Apropriação Econômica dos Recursos Hídricos no Brasil. In *Capitalismo Globalizado e Recursos Territoriais* (pp. 211–255). Rio de Janeiro: Lamparina.

Ioris, A. A. R. (2016a). Introduction: Underscoring Agribusiness Failures, Environmental Controversies, and Growing Food Uncertainties. In A. A. R. Ioris (Ed.), *Agriculture, Environment and Development. International Perspectives on Water, Land and Politics* (pp. 1–32). Basingstoke: Palgrave Macmillan.

Ioris, A. A. R. (2016b). Controversial Frontiers of Agricultural Development and Environmental Change. In A. A. R. Ioris (Ed.), *Agriculture, Environment and Development. International Perspectives on Water, Land and Politics* (pp. 221–250). Basingstoke: Palgrave Macmillan.

Ioris, A. A. R. (2017). Places of Agribusiness: Displacement, Replacement, and Misplacement in Mato Grosso, Brazil. *Geographical Review, 107*(3), 452–475.

Ioris, A. A. R. (2018). The Politics of Agribusiness and the Business of Sustainability. *Sustainability, 10*(5), 1648. https://doi.org/10.3390/su10051648.

Ivars, J. D. (2019). *Agua, Poder y Racionalizatión en la Provincia de Mendoza, Argentina*. Campina Grande: Eduepb.

López, R., Hoogendam, P., Vos, J., & Boelens, R. (2019). Transforming Hydrosocial Territories and Changing Languages of Water Rights Legitimation: Irrigation Development in Bolivia's Pucara Watershed. *Geoforum, 102*, 202–213.

Marchetto, M., Noquelli, L. H. M., Andrade, L. A. Z., Silva, M. A., Soares, S. R. A., & Evangelista, R. M. (2018). Avanços e Desafios da Política de Recursos Hídricos Quanto aos Instrumentos de Gestão. In D. M. Figueiredo, E. F. G. C. Dores, & Z. M. Lima (Eds.), *Bacia do Rio Cuiabá: Uma Abordagem Socioambiental* (pp. 498–536). EdUFMT: Cuiabá.

Martins, R. C. (2015). Fronteiras entre Desigualdade e Diferenças na Governança das Águas. *Ambiente & Sociedade, 18*(1), 221–238.

Mesquita, L. F. G. (2018). Os Comitês de Bacias Hidrográficas e o Gerenciamento Integrado na Política Nacional de Recursos Hídricos. *Desenvolvimento e Meio Ambiente, 45*, 56–80.

MMA (Ministério do Meio Ambiente). (2017). *CNRH Resolução No. 181 de 7 de dezembro de 2017. Aprova as Prioridades, Ações e Metas do Plano Nacional de Recursos Hídricos para 2016–2020*. Brasília: MMA.

Mohor, G. S., Rodriguez, D. A., Tomasella, J., & Siqueira Júnior, J. L. (2015). Exploratory Analyses for the Assessment of Climate Change Impacts on the Energy Production in an Amazon Run-of-River Hydropower Plant. *Journal of Hydrology, 4*, 41–59.

Moretto, E. M., Gomes, K. S., Roquetti, D. R., & Jordão, C. O. (2012). Histórico, Tendências e Perspectivas no Planejamento Espacial de Usinas Hidrelétricas Brasileiras: A Antiga e Atual Fronteira Amazônica. *Ambiente & Sociedade, 15*(3), 141–164.

Pahl-Wostl, C. (2019). The Role of Governance Modes and Meta-governance in the Transformation towards Sustainable Water Governance. *Environmental Science and Policy, 91*, 6–16.

Prado, J. R. S. (2018). *Qualidade da Água da Bacia do Rio Teles Pires na Área de Influência Direta da Futura UHE Sinop, MT*. MSc Dissertation, Federal University of Mato Grosso.

Ribeiro, P. E. A. M., & Hora, M. A. G. M. (2019). 20 anos da Lei n° 9.433/97: Percepções dos Comitês de Bacia Hidrográfica e dos Órgãos Gestores Acerca da Implementação da Política Nacional de Recursos Hídricos. *REGA, 16*(1). https://doi.org/10.21168/rega.v16e1.

Ribeiro, N. B., & Johnsson, R. M. F. (2018). Discussion on Water Governance: Patterns and Common Paths. *Ambiente & Sociedade, 21*. https://doi.org/10.1590/1809-4422asoc0125r2vu18l1ao.

Rollason, E., Bracken, L. J., Hardy, R. J., & Large, A. R. G. (2018). Evaluating the Success of Public Participation in Integrated Catchment Management. *Journal of Environmental Management, 228*, 267–278.

Rostirolla, S. L. (2018). *Análise Comparativa dos Estudos de Impacto Ambiental de Quatro Hidrelétricas no Rio Teles Pires*. MSc Dissertation, Federal University of Mato Grosso.

Schulz, C., & Ioris, A. A. R. (2017). The Paradox of Water Abundance in Mato Grosso, Brazil. *Sustainability, 9*(10), 1796. https://doi.org/10.3390/su9101796.

Schulz, C., Martin-Ortega, J., Glenk, K., & Ioris, A. A. R. (2017). The Value Base of Water Governance: A Multi-Disciplinary Perspective. *Ecological Economics, 131*, 241–249.

SEMA (Secretaria de Meio Ambiente). (2009). *Plano Estadual de Recursos Hídricos: Diagnóstico*. Cuiabá: SEMA.

SEMA (Secretaria de Meio Ambiente). (2017). *Relatório Técnico No. 001*. Cuiabá: COH/SURH/SEMA.

SEMA (Secretaria de Meio Ambiente). (2018). *Relatório de Monitoramento da Qualidade da Água Superficial de Mato Grosso*. Cuiabá: SEMA.

Siegmund-Schultze, M., Rodorff, V., Köppel, J., & Sobral, M. C. (2015). Paternalism or Participatory Governance? Efforts and Obstacles in Implementing the Brazilian Water Policy in a Large Watershed. *Land Use Policy, 48*, 120–130.

Silva, M. J., & Sato, M. T. (2012). Territórios em Tensão: O Mapeamento dos Conflitos Socioambientais do Estado de Mato Grosso – Brasil. *Ambiente & Sociedade, 15*(1), 1–22.

Sousa Junior, W. C., & Reid, J. (2010). Uncertainties in Amazon Hydropower Development: Risk Scenarios and Environmental Issues around the Belo Monte Dam. *Water Alternatives, 3*(2), 249–268.

Svampa, M. (2015). Commodities Consensus: Neoextractivism and Enclosure of the Commons in Latin America. *The South Atlantic Quarterly, 114*(1), 65–82.

Theodoro, H. D. (2017). *Análise da Gestão de Recursos Hídricos: Um Estudo de Caso do Comitê da Bacia Hidrográfica do Rio das Velhas*. PhD Thesis, Federal University of Minas Gerais.

Treib, O., Bähr, H., & Falkner, G. (2007). Modes of Governance: Towards a Conceptual Clarification. *Journal of European Public Policy, 14*(1), 1–20.

Veiga, L. B., & Magrini, A. (2011). Water Resources Management: Suggestions to the Brazilian Model Based on the American Experience. *WIT Transactions on Ecology and the Environment, 145*, 39–50.

Veiga, L. B., & Magrini, A. (2013). The Brazilian Water Resources Management Policy: Fifteen Years of Success and Challenges. *Water Resources Management, 27*, 2287–2302.

17

La Via Campesina's Agroecological Militancy at a Crossroads: New Research Avenues for Amazonian Studies

Claire Lagier

Acronyms

ABA *(National network)*	*Associação Brasileira de Agroecologia* (Brazilian Agroecology Association)
ANA *(National network)*	*Articulação Nacional de Agroecologia* (National Agroecology Articulation)
AS-PTA *(Brazilian NGO)*	*Assessoria e Serviços a Projetos em Agricultura Alternativa* (Assistance and Services to Alternative Agriculture Projects)
CLOC-LVC *(Regional network)*	*Coordinadora Latinoamericana de Organizaciones del Campo*—Via *Campesina* (Latin-American

C. Lagier (✉)
Rachel Carson Centre for Environment and Society, Ludwig-Maximilians-Universität München, Munich, Germany
e-mail: Claire.Lagier@rcc.lmu.de

A. A. R. Ioris (ed.), *Environment and Development*,
https://doi.org/10.1007/978-3-030-55416-3_17

	Coordination of Countryside Organisations of La Via Campesina)
CNBB *(Catholic Church)*	*Conferência Nacional dos Bispos do Brasil* (National Conference of Bishops of Brazil)
CNS *(Social movement)*	*Conselho Nacional dos Seringueiros* (National Council of Rubber Tappers)
CONAQ *(Social movement)*	*Coordenação Nacional de Articulação das Comunidades Negras Rurais Quilombolas* (National Coordination of Black Rural Quilombola Communities)
CPT *(Church network)*	*Comissão Pastoral da Terra* (Pastoral Land Commission)
INCRA *(Federal agency)*	*Instituto Nacional de Colonização e Reforma Agrária* (National Institute for Colonization and Agrarian Reform)
LVC *(Global network)*	*La* Via *Campesina*
MAB *(Social movement)*	*Movimento dos Atigindos por Barragens* (Movement of People Affected by Dams)
MAM *(Social movement)*	*Movimento pela Soberania Popular na Mineração* (Movement for Popular Sovereignty in Mining)
MMC *(Social movement)*	*Movimento das Mulheres Camponesas* (Movement of Peasant Women)
MPA *(Social movement)*	*Movimento dos Pequenos Produtores* (Movement of Small Agricultural Producers)
MPP *(Social movement)*	*Movimento dos Pescadores e Pescadoras Artesanais* (Movement of Artesanal Fisherfolk)
MST *(Social movement)*	*Movimento dos Trabalhadores Rurais Sem Terra* (Landless Rural Workers' Movement, or Landless Movement)
PAA *(Federal policy)*	*Programa de Acquisição de Alimentos* (Food Acquisition Programme)

PJR *(Social movement)*	*Pastoral da Juventude Rural* (Rural Youth Pastoral)
PLANAPO *(Federal plan)*	*Plano Nacional de Agroecologia e Produção Orgânica* (National Plan for Agroecology and Organic Production)
PNAE *(Federal programme)*	*Programa Nacional de Alimentação Escolar* (National School Food Programme)
PRONERA	Brazilian federal education program for land reform beneficiaries
SMO	Social Movement Organisation

Introduction and Context

The 2018–2019 global news cycle has been fairly apocalyptic when it comes to the Amazon. Reports of devastating fires in parts of the region under Brazilian and Bolivian control, as well as rising numbers of violent attacks and murders targeting activists in the region, have instilled a sense of urgency in observers. Above all, the rise to power of anti-environmentalist, pro-agribusiness far-right agitator Jair Bolsonaro and his allies in the 2018 Brazilian elections has raised justified fears that Amazonian ecosystems and marginalised human communities might be under threat of extermination, amidst heightened climate change reporting worldwide[1] as the global climate emergency escalates.[2] It is important to highlight the specificity of the current crisis, which involves the dismantlement of Brazilian environmental institutions and regulations; on top of causing terrible human rights violations, the degradation of Amazonian ecosystems has been linked to a rise in greenhouse gas emissions from carbon stored in forest biomass,[3] to the transformation of hydrological and climate cycles on a continental scale (Arraut et al. 2012; Fearnside 2017), to a drastic reduction of biodiversity (Fearnside 2008),

[1] See for example https://www.coveringclimatenow.org. All websites were accessed in March 2020.

[2] Current rates of deforestation in the Amazon are said to be at their highest rate in a decade, with a whopping 9762 km^2 of forest lost between August 2018 and July 2019, according to Brazilian space agency (INPE) data quoted in The Guardian on 19 November 2019 (Watts 2019).

[3] Some estimates claim that Amazonian biomass stores between 90 and 140 billion tons of carbon, the equivalent of over a century of human CO^2 emissions at early 2000s rates (Soares-Filho et al. 2006).

to the emergence of new modes of contagion in the context of infectious and parasitic diseases (Parente et al. 2012; Vasconcelos et al. 2001) as well as other public health issues such as mercury contamination throughout the food chain (Silva et al. 2009), and to lower soil quality and changed wildfire patterns (Foley et al. 2007). What is more, Brazilian Amazonian population has grown from 6 million people in 1960 to 25 million in 2010, transforming the dynamics of human occupation in the region (Davidson et al. 2012).

Nevertheless, it is important to highlight that Amazonia has played a central part in global politics and Brazilian state-building since the nineteenth century (Hecht 2013; Hecht and Cockburn 2011); rapid deforestation and other types of ecosystem degradation as well as agrarian conflicts in Amazonia have been steady since the 1960s, particularly in Brazil, which claims over half of the Amazonian rainforest (Malhi et al. 2008) over nine of its states—Acre, Amapá, Amazonas, Maranhão, Mato Grosso, Pará, Rondônia, Roraima and Tocantins. The Brazilian military regime (1964–1985) set it as a goal for internal control, economic growth and geopolitical strategy to 'develop' the Amazon through resource extraction, modernised agriculture, colonisation programmes overlooking the rights of existing communities and the development of infrastructure such as roads and railways (Hecht and Cockburn 2011; Hochstetler and Keck 2007; Pádua 2014). As Fearnside (2017) reminds us, rates of deforestation have gone up and down under the Workers' Party in the past two decades, following economic cycles, trade politics and export commodity prices levels, as well as global public opinion and institutionally fragile policy incentives (see also Waroux et al. 2019).

This clearly indicates that finding solutions to Amazonia's multiple crises is not a matter of searching for and implementing the right mix of economic 'development', ecological conservation and social acquiescence. As scholars of Brazilian environmental politics Kathryn Hochstetler and Margaret Keck put it, 'for all that the region's problems have been treated as socioeconomic, they are essentially political, requiring a struggle among competing visions of the region's future' (Hochstetler and Keck 2007: 141). One category of actors that have embodied and asserted the political nature of this complex situation are social movements organisations (SMOs), both Indigenous-identified and not. Some iconic references come to mind when one thinks about SMOs in the Brazilian Amazon, often linked to memorable episodes of violence: for instance, Indigenous struggles for land rights and cultural existence following assassinations such as the 2018 murder of Indigenous leader Jorginho Guajajara in Maranhão, the struggle and murder of iconic Acre-based rubber tapper political leader Chico Mendes in 1988 or the 17 April 1996 massacre of 19 activists of the Landless Workers' Movement (MST) in Eldorado dos Carajás, Pará[4] by the military police. This rightly attracts public international and national attention but also reinforces images of unchanging 'lawlessness' (Hochstetler and Keck 2007: 1), deflecting attention from dynamic alternatives and political transformations being enacted all around the region.

Indeed, relatively little is known empirically about the extent, practices and regional specificity of SMOs struggles and organising for alternative sociotechnical visions in the Amazon—different ways of imagining, debating, deciding, coordinating and collectively implementing priorities for important issues in the organisation of human life and environmental access. The aforementioned MST, a founding member of the global peasant alliance La Via Campesina (LVC), is one important organisation advocating for such a vision; along with the Movement of People Affected by Dams (MAB), the Movement of Peasant Women (MMC), the Movement of Small Agricultural Producers (MPA), the Rural Youth Pastoral (PJR), the Movement of Artisanal Fisherfolk (MPP)and the National Coordination of Black Rural Quilombola Communities (CONAQ), it forms the Brazil section of La Via Campesina's South American and Caribbean Coordination (CLOC-LVC), which struggles for social transformation and environmental justice in Brazil by clamouring for food sovereignty and agroecology as ways to organise rural life, food systems and ethical relations with the non-human world.

Broadly, agroecology emphasises promotion of soil fertility and ecosystem health, diversified cultures, locally adapted seeds, regionally- and culturally embedded markets, quality of life and labour for agriculturalists

[4] For a detailed description of the events that led to the deadly shooting of 19 *Movimento dos Trabalhadores Rurais Sem Terra* (Landless Rural Workers' Movement, or Landless Movement [MST]) activists by military police in Eldorado dos Carajás, in southeastern Pará, on 17 April 1996, see Branford and Rocha (2002: 138–142), see also Simmons (2005).

and attention to local ecologies, rather than standardisation, monoculture, transgenic seeds, large-scale commodity production and heavy use of synthetic inputs (Altieri 1995; Bell and Bellon 2018; Gliessman 2015, 2018).[5] Launched in 1996 at the World Food summit in Rome by Via Campesina activists, food sovereignty—conceptualised as a defence against the assaults of economic neoliberalism on national agricultural systems—was initially defined as 'the right of each nation to maintain and develop its own capacity to produce its basic foods, respecting cultural and productive diversity' (Via Campesina 1996, cited in Patel 2009: 665). It evolved in 2006 to 'The rights of peoples to define their own food and agriculture; to protect and regulate domestic agricultural production and trade in order to achieve sustainable development objectives; to determine the extent to which they want to be self-reliant' (Peoples Food Sovereignty Network 2002, cited in Patel 2009: 666). This formula has since expanded to involve higher attention to concerns of social position (gender, class, generation and ethnicity) *within* the food sovereignty movement, to privilege 'local and national' scales, and dropped any mention to sustainable development goals (SDGs) to instead claim justice and sovereignty over all natural resources beyond food and land. This resulted in an arguably more paradoxical and vaguer definition that possesses broader and more radical mobilising power, as a political cry for rural peoples', Indigenous peoples', small farmers' and peasants' autonomy, dignity and recognition as important political subjects and holders of rights, but is more difficult to operationalise on a policy level and to

[5] A review (Wezel et al. 2009) is widely cited for the three-fold definition of agroecology as a science, a set of practices and a social movement. Others argue that trying to define it chronologically from its first appearances in scientific literature in the first half of the twentieth century is politically reductive (Altieri and Nicholls 2012; Giraldo and Rosset 2017). It ignores that many so-called agroecological practices, while they can be enhanced through scientific and social scientific research in partnership with farmers themselves, are similar to traditional Indigenous and peasant practices that were well adapted to specific locales, and that farming techniques in themselves are not transformative of other problematic aspects of the current food system such as credit structures, social and cultural domination of certain groups by others and land access and market access (Mendéz et al. 2015). This characterises the posture of academics who are part of a more 'political' wing of academic agroecology and discursively support social movements' agroecological initiatives, as opposed to a more 'technical' wing which defines agroecology in ecological terms without taking a clear political stance (see Bellamy and Ioris 2017; Lamine 2017; Levidow et al. 2014 for analyses of this rift). For many authors and activists, agroecology is necessarily political because agroecosystems are the site of power relations and shaped by public decisions and institutions (Molina 2012).

build consensus around (Agarwal 2014; Edelman 2009; Patel 2009); as it is embodied in the Declaration of Nyéleni (La Via Campesina, 2007, cited in Patel 2009: 666):

> Food sovereignty is the right of peoples to healthy and culturally appropriate food produced through ecologically sound and sustainable methods, and their right to define their own food and agriculture systems. It puts those who produce, distribute and consume food at the heart of food systems and policies rather than the demands of markets and corporations. It defends the interests and inclusion of the next generation. It offers a strategy to resist and dismantle the current corporate trade and food regime, and directions for food, farming, pastoral and fisheries systems determined by local producers. Food sovereignty prioritises local and national economies and markets and empowers peasant and family farmer-driven agriculture, artisanal fishing, pastoralist-led grazing, and food production, distribution and consumption based on environmental, social and economic sustainability. Food sovereignty promotes transparent trade that guarantees just income to all peoples and the rights of consumers to control their food and nutrition. It ensures that the rights to use and manage our lands, territories, waters, seeds, livestock and biodiversity are in the hands of those of us who produce food. Food sovereignty implies new social relations free of oppression and inequality between men and women, peoples, racial groups, social classes and generations. (La Via Campesina 2007, cited in Patel 2009: 666)

Both agroecology and food sovereignty have gained momentum among international organisations such as the Food and Agriculture Organisation (FAO)[6] and, increasingly, national governments, including the federal government in Brazil, which adopted a tentative National Policy for Agroecology and Organic Production (Pnapo) in 2012.[7] This took place under pressure from Non-Governmental Organisations (NGOs) such as

[6] See, for example, the call to support agroecological food production by UN special rapporteur on the right to food (De Schutter 2010); see also growing support of the FAO, in the form on online declarations, centralisation of know-hows and opening of institutional space and funding for research into agroecology (FAO 2018).

[7] This policy strategy had access to two cycles of funding (2013–2015 and 2016–2019), including an operationalising plan (*Brasil Agroecológico*, or *Plano Nacional de Agroecologia e Produção Orgânica* (National Plan for Agroecology and Organic Production [PLANAPO]) from 2016 on, but is currently unfunded and has lost any political momentum within the federal government since Dilma Rousseff's impeachment in late 2016.

Assessoria e Serviços a Projetos em Agricultura Alternativa (Assistance and Services to Alternative Agriculture Projects [AS-PTA]) and countless local NGOs, national networks such as the *Associação Brasileira de Agroecologia* (Brazilian Agroecology Association [ABA]) and the *Articulação Nacional de Agroecologia* (National Agroecology Articulation [ANA]) and the earlier-mentioned LVC-affiliated SMOs, which have defended agroecology as a viable model to keep farmers on the land with dignity and produce quality food for the country's population while protecting rural ecosystems. The MST has been a vocal advocate for agroecology, and movement discourse describes it as something that is in the interest of all rural peoples, as a counter-hegemonic alternative to 'conventional' or pesticide-based agriculture, and as a foundation of its proposal for popular agrarian reform (MST 2014). However, my own recent ethnographic research (Lagier 2019), based on immersive fieldwork conducted over a year, as well as a corpus of earlier works by ethnographers working in the states of Santa Catarina, Pernambuco, Pará, Paraná, Rio Grande do Sul, Bahia, São Paulo and Rio de Janeiro states (e.g. Brenneisen 2002; Caldeira 2008; Delgado 2008, 2009; DeVore 2015; Gurr 2017; Flynn 2010; Wolford 2010) show us not to take a priori assumptions[8] for granted when it come to the local social and ecological effects of food sovereignty movements. Based on these reflections, I propose a three-fold research programme and a methodological reflection for new avenues of research on food sovereignty movements in Amazonian studies.

In the next section of this chapter, I give a short overview of my most recent research project and explain its relevance for Amazonian issues. Then, I briefly explain the emergence of LVC and notions of food sovereignty in Amazonia, based on the discourses of leadership. In the following section, I then present a methodological discussion on the importance

[8] For example, assumptions explored by the works I cite include the notions that all people who live in areas affiliated with rural social movements organisations (SMOs) consider themselves active members; that all members of food sovereignty SMOs fully understand, agree with and act on their movements' programme; that because SMOs struggle against historical oppression, they always enhance local and participatory democracy; that democratic and ecological choices are always compatible; or that because movements campaign for agroecological transitions in food systems, their actual local effects always enhance sustainability transitions.

of ethnography, participant observation and case studies. I then move on to my three-fold research suggestion. Firstly, I point out the necessity to study what agroecology means within interactions between humans and non-humans in Amazonian contexts, especially when it comes to the political significance of (agro)forestry, wetland activities and small-scale extractivist practices in places where frontal activism can be significantly more dangerous than in other contexts. I then ask to what extent a series of institutional partnerships and articulatory spaces set up in the 2010s to foster agroecological training in Amazonia have created conditions for young activists to identify with agroecology as a political project and to convince others to join it. Finally, it shall be suggested that studying agroecology's social and intergenerational sustainability is crucial to countering current devastating ecological and land-use patterns, which means paying attention to Amazonian youth and women's preoccupations in research.

La Via Campesina's Complex and Contradictory Effects on Ecological Transitions

As we have seen, LVC movements clamour for agroecology and food sovereignty. In order to enable this alternative model, which runs counter to the ways technical assistance, market access and agronomic research are currently organised (Buttel 2006; Goodman et al. 2012), proponents of agroecology have sought to organise alternative institutions. This is what academic agroecologist Manuel Gonzalez de Molina argues for when he points out the necessity for a 'political agroecology' capable of effecting concrete policy change, to ensure that it is not reduced to a set of depoliticised, 'greener' agricultural practices, or to an archipelago of isolated micro-initiatives. Critically praising the actions of local agroecological initiatives, the author claims that:

> Without a profound change in the institutional framework in force it will not be possible for successful agroecological experiences to spread and for the ecological crisis in the field to be combated effectively. Consequently, political agroecology examines the most suitable way to participate in these

> movements and to use those tools that render institutional change possible. Such a change, in a world still organised around nation states, is only possible through political mediation. In democratic systems, for example, it implies collective action through social movement, electoral political participation, the game of alliance between different social forces to build government majorities, etc. In other words, it calls for the creation of essentially political strategies. (Molina 2012: 51)

Consequently, rural social movements, such as those which make up LVC, have been central actors in the politicisation of agroecology throughout Brazil and Latin America (as well as globally). The MST in particular has invested substantial activist energy and resources into fostering training spaces to broaden the reach of agroecological knowledge and expertise, lobbied governments to bring forth public policies designed to sustain it and defends it in public space and the media. Educational initiatives for adults, youth and children have been particularly important to this end (Meek and Tarlau 2016; Meek et al. 2017; Pahnke 2014, 2015); indeed, rural social movements organisers know very well that agricultural models gain legitimacy, in part, through the discourse and actions of professionally trained agronomists, extension agents and individuals seen as holders of differentiated knowledge within communities (Delgado 2008, 2009; Lagier 2019; Meek 2014, 2015).

On the other hand, it is fair to say that within the existing literature, the extent of LVC movements' successes in the 'agroecological turn' has been exaggerated (Pahnke 2015) in some cases and romanticised in others (see Edelman 2009). Discursively, it is rather simple to assert that agroecology can be scaled 'up' and scaled 'out'[9] through farmers' 'active participation in the process of technological innovation and dissemination through models that focus on sharing experiences, strengthening local research, and problem-solving capacities' (Altieri and Nicholls 2012: 472). Such activities have led authors to suggest that a peasant-led 'agroecological revolution' (Altieri and Toledo 2011) has swept across Latin

[9] Scaling 'up' refers to institutionalising agroecology in state structures and market processes, while scaling 'out' means enlisting more farmers and territories in agroecological production (Altieri and Nicholls 2012; Cacho et al. 2018; Dalgaard et al. 2003; Rosset and Martinez-Torres 2012; Varghese and Hansen-Kuhn 2013).

America. This elides the actual structural constraints that a new generation of farmers, peasants and forest dwellers must face as they attempt to build a model of alternative food production that runs counter to established credit structures and markets, as well as the multifarious ways the agribusiness sector has dramatically degraded Brazilian landscapes[10] and shaped both politics and the economy in the past 60 years. Most importantly, it leaves unopened the black box of social movements' internal workings and does not address the obstacles farmers seeking to 'ecologise' their practices (e.g. by cutting back on pesticides, building local seed and seedling banks or accessing different markets) and young activists training towards professional practice in agroecology may experience within their organisations. In the case of the MST, for example, according to a recent, nationwide estimate, only 5–15% of settler families associated with the movement have adopted agroecological practices (Pahnke 2015), thus adhering to its official agroecological mission.

My recent research draws on ethnographic fieldwork conducted in a settlement affiliated with the MST, which hosts a transnational LVC training centre for agroecology in Paraná, southern Brazil (Lagier 2019). This research suggests that even within a 'model' agroecological space in Brazil's south,[11] agroecology is hotly contested and a source of social polarisation within the MST's rank-and-file bases—not least of all because of centralised internal power, which sidelines those without close ties to leadership and resources to transition to agroecology relatively unsupported from community social capital. What is more, young LVC activists from Brazil, Bolivia, Chile, Argentina, the Dominican Republic and Paraguay enrolled in a bachelor of agroecology programme born of a state-movements partnership under the federal agrarian reform education programme (PRONERA) reported complex obstacles in their agroecological endeavours, coming from both inside and outside of their

[10] For striking descriptions of Brazilian agribusiness-moulded landscapes in the context of MST ethnographic research, see for example Gurr (2017: 154–156) and Meek (2014: 1–2).

[11] This is where the MST has been active for the longest and formed its organisational methods and ethical conception of land justice (Wolford 2010), which permeate large sectors of La Via Campesina (LVC) Brazil since the MST has helped consolidate strategies and train many cadres of other LVC movements in Brazil.

social movements—sometimes leading to confusion, disengagement and disappointment in movement support for young activists, especially female ones. While my study's conclusions cannot be easily generalised to other settings, they show that contrary to what is often taken for granted by the more celebratory literature on rural social movements and agroecological transitions, these organisations produce complex and contradictory effects on both local ecological transitions and young activists' capacity to promote agroecology as a politicised alternative to agribusiness and a *socially* sustainable way of life—even in so-called model spaces. Therefore, I claim here that it is necessary to ask similar questions of Amazonian activist spaces, which have been the focus of a relatively small amount of social scientific literature compared to Brazil's widely studied southern, southeastern and northeastern rural activist spaces in spite of spectacular cultural, ecological and territorial specificities. Indeed, LVC SMOs' effects on ecological transitions and social emancipation in the Amazon cannot be assumed a priori because of a pre-existing sympathy for social movements' goals, and understanding their dynamics is potentially key to transforming socio-ecological realities in disputed frontier spaces.

La Via Campesina in the Amazon

This section briefly traces the emergence and presence of La Via Campesina in Amazonia, a particularly dangerous region to be a rural political activist in Brazil (Wright and Wolford 2003: 180), and raises further research questions based on what is known on the subject. One interesting aspect of reviewing the literature on LVC movements in the Amazon is that there is very little interaction between literatures addressing environmentalism and social change in the Amazon (e.g. Hochstetler and Keck 2007; Hecht and Cockburn 2011; Issberner and Léna 2016[12])

[12] Within Issberner and Léna's edited volume, some contributions (e.g. May et al. 2017) address the intersection of socio-ecological transformation and land right movements in the Amazon, but approach the topic from a geographical and policy analysis perspective—this is highly relevant and useful, but does not help analyse the social effects of movements on local cultures, everyday lives and locally specific conceptions of agroecology.

and empirical literature on LVC, even though the MST has been present in the Amazon region since the early 1990s (Branford and Rocha 2002; Wright and Wolford 2003). This suggests that there has been very little immersive socio-anthropological research on the actions of these movements in the Amazon and their effects on socio-ecological transitions. One exception to this is David Meek's anthropological study on the MST and the political ecology of education in southern Pará (Meek 2014, 2015), which I shall return to later.

Violent land conflicts in the Amazon have been going on for a long time (see e.g. Simmons 2004; Hochstetler and Keck 2007), and LVC movements are rather late joiners in the complex network of localised, small-scale struggles by marginalised land dwellers (*posseiros*), rubber tappers' organisations, rural unions and church-led groups leading the land right campaign in the region; what is more, pre-existing groups often saw the MST's arrival, by far the earliest LVC movement to organise in the Amazon, as a possible threat to their ongoing struggles (Ondetti et al. 2015). These authors, who trace the emergence of MST activism in Pará since the late 1980s, remind us that the struggle for MST implantation in the region has been long, contentious and not always seen in a positive light by other political organisations struggling for land rights in the region, such as rural trade unions and the National Council of Rubber Tappers (*Conselho Nacional dos Seringueiros* [CNS]). The MST was initially ambivalent about organising in this region as one of the realities that initially sparked the movement was the unwillingness of southern agriculturalists to be transferred to the Amazon under the dictatorship's policy of using the region as an empty zone of resettlement.

Ondetti and his co-authors claim that the main contributions of the MST in the region have been two. The first one is an expansion and a transformation of what social movements scholar Charles Tilly calls the 'repertoire of contention' (shared knowledge on how to successfully make demands on governments and other actors), from older tactics associated with local *posseiro* struggles ('attempts of local groups to gain control of particular pieces of land, sometimes using armed force' (Ondetti et al. 2015: 220), without much attention to long-term strategy and appealing to public opinion) to the MST's classic tactics of highly public, disciplined protests and flexible alliance-building as part of a highly defined long-term strategy. The second contribution is an important quantitative increase in the number of agrarian reform settlements (ARS) in the Amazon. Indeed, today ARS cover 5% of the Amazon Biome in Brazil (which represents a very large amount of land), with hundreds of thousands of families having gained access to land in the region since 1995, although this has also been linked to low federal, state and local investment in quality of life and infrastructure in these areas, the migration of people with little knowledge of locally adapted agricultural and resource management practices, public perceptions that this policy does not truly support economic and social development in the Amazon, and a reduction of forest cover in ARS areas (see also Le Tourneau and Bursztyn 2010), with heightened resource management challenges arising in areas under collective management (May et al. 2017). Given these challenges, researchers can obviously not construct a narrative about how agroecological alternatives are being built in these areas, how successful they have been or even what criteria should be used to evaluate what success means without further substantial empirical research.

Besides the MST, which is organised in almost all Amazonian states and articulates a public discourse linking rural and forest peoples' human rights, an international peasant identity rooted in solidarity and taking care of specific territories, and agroecology, the MMC, which officialised its existence as a nationwide rural women's movement in 2005 (from a regional movement born in Santa Catarina in the 1980s), is organised in Pará, Amazonas, Acre and Tocantins. The MPA, founded in 1996 to advocate for the rights of small landed agriculturalists to remain on the land, is present in 17 states of Brazil; I have not been able to find the exact list of Amazonian states where it is organised, but it seems to have a particularly strong presence in Rondônia. The MAB, which became a national organisation in 1991 after an early start in southern Brazil in the 1980s supported by the Pastoral Land Commission (*Comissão Pastoral da Terra* [CPT]) like the MST, is particularly well organised in Pará (where it has responded to several mega-dam projects such as the Belo Monte dam by organising the terms of resettlement with federal authorities and organising people around a rights-based discourse), in Amapá and in Rondônia, as well as in Mato Grosso.[13]

This leads me to my last point, which is the emergence of a declared will to articulate a specifically *Amazonian* struggle [*luta amazônica*] on the part of LVC leadership in recent years; for example, Charles Trocate, who used to be a member of the MST's national coordination committee and helped found the Movement for Popular Sovereignty in Mining (*Movimento pela Soberania Popular na Mineração* [MAM][14]) in 2012–2014, writes in 2014 in a book that explains LVC-Brazil's vision for the Amazon in the MST's characteristic style:

> The Amazonian Peasantry has a double task coming ahead; it must recognise itself as a political subject antagonistic to big capital's systemic project (to seek alternatives to the commodification of nature in a struggle of knowledges and power) as well as situate itself radically in the urban Amazon, striving to work with the oppressed in cities and private companies, to construct a new popular field, feeding on all the processes of struggle and resistance as well as producing its own cultural world, organisational science and strategic programme. These are implications of the current pattern of class struggle in the Amazon, and they must be taken on. (Trocate 2014: 13, my translation)

This echoes the public discourse of several LVC movements' leadership in 2018–2019, be it MAB's participation in the regional articulation event (strongly supported by Indigenous organisations) *Amazonia Centre of the World* in November 2018,[15] the MST national leader Alzerina Montello's declaration that 'We will organise popular assemblies in all of the Amazon's main cities to discuss the sovereignty of this region' at the MST's 18th state meeting in Maranhão, in December 2019[16] and the

[13] In addition, the Rural Youth Pastoral (PJR) is present in Pará, Mato Grosso and Tocantins. The Movement of Artesanal Fisherfolk (MPP) and the National Coordination of Quilombola Communities (CONAQ) have more prominence in the Northeast and relatively little presence in the Amazon, where CONAQ is present in Maranhão, Pará and Tocantins and MPP in Pará.

[14] Movement for Popular Sovereignty in Mining (*Movimento pela Soberania Popular na Mineração* [MAM]) is not a member of La Via Campesina Brazil at this time.

[15] https://www.mabnacional.org.br/noticia/defender-amaz-nia-uma-s-voz

[16] https://www.instagram.com/p/B6By245hWW_/

emphasis on qualitative progress and urban-rural alliances at the MPA's most recent state meeting in Rondônia, also in December 2019.[17] Last but not least, one must here mention the Synod for Bishops for the Pan-Amazon Region [*Sínodo pela Amazônia*], which took place in Rome in October 2019 with strong environmental undertones, and has had great resonance within the PJR, a movement linked to the youth branch of Brazil's National Bishop's Conference (CNBB) as well as to LVC, and within the CPT, which has strong links to MST leadership and urban activists working to make agroecology more widespread and accepted in cities.

Therefore, it is clear that LVC movements are widely organised in the Amazon region, with intense collective fora and alliance-building taking place in 2018–2019—at the same time as the new federal government has announced a crackdown on NGOs in the Amazon and dismantled environmental institutions within the Brazilian federal state. However, this leaves largely unaddressed the question of how these movements interact with each other at different levels, outside of large events; how their styles of organisation and specific goals guide their support for specific communities (and specific individuals within these communities); how their training initiatives for agroecological militants are organised and experienced; how they work in specific contexts with scientists and educators to transform what knowledge is considered relevant in agroecosystem management; and how they interact with the broader public through markets, fairs and other alternative market channels, social media, international partnerships, university partnerships and urban-rural partnerships—which Trocate (2014) rightly highlights as crucial to LVC's struggle in the Amazon, given the fast rate of urbanisation in the region.

Given the MST's long history of organised struggle in the Amazon and in Brazil, and the preponderance of its current and past leadership in setting the agenda for LVC Brazil in many places, it is fair to think that the MST's national leadership may have a good degree of influence over LVC organising and institutional partnership building in the Amazon. To what degree is this the case, and what are the implications for local struggles and democracy, for relations between LVC organisations in the

[17] https://mpabrasil.org.br/noticias/mpa-realiza-seu-xx-encontro-em-rondonia

region and for socio-environmental change in local contexts? What kind of production and social organisation alternatives are being promoted, constructed and influenced by LVC movements in local contexts? How are agroecology and food sovereignty's meaning understood and constructed in specific communities, and are these conceptions *socially* sustainable within community and intergenerational relations? Before moving on to proposed directions for future research in the Amazon, in the next section, I conduct a brief methodological reflection on some ways *how* future research should be conducted.

Methodological Considerations

As the literature points out, LVC and its member organisations are heterogeneous organisations acting at different scales (DeVore 2015; Lagier 2019; Rubbo 2016; Wolford 2010), and its levels of organisation do not always smoothly combine or have the same perceived interests (Lagier 2019; Wolford 2010); for example, the MST consists of many different fields that rely on the support of different organisations, social bases with different conceptions of land and farming, regional political configurations, government programmes and local alliances to carry out particular actions. Wolford (2010: 11) explains about the MST membership[18] that

> people join movements for different reasons, and the act of joining does not preclude questioning, rejecting or even deliberately misunderstanding the organisation's ideology, tactics and ultimate goals. But this is not the information that is presented at movements meetings, or in movement newsletters. It is the stuff of mobilisation, but it is not the stuff that activists generally like to talk about. To see the full continuum of MST members, one has to focus on the banal, on the everyday, and on the subalterns among subalterns. This continuum is rarely present in social movement

[18] As Gurr (2017: 93) reminds us, in great resonance with my own experiences and observations, the notion of MST membership is a subject of scholarly debate as the MST does not officially keep a register of its members and conceptions of movement belonging and identity change fluidly in time and space (Issa 2014), with residence in an MST-affiliated space (Sigaud 2015), with evolving movement discourse (Vergara-Camus 2013) and according to local and regional dynamics of mobilisation responding to structural factors (Wolford 2010).

> analyses because in social movement research by activists and scholars alike, model cases and model members tend to be singled out. We tend to study the ideal members, the coherent messages and the brightest media stars. We do not focus on the ambivalent or the half-hearted members; social movements are often read sympathetically as organisations of ideologically committed members for whom the act of joining the movement and participating in movement activities.

In the same way, people and communities' decisions to turn to agroecology and participate in food sovereignty activism, and the broader conditions that enable or prevent them to do so, cannot be explained without looking at specific socio-environmental configurations in depth, through immersive research, and as embedded within institutions, markets, power relations and local alliances (for an earlier formulation of this methodological suggestion in the context of SMOs, see Burdick 1995). This includes participant observation, detailed note-taking of everyday settings and interactions, building trust with participants over time, experiencing everyday life, life cycles and important events along participants as an interactive process and sometimes in-depth interviews and visual methods (photo and video recordings). Studying the context of emergence and public discourse of social movements tells us little about how alternatives are constructed in specific contexts, nor how these alternatives are subjectively experienced by SMO members and other relevant actors. Going beyond Wolford's above citation and building on Vergara-Camus's critical review of her book (2013), it is important not to maintain a rigid methodological divide between 'rank-and-file members' and 'leaders' or between 'model' and 'non-model places': including the discourse of prominent leaders and critically exploring well-known activist spaces in research can be productive of new knowledge on the complex local effects of SMOs, as I have shown in my own research.

Moreover, limiting research on agroecology within LVC movements to official movement discourse and communication, interviews with leaders and political economic, geographic or historical factors has several epistemic consequences. The first one is that knowledge produced this way is normative (showing why agroecology and food sovereignty are the way forward and must be supported) and therefore will be insufficient to explain why some SMO members have not wanted to or have been unable to adopt them. The second consequence, as Wolford (2010: 22) clearly explains, is that SMOs and their members end up being represented in problematic ways. Movements operate from the most intimate personal transformation to national and transnational politics. Conflating personal motivations and local dynamics with a fictional universal social movement subjectivity (the 'populist' external identity carefully crafted by social movement leaders for strategic reasons, see Wolford 2010; Flynn 2010) creates subject elision (Wolford 2010) and lack of empirical complexity (see also DeVore 2015; Gurr 2017; Meek 2014). Thirdly, addressing the ways that food sovereignty is constructed from a non-immersive perspective does not help understand the relational, qualitative aspects of political organising, either in transnational political spaces where organisations articulate their networks, in partnerships being built with non-LVC organisations and state actors, or in meanings constructed with urban consumers around sales of the food produced in ways deemed 'agroecological' (Lagier 2019).

Therefore, it is crucial to address them through thorough empirical research, based on trust with participants, taking into account everyday life in a community, a specific local context and subjective understandings of agroecology and food sovereignty. Participant observation, ethnography and comparative case studies based on thorough immersive research are excellent methods in order to gather this kind of information, in ways that are irreplaceable by short visits, attending only public events and activist meetings, and decontextualised interviews. Because of this necessity for thorough empirical research, a broad research programme on LVC in Amazonia can obviously not be conducted alone (or even in a small team) and ideally requires a coordinated and multidisciplinary research team for at least 4–5 years in order to formulate specific questions and objectives, identify possible fieldwork sites, negotiate access ethically with local communities and SMO leadership, perform fieldwork and data gathering and analyse data without rushing. For this reason, it would be interesting for a team of professional researchers to organise research projects to this effect and involve graduate students with several years of funding and time ahead of them in order to gather both in-depth and multi-sited data.

It is easy to notice that a lot of both immersive and non-immersive research on LVC SMOs in Brazil focuses on the MST, possibly because the MST is the most globally well-known of these movements, is present in almost all Brazilian states, is well connected among global academia, and generally welcomes research attention as long as it is friendly to the movement. In the years to come, it is important to develop in-depth research projects focusing on non-MST LVC movements, as well as on spaces of articulation where LVC holds courses, meetings and events, which is where researchers are most likely to be able to observe the dynamics of decision-making, the making of a multi-movement activist culture and the negotiation of a common public voice. Finally, heightened care should be taken to protect research participants' privacy and anonymity, as such research is known to reveal uncomfortable truths and expose power dynamics. These methods can be complemented by quantitative and GIS methods such as analysis of satellite data on land-use change over time, surveys to sketch the extent of SMO membership engaging with practices deemed 'agroecological', and of transformations in the circulation and destination of 'agroecological' production, archival data and oral history narratives to retrace the arrival of social movements in specific regions, etc.

Avenues for Future Research

This last part of my discussion will be brief, as it is meant to raise further research questions to be addressed empirically by future research. I mainly focus on three points, namely: (1) the dynamic interactions of LVC actions and tactics in the Amazon with pre-existing conceptions of land, ecological relations and rural economic activities; (2) the ways that agroecological and environmental education fostered by LVC movements play out 'on the ground'; and (3) the extent to which (and ways that) LVC campaigning fosters women and youth's inclusive and active participation, emancipation and leadership.

First, any enquiry into LVC in the Amazon must question how different movements' strategies and local tactics relate to pre-existing socio-environmental practices, conceptions of land and resource management.

This is all the more important because, as mentioned earlier, the Amazon, and particularly the state of Pará (where Ondetti et al. 2015: 203, claim that one-fifth of all agrarian reform beneficiaries nationwide are settled), has been the stage of the creation of large numbers of agrarian reform by Brazil's federal agency for agrarian reform (INCRA), in the past two decades, but has not necessarily benefited from infrastructural help, market structuring and educational policies in proportion (Le Tourneau and Bursztyn 2010; Ondetti et al. 2015) resulting in a settled population with little previous local and agricultural knowledge in some areas (see also Branford and Rocha 2002). How has this played out in the construction of what agroecology means in different areas of Amazonia, where rural activities are also heavily influenced by forest and water cycles (e.g. in riverside *ribeirinho* communities and communities living in *várzea* seasonal floodplains), as well as small-scale extractivist activities such as rubber tapping, nut gathering and babaçu coconut artisanal processing—activities which depend on vastly different relations with ecosystems than do planting and harvest? How have LVC movements impacted the construction of and access to alternative markets, in an era of rapidly unravelling policy frameworks (such as the Zero Hunger—*Fome Zero* policy programme) and consolidation of agribusiness control over the entire food system?

What is more, agroforestry—the practice of including trees, bamboos and shrubs within agroecosystems, which has been practised under many forms by Indigenous populations around the world for long periods of time—has been identified by scientific research (e.g. Tremblay et al. 2015; see Bezerra 2011 for a case study in an agrarian reform settlement in Pará) as a promising alternative for the Amazon, in order to rebuild tree cover and slow down deforestation and the use of fire in agroecosystems, integrate human extractive activities, agriculture and conservation, and generate stable incomes and varied diets for rural families. It is a category of agroecological practices that has been adopted in LVC communities in many regions of Brazil (e.g. Paraná, São Paulo, Bahia), often through the initial intervention of partnerships between non-SMO actors (scientists or non-governmental organisations) and social movement leadership. How have these practices been introduced and developed in

LVC Amazonian communities?[19] How widespread are these practices in areas affiliated with LVC movements today, have they taken hold in communities where they were introduced beyond pilot projects, and are they becoming important income-generating and autonomy-building strategies for families, as prescribed by food sovereignty?

Moreover, as Wolford (2010) shows, conceptions of land—as a place to live and work as independent agriculturalists, or as a place to work for a wage and live as a salaried worker in the case of her comparative work based on research conducted in Santa Catarina and Pernambuco states—vary widely within Brazil, based on regional agrarian and environmental histories, historical land uses and different configurations of population formation based on migration, slavery, plantation systems and rural political economic agro-industrial transformations. Her work highlighted how these pre-existing conceptions, which made MST goals, politics of land redistribution and ethics seem appropriate in Brazil's south where they originated, were not adequate in the wage labour-based sugarcane economy of Pernambuco. How would this comparison play out in different Amazonian regions? For example, how has the MST's ideal for collectively organised 'peasant' agriculturalists who derive their livelihood from harvests and small agro-industrial cooperatives been understood, negotiated (or imposed), transformed and materialised in the Amazon, and with what effects on local ecological transitions? How have LVC leaders adapted their discourse and practices to local conceptions of land, and how have priorities been defined when local democracy has clashed with national-level or transnational-level strategic orientations? Has the advance of these social movements impacted relations within communities, and brought along new social distinctions and hierarchies? In sum, how have LVC movements influenced socio-ecological transformations, production relations and subjective understandings of agroecology, autonomy and sovereignty in the Amazon region?

[19] See Costa (2017) for a Masters' dissertation by a Brazilian researcher conducting this type of research in an agrarian reform settlement in Rondônia, which shows that the construction of relations with urban consumers, within and outside of public policies designed to support school attendance, proper school-aged child nutrition and smallholder income such as the Food Acquisition Programme (*Programa de Aquisição de Alimentos,* PAA) and the National School Food Programme (*Programa Nacional de Alimentação Escolar,* PNAE) is a crucial factor in how successful agroecological projects can become on settlements located close to urban centres.

Second, earlier in this chapter, I claimed that LVC—most notably through MST leadership (Trocate 2014) but also through other movements such as MAB, MPA and MMC—have articulated a specific discourse on the alliance's goals and achievements in Amazonia. An important aspect of this is educational, as the formidable challenges of bringing literacy, fostering place-based understandings of structural transformations (Meek 2014) and alternative agricultural practices and conceptions of agriculture and new, emancipatory ways of generating knowledge to new territories call for much innovation in training. In May 2007, at the 3rd Regional Conference for Countryside Education in Marabá (Pará), LVC leadership from the region issued a declaration stating their intention to pressure municipal, state and federal governments to support new educational partnerships at all levels to support countryside and food sovereignty education (Trocate 2014: 119–123). In 2009, LVC's Amazonian Latin American Institute of Agroecology (*IALA Amazônico*) was also built in an agrarian reform settlement in southern Pará. It was the fourth agroecology training school built by the coalition in Latin America, making it an important hub in the alliance's Latin American strategy for food sovereignty. Twelve years on, how have these partnerships been built and mobilised across federal, state and municipal levels, but also on a transnational scale across the Amazon? How have *IALA Amazônico*'s training programmes, some of them notably financed through the Brazilian federal programme for agrarian reform education, PRONERA, been set up, taught and experienced by young activists? Have graduates gone on the successfully organise, model and research agroecological initiatives in their communities? How have different movements within LVC related to this space and other training spaces?

Significantly, this is one area of research where future researchers can build upon earlier immersive research. Meek's 2014 ethnographic doctoral dissertation, based on 17 months of fieldwork in southeastern Pará, examines the institutional and interpersonal construction of agroecological and environmental education initiatives in the MST. Meek found that educators taking part in these partnerships were strategic intermediaries between movement and state, making them key participants to interview in future research. He insists that 'agroecological education has the *potential* to maintain MST members' political participation, identification with the movement, utilisation of the agricultural principles it advocates, and role as individuals committed to a longer process of emancipatory social change' (Meek 2014: 256), showing that educational institutionalisation is a major opportunity for the movement, and that the MST's initiatives in this region have uniquely opened up space within Pará's federal universities, constructed a new vision of agroecological education and promoted anti-neoliberal education in the state. However, his research also shows that training centres have created unique dynamics within the settlements where they have been implanted due to the uneven political participation of educators and rank-and-file settlers in the MST, and pre-existing land uses and conceptions of agriculture in direct opposition to LVC's agroecology. His claim that 'movement educational spaces are key incubators for larger scale processes of educational, social, and environmental change' (Meek 2014: 272) and that '[one of the schools he studied, on the 17 de Abril settlement] would be a productive site for extended ethnographic research to assess the success of its movement-affiliated educators in integrating MST pedagogy and ideals into the curriculum. Scholars would undoubtedly learn important lessons by continuing to pay attention to the daily politics of educators' interrelations, and their interactions with students' parents concerning the role of politics in the school' invites further research on this theme, now that five years have passed and federal politics for agrarian reform have been radically transformed by the rise of the far right within Brazilian institutions.

Finally, an important area of further research into LVC initiatives in the Amazon is the extent to which local initiatives have fomented inclusive and emancipatory participation of women and youth and transformed gender relations. Indeed, even though official LVC discourse has emphasised feminist discourse for years,[20] partly under the impulse of female leaders and feminist-minded LVC member organisations (such as the MMC[21]), recent works (e.g. Caldeira 2008, 2009; Deere 2003; Flynn 2010; Gurr 2017; Lagier 2019) show that realities have only slowly

[20] Desmarais (2007), already emphasises the aspect in the first broad work on LVC, although not drawing on immersive research.

[21] It is interesting to note that the MMC was partly born out of the will of young women who separated from the early MST in Rio Grande do Sul and Santa Catarina states in the 1980s to form their own movement, one of their concerns being that MST leaders wouldn't listen to women and

started to change at the local level—with female and young members of SMOs often facing a lot of resistance to their political, economic, bodily and reproductive autonomy within social movement leadership networks, families and extended communities. Researching this topic is crucial, given the current global crisis in rural youth unemployment, underemployment and outmigration (White 2012); the willingness and ability of women and youth to stay in rural areas and make a living there will be a crucial factor in the renewal of small-scale, community-based food sovereignty all over the world, as Pierre Bourdieu (1962) noted about women' willingness to marry and remain in rural France.

A report on gender and forests in the Amazon was recently conducted by the Centre for International Forestry Research (Schmink and García 2015). Highlights of this report's findings include similar dynamics to those observed in other regions of Latin America, such as the prioritisation of community unity over women's autonomy, patriarchal cultural values preventing women from accessing decision-making in forest and ecosystem management, markets and autonomous property rights; a gender-segregated division of labour; the invisibilisation of the value and extent of women's productive and reproductive labour; physical isolation of women, making access to help and social services difficult in case of violence; and adverse effects of recent timber development projects on women's participation and financial autonomy (Schmink and García 2015). Significantly, the report also points to a lack of systematic research 'on the diverse forms of gender relations among forest populations of the Amazon region, despite the importance of this topic for forest management, community food security, sustainable livelihoods and the capacity of Amazonian people to respond to external pressures and changing climates' (Schmink and García 2015: 26). The 2013 publication of an edited book coordinated by Brazilian sociologists and supported by the MPA on peasant women and their specificities, which includes several chapters on Amazonian women (Neves and Medeiros 2013), is an important step in this direction.

---

respect them as political subjects (Rubin and Sokoloff-Rubin 2013). A social history and sociology of the MMC in the Amazon remains to be written.

However, it is crucial that further research pay attention to these dynamics in the context of LVC-affiliated communities and organisational practices, as the future of social movements (along that of food sovereignty and agriculture itself) is at stake. Have LVC movements actually worked towards the recognition and valuation of women and youth's (re)productive work within rural and extractive communities, as an integral part of political agroecology? Have new practices introduced and strengthened by these organisations in the Amazon transformed the dynamics of income generation and economic autonomy for them? How have rural development policies aiming to support women's income (such as the PAA, or new legal protection for babaçu coconut reserves, where women traditionally gather and manually extract seeds) impacted dynamics within families and communities? How has movements' insistence on the category of family agriculture [*agricultura familiar*] worked for women and youth organising to free themselves of patriarchal authority *within* families? Importantly, are sexual assaults and reproductive rights taken and treated seriously, with accountability demanded of perpetrators, within LVC movement structures?

Conclusions

Building on some of the conclusions of my recent research on LVC and agroecological transitions in southern Brazil, as well as other immersive socio-anthropological work exploring similar topics, this exploratory text aimed to suggest new potential avenues of research for exploring what food sovereignty has meant and what its future could be in the Amazon region of Brazil. Alternative sociopolitical visions for the organisation and distribution of power, resources and priorities in rural areas and food systems, such as agroecology-based food sovereignty as advocated by LVC, have important potential to durably transform the Brazilian Amazon, slow down the advances of agribusiness in the region and enhance the well-being of its human and non-human communities. LVC movements in Brazil, such as the MST, MMC, MPA, MAB, PJR,

CONAQ and MPP, have together campaigned all over the Amazon region, with special intensity in Southern Pará (where the MST has campaigned since the early 1990s and MAB campaign against several dam projects), Mato Grosso, Acre and Rondônia.

However, I have pointed out that little is known empirically about the on-the-ground and local effects of this campaigning on social cohesion, ecological transitions in food systems and development of critical environmental education initiatives, all crucial parts of rural development alternatives' social sustainability. In particular, I have suggested that further research focus on LVC's interactions with specific ecological practices related to Amazonian practices such as agroforestry, small-scale extractivism, rubber tapping, babaçu coconut artisanal processing and Brazil nut gathering, as well as local understandings of the meaning of land ownership and regional food markets, on the continued effects of a series of state-movements educational partnerships started in the early 2010s, and on women and youth's emancipatory inclusion within SMOs' rank-and-file bases and militancy at all levels.

I have claimed that such research, in order to deploy its critical potential and comprehend ambiguity and local dynamics, should include important immersive components such as long-term fieldwork, ethnography and in-depth interviews with social movement rank-and-file members, activists, and ex-members, as well as state actors and other relevant actors (e.g. scientists, technicians and educators involved in agroecological projects as well as non-Brazilian SMO actors active in the Amazon), and suggested other complementary methods. By no means do I wish to suggest that these are the only promising areas of research when it comes to LVC SMOs in the Brazilian Amazon; this chapter aimed to raise further questions. However, similarly to Burdick (1995) and Edelman (2009), I believe that the best way for critical ally researchers to help SMOs better attain their objectives is by engaging in this kind of research and contributing insights, rather than by repeating social movement discourse in academic jargon.

References

Agarwal, B. (2014). Food Sovereignty, Food Security and Democratic Choice: Critical Contradictions, Difficult Conciliations. *The Journal of Peasant Studies, 41*, 1247–1268.

Altieri, M. (1995). *Agroecology: The Science of Sustainable Agriculture* (2nd ed.). Colorado/London: CRC Press Boulder.

Altieri, M., & Nicholls, C. I. (2012). Agroecology Scaling up for Food Sovereignty and Resiliency. In E. Lichtfouse (Ed.), *Sustainable Agriculture Reviews* (pp. 1–29). Dordrecht: Springer.

Altieri, M., & Toledo, V. M. (2011). The Agroecological Revolution in Latin America: Rescuing Nature, Ensuring Food Sovereignty and Empowering Peasants. *The Journal of Peasant Studies, 38*, 587–612.

Arraut, J. M., Nobre, C. A., Barbosa, H. M., Obregon, G., & Marengo, J. A. (2012). Aerial Rivers and Lakes: Looking at Large-scale Moisture Transport and Its Relation to Amazonia and to Subtropical Rainfall in South America. *Journal of Climate, 25*(2), 543–556.

Bell, M. M., & Bellon, S. (2018). Generalization Without Universalization: Towards an Agroecology Theory. *Agroecology and Sustainable Food Systems, 42*, 605–611.

Bellamy, A. S., & Ioris, A. A. R. (2017). Addressing the Knowledge Gaps in Agroecology and Identifying Guiding Principles for Transforming Conventional Agri-Food Systems. *Sustainability, 9*(3), 330. https://doi.org/10.3390/su9030330.

Bezerra, N. F. C. (2011). A Amazônia e os Novos Paradigmas de Desenvolvimento Rural: Uma Breve Reflexão Teórica. *Revista Brasileira de Agroecologia, 6*(2), 40–54.

Bourdieu, P. (1962). Célibat et Condition Paysanne. *Etudes Rurales, 5–6*, 161–176.

Branford, S., & Rocha, J. (2002). *Cutting the Wire. The Story of the Landless Movement in Brazil*. London: Latin America Bureau.

Brenneisen, E. C. (2002). *Relações de Poder, Dominação e Resistência: O MST e os Assentamentos Rurais*. Cascavel: Edunioeste.

Burdick, J. (1995). Uniting Theory and Practice in the Ethnography of Social Movements: Notes Toward a Hopeful Realism. *Dialectical Anthropology, 20*(3–4), 361–385.

Buttel, F. (2006). Sustaining the Unsustainable: Agro-Food Systems and Environment in the Modern World. In *The Handbook of Rural Studies* (pp. 213–229). London: Sage.

Cacho, M. M. T., Giraldo, O. F., Aldasoro, M., Morales, H., Ferguson, B. G., Rosset, P., Khadse, A., & Campos, C. (2018). Bringing Agroecology to Scale: Key Drivers and Emblematic Cases. *Agroecology and Sustainable Food Systems, 42*(6), 637–665.

Caldeira, R. (2008). 'My Land, Your Social Transformation': Conflicts Within the Landless People Movement (MST), Rio de Janeiro, Brazil. *Social Movements and Rural Politics, 24*(2), 150–160.

Caldeira, R. (2009). The Failed Marriage Between Women and the Landless People's Movement (MST) in Brazil. *Journal of International Women's Studies, 10*(4), 237–258.

Costa, L. M. (2017). *Agroecologia na Amazônia Desafios e Perspectivas no Contexto da Reforma Agrária: Um Estudo de Caso em Ariquemes – Rondônia.* MSc Dissertation, Universidade Federal de Santa Catarina.

Dalgaard, T., Hutchings, N. J., & Porter, J. R. (2003). Agroecology, Scaling and Interdisciplinarity. *Agriculture, Ecosystems & Environment, 100*(1), 39–51.

Davidson, E. A., Araujo, A. C., Artaxo, P., Balch, J. K., Brown, F., Bustamante, M. M. C., Coe, M. T., DeFries, R. S., Keller, M., Longo, M., Munger, J. W., Schroeder, W., Soares-Filho, B. S., Souza, C. M., Jr., & Wofsy, C. S. (2012). The Amazon Basin in Transition. *Nature, 481*, 321.

De Schutter, O. (2010). *Report Submitted by the Special Rapporteur on the Right to Food.* Geneva: United Nations General Assembly, Human Rights Council.

Deere, C. D. (2003). Women's Land Rights and Rural Social Movements in the Brazilian Agrarian Reform. *Journal of Agrarian Change, 3*(1–2), 257–288.

Delgado, A. (2008). Opening Up for Participation in Agro-Biodiversity Conservation: The Expert-Lay Interplay in a Brazilian Social Movement. *Journal of Agricultural and Environmental Ethics, 21*(6), 559.

Delgado, A. (2009). Activist Trust: The Diffusion of Green Expertise in a Brazilian Landscape. *Public Understanding of Science, 19*(5), 562–577.

Desmarais, A. A. (2007). *La Vía Campesina. Globalization and the Power of Peasants.* Halifax: Fernwood Publishing.

DeVore, J. (2015). The Landless Invading the Landless: Participation, Coercion, and Agrarian Social Movements in the Cacao Lands of Southern Bahia, Brazil. *The Journal of Peasant Studies, 42*(6), 1201–1223.

Edelman, M. (2009). Synergies and Tensions Between Rural Social Movements and Professional Researchers. *The Journal of Peasant Studies, 36*(1), 245–265.

Fearnside, P. M. (2008). The Roles and Movements of Actors in the Deforestation of Brazilian Amazonia. *Ecology and Society, 13*(1), 23.

Fearnside, P. M. (2017). Deforestation of the Brazilian Amazon. In H. Shugart (Ed.), *Oxford Research Encyclopaedia of Environmental Science.* New York:

Oxford University Press. https://doi.org/10.1093/acrefore/9780199389414.013.102.

Flynn, A. (2010). *Pluralist Identities and Empowering 'the People': Brazil's Landless Workers' Movement (MST) at the Crossroads*. PhD Thesis, University of Manchester.

Foley, J. A., Asner, G. P., Costa, M. H., Coe, M. T., DeFries, R., Gibbs, H. K., Howard, E. H., Olson, S., Patz, J., Ramankutty, N., & Snyder, P. (2007). Amazonia Revealed: Forest Degradation and Loss of Ecosystem Goods and Services in the Amazon Basin. *Frontiers in Ecology and the Environment, 5*(1), 25–32.

Food and Agriculture Organisation of the United Nations. (2018). *Understanding Agroecology*. http://www.fao.org/publications/highlights-detail/en/c/1113542. Accessed 15 Nov 2019.

Giraldo, O. F., & Rosset, P. M. (2017). Agroecology as a Territory in Dispute: Between Institutionality and Social Movements. *The Journal of Peasant Studies, 43*(3), 545–564.

Gliessman, S. (2015). *Agroecology: The Ecology of Sustainable Food Systems* (3rd ed.). Boca Raton: CRC Press.

Gliessman, S. (2018). Defining Agroecology. *Agroecology and Sustainable Food Systems, 42*(6), 599–600.

Goodman, D., DuPuis, E. M., & Goodman, M. K. (2012). *Alternative Food Networks: Knowledge, Practice, and Politics*. Oxon/New York: Routledge.

Gurr, M. (2017). *The Limits of Liberation: Youth and Politics in Brazil's Landless Rural Workers' Movement (MST)*. PhD Thesis, Syracuse University.

Hecht, S. B. (2013). *The Scramble for the Amazon and the 'Lost Paradise' of Euclides da Cunha*. Chicago: University of Chicago Press.

Hecht, S. B., & Cockburn, A. (2011). *The Fate of the Forest. Developers, Destroyers and Defenders of the Amazon*. Chicago: University of Chicago Press.

Hochstetler, K., & Keck, M. E. (2007). *Greening Brazil. Environmental Activism in State and Society*. Durham/London: Duke University Press.

Issa, D. (2014). Praxis of Empowerment: Mística and Mobilisation in Brazil's Landless Rural Workers' Movement (MST). In R. Stahler-Sholk, H. E. Vanden, & M. Becker (Eds.), *Rethinking Latin American Social Movements: Radical Action from Below* (pp. 85–100). London: Rowman & Littlefield.

Issberner, L.-R., & Léna, P. (Eds.). (2016). *Brazil in the Anthropocene, Conflicts between Predatory Development and Environmental Politics*, Routledge Environmental Humanities series. Taylor and Francis.

Lagier, C. (2019). *Constructing Legitimacy? Agroecology Within and Beyond the Brazilian Landless Workers' Movement (MST)*. PhD Thesis, LMU Munich.

Lamine, C. (2017). *La Fabrique Sociale de l'Ecologisation de l'Agriculture*. Marseille: La Discussion.

Le Tourneau, F. M., & Bursztyn, M. (2010). Assentamentos Rurais na Amazônia: Contradições Entre a Política Agrária e a Política Ambiental. *Ambiente & Sociedade, 13*(1), 111–130.

Levidow, L., Pimbert, M., & Vanloqueren, G. (2014). Agroecological Research: Conforming – or Transforming the Dominant Agro-Food Regime? *Agroecology and Sustainable Food Systems, 38*(10), 1127–1155.

Malhi, Y., et al. (2008). Climate Change, Deforestation, and the Fate of the Amazon. *Science, 319*, 169–172.

May, P., Davenport, R., Nogueira, P., & Nunes, P. C. (2017). Collective Forest Reserves in Agrarian Reform Settlements. In L. Issberner & P. Léna (Eds.), *Brazil in the Anthropocene* (pp. 327–349). Oxon/New York: Routledge.

Meek, D. (2014). *Movements in Education: The Political Ecology of Education in the Brazilian Landless Workers' Movement*. PhD Thesis, University of Georgia.

Meek, D. (2015). Learning as Territoriality: The Political Ecology of Education in the Brazilian Landless Workers' Movement. *The Journal of Peasant Studies, 42*, 1179–1200.

Meek, D., & Tarlau, R. (2016). Critical Food Systems Education (CFSE): Educating for Food Sovereignty. *Agroecology and Sustainable Food Systems, 40*, 237–260.

Meek, D., Bradley, K., Ferguson, B., Hoey, L., Morales, H., Rosset, P., & Tarlau, R. (2017). Food Sovereignty Education Across the Americas: Multiple Origins, Converging Movements. *Agriculture and Human Values, 36*, 611.

Mendéz, E., Bacon, C., & Cohen, R. (2015). Introduction: Agroecology as a Transdisciplinary, Participatory and Action-oriented Approach. In E. Méndez, C. Bacon, R. Cohen, & S. Gliessman (Eds.), *Agroecology: A Transdisciplinary, Participatory and Action-Oriented Approach*. Boca Raton: CRC Press, Taylor & Francis Group.

Molina, M. G. (2012). Agroecology and Politics. How to Get Sustainability? About the Necessity for a Political Agroecology. *Agroecology and Sustainable Food Systems, 37*(1), 45–59.

MST. (2014). *Programma Agrário do MST – Texto em construção para o VI Congresso Nacional*. São Paulo: Secretaria Nacional do MST. https://mst-brasilien.de/wp-content/uploads/2014/02/Cartilha-Programmea-agr%C3%A1riodoMST-FINAL.pdf. Accessed 15 Nov 2019.

Neves, D. P., & Medeiros, L. S. (Eds.). (2013). *Mulheres Camponesas: Trabalho Produtivo e Engajamentos Políticos*. www.icmbio.gov.br/educacaoambiental/images/stories/biblioteca/educacao_ambiental/livro_mulheres_camponesas.pdf. Accessed 10 Nov 2019.

Ondetti, G., Wambergue, E., & Afonso, J. B. G. (2015). From *Posseiro* to *Sem Terra*. The Impact of MST Land Struggles in the State of Pará. In M. Carter (Ed.), *Challenging Social Inequality. The Landless Rural Workers' Movement and Agrarian Reform in Brazil* (pp. 202–226). Durham/London: Duke University Press.

Pádua, J. A. (2014). Brazil in the History of the Anthropocene. In L. Issberner & P. Léna (Eds.), *Brazil in the Anthropocene* (pp. 19–40). Oxon/New York: Routledge.

Pahnke, A. (2014). The MST and Its Educational Programme on the World Stage. In N. P. Wood (Ed.), *Brazil in 21st Century Popular Media: Culture, Politics and Nationalism on the World Stage* (pp. 157–184). Lanham: Lexington Books.

Pahnke, A. (2015). Institutionalising Economies of Opposition: Explaining and Evaluating the Success of the MST's Cooperatives and Agroecological Repeasantisation. *The Journal of Peasant Studies, 42*(6), 1087–1107.

Parente, A. T., Souza, E. B., & Ribeiro, J. B. M. (2012). Occurrence of Malaria in Four Cities in the State of Pará During 1988 to 2005 and its Relationships with Deforestation. *Acta Amazonica, 42*(1), 41–48.

Patel, R. (2009). Grassroots Voices: What Does Food Sovereignty Look Like? *Journal of Peasant Studies, 36*(3), 663–706.

Rosset, P., & Martinez-Torres, M. E. (2012). Rural Social Movements and Agroecology: Context, Theory, and Process. *Ecology and Society, 17*(3), 17.

Rubbo, D. I. A. (2016). *Párias da Terra: o MST e a Mundialização da Luta Camponesa*. São Paulo: Alameda.

Rubin, J. W., & Sokoloff-Rubin, E. (2013). *Sustaining Activism. A Brazilian Women's Movement and a Father-Daughter Collaboration*. Durham/London: Duke University Press.

Schmink, M., & García, M. A. G. (2015). *Under the Canopy: Gender and Forests in Amazonia* (Occasional Paper 121). Bogor: Centre for International Forestry Research (CIFOR).

Sigaud, L. M. (2015). Under the Black Tarp: The Dynamics and Legitimacy of Land Occupations in Pernambuco. In M. Carter (Ed.), *Challenging Social Inequality: The Landless Workers Movement and Agrarian Reform in Brazil* (pp. 182–201). Durham: Duke University Press.

Silva, D. S., Lucotte, M., Paquet, S., & Davidson, R. (2009). Influence of Ecological Factors and of Land Use on Mercury Levels in Fish in the Tapajós River Basin, Amazon. *Environmental Research, 109*, 432–446.

Simmons, C. S. (2004). The Political Economy of Land Conflict in the Eastern Brazilian Amazon. *Annals of the Association of American Geographers, 94*(1), 183–206.

Simmons, C. S. (2005). Territorialising Land Conflict: Space, Place, and Contentious Politics in the Brazilian Amazon. *GeoJournal, 64*(4), 307–317.

Soares-Filho, B. S., Nepstad, D., Curran, L., Cerqueira, G., Garcia, R., Ramos, C., Voll, E., MacDonald, A., Lefebvre, P., & Schlesinger, P. (2006). Modelling Conservation in the Amazon Basin. *Nature, 440*, 520–523.

Tremblay, S., Lucotte, M., Revéret, J. P., Davidson, R., Mertens, F., Passos, C. J. S., & Romaña, C. A. (2015). Agroforestry Systems as a Profitable Alternative to Slash and Burn Practices in Small-scale Agriculture of the Brazilian Amazon. *Agroforestry Systems, 89*, 193–204.

Trocate, C. (2014). *Ideias para o Pensamento Político da Via Campesina na Amazônia: Pela Construção de um Bloco Camponês e Popular*. Marabá: Editorial iGuana.

Varghese, S., & Hansen-Kuhn, K. (2013). *Scaling Up Agroecology: Toward the Realisation of the Right to Food*. Minneapolis: Institute for Agriculture and Trade Policy.

Vasconcelos, P. F. C., Travassos da Rosa, A. P. A., Rodrigues, S. G., Travassos da Rosa, E. S., Dégallier, N., & Travassos da Rosa, J. F. S. (2001). Inadequate Management of Natural Ecosystem in the Brazilian Amazon Region Results in the Emergence and Reemergence of Arboviruses. *Cadernos de Saúde Pública, 17*, 155–164.

Vergara-Camus, L. (2013). This Land is Ours Now: Social Mobilization and the Meanings of Land in Brazil, Wendy Wolford, Durham, NC.: Duke University Press, 2010. *Historical Materialism, 21*, 169–178.

Waroux, Y. P., Garrett, R. D., Graesser, J., Nolte, C., White, C., & Lambin, E. F. (2019). The Restructuring of South American Soy and Beef Production and Trade Under Changing Environmental Regulations. *World Development, 121*, 188–202.

Watts, J. (2019, November 18). Amazon Deforestation 'at Highest Level in a Decade'. *The Guardian*. https://www.theguardian.com/environment/2019/nov/18/amazon-deforestation-at-highest-level-in-a-decade. Accessed 15 Nov 2019.

Wezel, A., Bellon, S., Doré, T., Francis, C., Vallod, D., & David, C. (2009). Agroecology as a Science, a Movement and a Practice: A Review. *Agronomy for Sustainable Development, 29*(4), 503–515.
White, B. (2012). Agriculture and the Generation Problem: Rural Youth, Employment and the Future of Farming. *IDS Bulletin, 43*(6), 9–19.
Wolford, W. (2010). *This Land Is Ours Now: Social Mobilisation and the Meanings of Land in Brazil.* Durham: Duke University Press.
Wright, A., & Wolford, W. (2003). *To Inherit the Earth. The Landless Movement and the Struggle for a New Brazil.* Oakland: Food First Books.

18

Oxford Letter for the Amazon

Antonio Augusto Rossotto Ioris

The Amazon rainforest and its people are harbingers of life, knowledge and creativity. The numerous and highly diverse communities that live in the Amazon region demonstrate every day that it is possible to align the values of nature with those of one's labour so as to form a seamless whole of well-being.

Today, however, the Amazon is under an assault that has provoked the displacement of communities, the expropriation of land, resources and livelihoods, the privatisation of collectively managed ecosystems and an increase in assassinations of rural and indigenous Amazonian leaders. This assault is being led by a narrow, unethical business and political elite seeking rapid enrichment. With the support of authoritarian state planning, this unscrupulous minority has transformed the Amazon into one of the most disputed frontiers of Western modernity.

We, students, indigenous leaders, peasants, workers, researchers, academics, activists and civil society members from different parts of the world, gathered at Oxford University for the International Colloquium AMAZON'S RISING VIOLENCE AND DISTURBING TRENDS,

A. A. R. Ioris (✉)
School of Geography and Planning, Cardiff University, Cardiff, UK
e-mail: IorisA@cardiff.ac.uk

A. A. R. Ioris (ed.), *Environment and Development*,
https://doi.org/10.1007/978-3-030-55416-3_18

from 31 January to 02 February 2020, the fifth main event of the Agrocultures International Network, are living proof of the mounting opposition against such perverse forms of economic development based on systematic socio-ecological exploitation.

We believe that the Amazon is more precious and has a much brighter future than the initiatives and strategies of national governments and their geopolitical and economic partners. Working with the forest and other ecosystems, rather than against them, constitutes a crucial element of the concrete alternative path that will lead us safely away from the destructive trends of mainstream development.

We demand that the political and economic rights of Amazonian people be honoured and respected. In full solidarity with other movements, demonstrations and initiatives to save the Amazon, we will do everything in our power to help humanity prevent catastrophic climate change and rising inequalities associated with market-based globalisation in the Amazon and elsewhere on the planet. We are part of a growing awareness throughout the world that social exclusion and failed economic thinking are putting us at risk.

Based on the above, we want to communicate the following demands and urgent messages:

> We denounce the systematic and increased number of killings of human rights activists, community leaders, peasants, members of indigenous groups and environmental campaigners in the Amazon, particularly in Colombia and Brazil;
>
> We call for the strict enforcement of the national constitutions and related legislation in order to demarcate indigenous lands, stop mining in the Yanomami territory, to prevent the discharge of mercury and other toxins into rivers and bring to justice those responsible for atrocities associated with drug trafficking (a growing and disturbing phenomenon that is reshaping the region) and the invasions and violent assaults on Amazonian communities;
>
> We condemn the regional governments, especially the current Brazilian administration, for dismantling public policies, for enflaming racial and ethnic tensions, for undermining and inhibiting the work of researchers and scientists, for the gross mismanagement of natural resources, for dishonestly encouraging further environmental devastation (wildfires, defor-

estation and pollution, among other impacts) and for the vicious, escalating attacks on indigenous territories, peasant and *quilombola* communities, shanty towns, and other lived spaces;

We demand a profound and meaningful reconsideration of economic plans and development programmes that have caused major disruption in the region, mainly for the benefit of external economic agents and social groups, including hydropower dams, large-scale navigation, mining operations, logging, ranches and agribusiness farms;

We urge governments, businesses and individuals to boycott agricultural goods (soy and meat in particular), timber, gold and minerals from farmers, illegal goldminers, businesses and corporations who clear the rainforest, use slave labour or threaten the basic civil, labour and environmental rights;

We accuse investment banks and sovereign debt funds of significantly aggravating deforestation and the emission of greenhouse gases, of further encouraging the exploitation of low-paid workers and facilitating the perpetuation of genocidal policies;

We repudiate the inadequacy and the warped consequences of market-based environmental conservation schemes, such as carbon markets and biodiversity commodification, that end up creating new forms of alienation and conflict;

We call on the Inter-American and United Nations Human Rights Commissions to investigate the violations of human and ethnic rights occurring throughout the Amazon;

We expect dedicated, genuine actions from academics, civil servants, politicians and the business community to revert these trends, promote critical and responsible science and prioritise the lives, the heritage, the needs and the autonomous contribution of the peoples of the Amazon.

We, the participants of this event, jointly raise the voice of the majority of social groups in the region, who know and passionately believe that the Amazon has had a troubling past and is undergoing a turbulent present but, because of its people, can also nourish a much better, and fairer, future.

Oxford, 31 January 2020, signed by indigenous and peasant representatives, politicians, civil society members, students, artists, activists, researchers and academics (Fig. 18.1).

Fig. 18.1 Poster of the agrocultures international colloquium Amazon's rising violence and disturbing trends

Index[1]

[1] Note: Page numbers followed by 'n' refer to notes.

A. A. R. Ioris (ed.), *Environment and Development*, https://doi.org/10.1007/978-3-030-55416-3

The manufacturer's authorised representative in the EU is Springer Nature Customer Service Centre GmbH, Europaplatz 3, 69115 Heidelberg, Germany. If you have any concerns regarding our products, please contact ProductSafety@springernature.com

Printed and bound by CPI Group (UK) Ltd, Croydon, CR0 4YY

23/09/2026

02221529-0001